An Introduction to Organic Geochemistry

Other titles in the **Longman Geochemistry Series**:

Chemical Thermodynamics for Earth Scientists
  Philip Fletcher

Using Geochemical Data: Evaluation, Presentation, Interpretation
  Hugh Rollinson

An Introduction to
Organic Geochemistry

S D Killops   V J Killops

Series Advisory Editor:

Dr R C O Gill
Royal Holloway, University of London

Longman
Scientific &
Technical

Copublished in the United States with
John Wiley & Sons, Inc., New York

Longman Scientific & Technical
Longman Group UK Limited
Longman House, Burnt Mill, Harlow,
Essex CM20 2JE, England
and Associated Companies throughout the world.

Copublished in the United States with
John Wiley & Sons, Inc., 695 Third Avenue, New York, NY 10158

© Longman Group UK Ltd 1993

First published 1993

British Library Cataloguing in Publication Data
A catalogue record for this book is available from the British Library

ISBN 0 582 08040 1

Library of Congress Cataloging-in-Publication Data
A catalogue record for this book is available from the Library of Congress

ISBN 0 470 22073 2

Set by 6 in 10/12 Ehrhardt Ro.

Printed in Hong Kong
WP/01

Contents

Preface xiii
Acknowledgements xiv

Chapter 1 **Production and fate of organic matter** **1**

1.1 **Global carbon cycle** 1
1.1.1 Organic geochemistry and the carbon cycle: an introduction 1
1.1.2 Carbon reservoirs and fluxes between them 1
　　　Organisms and carbon cycling 2
　　　Accumulation of sedimentary organic matter 3
　　　Deviation from the steady state system 4
　　　Marine primary production and sedimentation 4
1.1.3 Biochemical and geochemical subcycles 5
　　　Organic-rich sediments and primary production 6
1.2 **Photosynthesis and the evolution of life** 6
1.2.1 Atmospheric oxygen, photosynthesis and the first organisms 6
1.2.2 Evolution of marine life 13
1.2.3 Evolution of terrestrial life 14
1.2.4 Regional variations in ecosystems 15
1.3 **Major contributors to sedimentary organic matter** 16
1.3.1 Major present-day contributors 16
1.3.2 The fossil record of major contributors 17
1.4 **Photosynthesis and stable isotopes of carbon** 18

Chapter 2 **Chemical composition of biogenic matter** **22**

2.1 **Structure of natural products** 22
2.1.1 Introduction 22
2.1.2 Bonding in organic compounds 23
2.1.3 Stereoisomerism 25
2.2 **Carbohydrates** 28
2.2.1 Composition 28
2.2.2 Occurrence and function 31
2.3 **Proteins** 34
2.3.1 Composition 34
2.3.2 Occurrence and function 37
2.4 **Lipids** 39

2.4.1   Glycerides   39
         Fats   39
         Phospholipids, glycolipids and ether lipids   42
2.4.2   Waxes and related compounds   43
         Waxes   43
         Cutin and suberin   45
2.4.3   Terpenoids   45
         Monoterpenoids   47
         Sesquiterpenoids   47
         Diterpenoids   47
         Triterpenoids   47
         Steroids   50
         Tetraterpenoids   53
2.4.4   Tetrapyrrole pigments   56
2.5   **Lignin, tannins and related compounds**   58
2.5.1   Lignin   58
2.5.2   Tannins and other hydroxy-aromatic pigments   60
2.6   **Geochemical implications of compositional variation**   61
2.6.1   Compositional variation of organisms   61
2.6.2   Variations throughout geological time   62

Chapter 3   **Conditions for the accumulation of organic-rich sediments**   **63**

3.1   **Introduction**   63
3.2   **Factors affecting primary production**   63
3.2.1   Stratification of the water column   65
3.2.2   Light   68
3.2.3   Nutrients   70
         Low latitude oceans   71
         Middle latitude oceans   72
         High latitude oceans   73
         Stratified lakes   73
3.2.4   Spatial variation in marine primary production   74
3.2.5   Variation in phytoplankton populations   76
3.3   **Preservation and degradation of organic material**   77
3.3.1   Fate of primary production in the water column   77
3.3.2   Sedimentary fate of organic material   77
         Aerobic decomposition   77
         Anaerobic decomposition   78
         Bacterial communities and their interactions   80
3.3.3   Factors affecting sedimentary preservation of organic matter   82
3.4   **Depositional environments**   84
3.4.1   Lacustrine environments   84
         Open lakes   85
         Closed lakes   86
3.4.2   Peat swamps and coal formation   87
         Okefenokee swamp   88

3.4.3	Marine environments	89
	Marine shelf deposits	89
	Enclosed and silled basins	90
	Production and preservation of organic matter in the Black Sea	91
	Cretaceous anoxic events	91

Chapter 4   Formation of humic material, coal and kerogen   93

4.1	**Diagenesis**	93
4.1.1	Introduction	93
4.1.2	Microbial degradation of organic matter during diagenesis	93
4.1.3	Geopolymer formation	94
4.2	**Humic material**	95
4.2.1	Occurrence and classification	95
4.2.2	Composition and structure	95
4.2.3	Formation of humic substances	98
4.3	**Coal**	98
4.3.1	Classification and composition	98
	Classification	98
	Petrological composition	99
	Chemical composition	100
4.3.2	Formation	100
	Peatification	101
	Biochemical stage of coalification	101
	Geochemical stage of coalification	102
	Structural changes during coal formation	103
4.4	**Kerogen**	106
4.4.1	Formation	106
	Geopolymer formation during diagenesis	106
	Biomarkers	109
	Sulphur incorporation	110
4.4.2	Kerogen composition	112
4.4.3	Kerogen classification	113
	Type I kerogen	113
	Type II kerogen	114
	Type III kerogen	114
	Type IV kerogen	115
	Improved kerogen typing	115
4.4.4	Thermal evolution of kerogen	115
	Structural changes	115
	Changes in chemical composition	116

Chapter 5   Generation and composition of petroleum   119

5.1	**Petroleum generation**	119
5.1.1	Introduction	119

5.1.2	Hydrocarbons from coal	120
5.1.3	Variation in hydrocarbon composition with kerogen maturity	120
5.1.4	Reactions involved in hydrocarbon generation	122
	Isotopic fractionation	123
5.2	**Importance of time and temperature in petroleum formation**	124
5.2.1	Effects of time and temperature on hydrocarbon generation	124
	Temperature	124
	Time	125
5.2.2	Kinetic models of petroleum formation	125
	Time–temperature index	125
	More complex models	128
5.3	**Migration of hydrocarbons**	129
5.3.1	Primary migration	129
	Mechanisms of primary migration	129
	Expulsion efficiency	131
5.3.2	Secondary migration	132
5.3.3	Traps and reservoirs	133
5.4	**Petroleum composition**	135
5.4.1	Gross composition of oils	135
5.4.2	Hydrocarbons in petroleum	136
	Major hydrocarbons	136
	Biomarkers	138
5.4.3	Comparison of crude oil composition with source rock bitumen	140
	Migration	140
	Water washing	141
	De-asphalting	141
	Biodegradation	141
	Thermal alteration	141
5.5	**Occurrence of fossil fuels**	141
5.5.1	Temporal distribution of fossil fuels	141
5.5.2	Oil reserves	144
	Conventional oils	144
	Heavy oils	145
	Oil shales	145
5.5.3	Coal	146
5.5.4	Gas	146
5.6	**Assessment of petroleum source rocks**	146
5.6.1	Amount and type of organic matter	146
	Amount of organic matter	147
	Type of organic matter–optical methods	147
	Type of organic matter–physico-chemical methods	148
5.6.2	Maturity of organic matter	148
	Optical measurements of maturity	148
	Pyrolytic measurements of maturity	150
	Effect of maturity on identification of organic matter type	150
	Chemical measurements of maturity based on bitumen	150
5.6.3	Stable isotopes and correlation of petroleums with source rocks	151

Chapter 6   **Molecular evaluation of Recent sediments**   **153**

6.1   **Biomarker distributions inherited from organisms**   153
6.1.1   Introduction   153
6.1.2   General differences between major groups of organisms   154
6.1.3   Factors affecting the lipid composition of organisms   155
6.2   **Examples of source indicators in Recent sediments**   157
6.2.1   Fatty acids   157
         Monounsaturated fatty acids   157
         Polyunsaturated fatty acids   158
         Iso and *anteiso* methyl-branched fatty acids   158
         Internally branched and cycloalkyl fatty acids   159
         Hydroxy fatty acids   159
6.2.2   Sterols   159
6.2.3   Carbohydrates   161
6.2.4   Lignins   162
6.2.5   Carbon isotopes   162
6.3   **Diagenesis at the molecular level**   165
6.3.1   General diagenetic processes   165
         Carbohydrates and lignins   165
         Biomarkers   166
6.3.2   Lipid diagenesis in the water column   168
6.3.3   Sedimentary diagenesis of lipids   170
         Fatty acids   170
         Photosynthetic pigments   172
         Steroids   175
         Terpenoids   179
6.4   **Palaeotemperature measurement**   187
6.4.1   Amino acid epimerisation   187
6.4.2   Degree of unsaturation in long-chain ketones   187

Chapter 7   **Molecular assessment of ancient sediments and petroleum**   **190**
             formation

7.1   **Source indicators**   190
7.1.1   Introduction   190
7.1.2   Hydrocarbons   190
         Normal and methyl-branched alkanes   190
         Acyclic isoprenoids   191
         Cycloalkanes   192
7.1.3   Carbon isotopes   193
7.2   **Indicators of depositional environment**   194
7.2.1   Hypersalinity   194
7.2.2   Redox conditions   195
         Phytol diagenesis   195
         Nickel and vanadium distributions   197

7.2.3   Recognition of different types of marine and lacustrine
        environments   198
  7.3   **Thermal maturity and molecular transformations**   199
7.3.1   Configurational isomerism   199
        Acyclic isoprenoidal alkanes   199
        Steranes   199
        Triterpanes   200
7.3.2   Aromatisation   201
7.3.3   Enrichment of short-chain hydrocarbons and cracking processes 201
        Steroids   201
        Porphyrins   201
  7.4   **Molecular maturity and source parameters in petroleum   202
        exploration**
7.4.1   Molecular maturity parameters   202
        Light hydrocarbons   202
        Carbon preference index   203
        Pristane formation index   203
        Biomarker transformations   204
        Methyl group isomerism in aromatic hydrocarbons   205
7.4.2   Effect of geothermal gradient on molecular maturity parameters 208
7.4.3   Correlation of oils and source rocks   210
  7.5   **Biomarker hydrocarbon analysis**   211
7.5.1   Introduction   211
7.5.2   Gas chromatography–mass spectrometry   211
7.5.3   Evaluation of biomarker distributions   213

Chapter 8   **Environmental behaviour of anthropogenic organic   217
            compounds**

  8.1   **Introduction**   217
  8.2   **Human influence on the carbon cycle**   217
8.2.1   Carbon dioxide and the greenhouse effect   217
8.2.2   Effects of other trace gases on global warming   220
        Methane   221
        Carbon monoxide   223
        Dimethyl sulphide   223
8.2.3   Eutrophication   224
  8.3   **Halocarbons and ozone depletion**   225
  8.4   **Hydrocarbon pollution in aquatic environments**   227
8.4.1   Fossil fuel combustion   227
        Polycyclic aromatic hydrocarbons in Recent sediments   227
        Polycyclic aromatic hydrocarbons in ancient sediments   228
8.4.2   Oil spills   230
        Effects of oil pollution   230
        Oil pollution monitoring   231
  8.5   **Some xenobiotic organic substances**   232
8.5.1   DDT and related compounds   233

8.5.2   Polychlorinated biphenyls   234
 8.6   **Factors affecting the fate of anthropogenic inputs**   237
8.6.1   General considerations   237
8.6.2   Humic substances and pollutants   238

        References   239
        Further reading   249
        Index   253

Preface

To begin with, a brief statement of what constitutes organic geochemistry is probably called for. It is the study of the transformations undergone by all organic matter from its origin, whether biological or man-made, in the geosphere. The term geosphere is usually interpreted in its widest sense, encompassing the range of possible environments (earth, air, water and sediment). The transformations involved vary from those mediated by biological agents during recycling of the constituents of living organisms to those controlled by temperature and pressure at depth in the crust on the way to long-term preservation of organic matter in sedimentary rock.

Our knowledge of organic geochemistry has expanded so greatly in recent years that a comprehensive text on the subject would fill many books of this size. To a newcomer, the bulk of information and the terminology adopted from a range of disciplines, such as chemistry, geology, ecology, biochemistry, botany and oceanography, can be quite daunting. However, to those who are not readily deterred, the fascination of the subject soon becomes apparent. If only the basics of organic geochemistry could be found readily at hand and not scattered through the textbooks and journals of a number of disciplines! These were our thoughts when we first came to the subject some ten years or so ago and which provided the stimulus for this book when one of us (S.D.K.) came to teach the subject to both undergraduates and postgraduates.

This book is an attempt to present a readily accessible, up-to-date and integrated introduction to organic geochemistry, at a reasonable price. It does not assume any particular specialist knowledge other than some basic chemistry and is intended to serve as a text for both undergraduate and postgraduate courses in which organic geochemistry is an important component. It may also be found a useful companion by experienced scientists from other disciplines who may be moving into the subject for the first time.

Naturally, in a book of this size it is not possible to discuss all aspects of organic geochemistry, and emphasis has been placed on the formation of organic-rich sedimentary deposits. While this approach will be of particular relevance to those interested in petroleum exploration, the important area of environmental geochemistry has also received consideration. In addition, boxes have been used to explain concepts introduced from other disciplines. We hope that this book will stimulate the reader to continue studying organic geochemistry.

Acknowledgements

The authors wish to thank the following people for their advice during the preparation of this book: Prof. W. G. Chaloner, F.R.S. (Biology Dept., Royal Holloway and Bedford New College, University of London); Dr. P. A. Cranwell (Institute of Freshwater Ecology, Windermere); Dr. P. Finch (Chemistry Unit, Royal Holloway and Bedford New College, University of London); Dr. A. Fleet (Geochemistry Section, BP Research Centre, Sunbury); Prof. S. Larter (Newcastle Research Group in Fossil Fuels and Environmental Geochemistry, University of Newcastle); Dr. J. McEvoy (National Rivers Authority, Warrington); Dr. R. J. Parkes (Geology Dept., University of Bristol); Prof. J. B. Pridham (Biochemistry Dept., Royal Holloway and Bedford New College, University of London); Dr. D. J. Smith (Water Research Centre, Medmenham); and Dr. G. Wolff (Earth Sciences Dept., University of Liverpool).

Chapter 1   Production and fate of organic matter

1.1   Global carbon cycle

1.1.1   Organic geochemistry and the carbon cycle: an introduction

Organic geochemistry concerns the fate of all organic compounds in the geosphere as a whole. From chiefly biological origins, organic compounds can be incorporated into sedimentary rocks and preserved for geological periods, but they are ultimately returned to the Earth's surface, either by natural processes or by human action, to participate again in biosynthetic reactions. This cycle involves various biochemical and geochemical transformations, which form a central part of any consideration of organic geochemistry.

The chemistry of organic compounds is based on carbon, the twelfth most abundant element in the Earth's crust. Carbon accounts for only ca. 0.08% of the combined lithosphere, hydrosphere and atmosphere but it is, nevertheless, an extremely important element and its compounds form the basis of all life. Carbon-rich sedimentary deposits are of great importance to humans. They comprise diamond and graphite (the native forms of carbon), calcium and magnesium carbonates (calcite, limestone, dolomite, marble and chalk) and hydrocarbons (gas, oil and coal). The last category is the main consideration of this book. The ability of carbon to form an immense variety of naturally occurring compounds, primarily with the elements hydrogen, oxygen, sulphur and nitrogen, with an equally wide range of properties is unparalleled by other elements.

1.1.2   Carbon reservoirs and fluxes between them

A useful starting point for the study of organic carbon compounds in the biosphere and geosphere is the global carbon cycle. A much simplified summary of this cycle is shown in Fig. 1.1, which gives an idea of the sizes of the various compartments (or **reservoirs**) in which carbon is located, the exchange rates (or **fluxes**) between these reservoirs and the main forms in which carbon exists in each reservoir. The metric tonne (t or 10^3 kg) is a convenient unit for large masses and is used in Fig. 1.1 and throughout the text. All reservoir and flux values in Fig. 1.1 are approximate, as many cannot be measured directly but are inferred from other measurements. It is likely that many of the estimates used will be revised, some by significant amounts, as research progresses. It can be seen that by far the largest reservoir of carbon, accounting for ca. 99.9% of the total, is sedimentary rock, mainly in the form of carbonates. In addition, inputs to and outputs from the various carbon reservoirs are broadly in balance, resulting in what can be considered a steady state system.

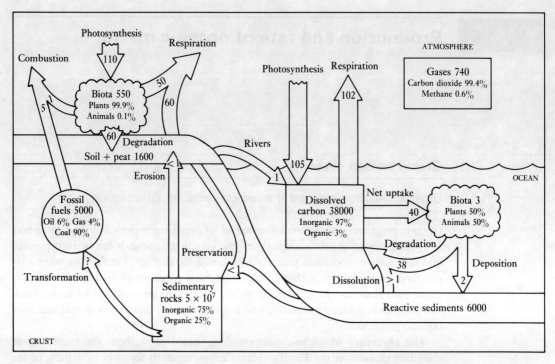

Figure 1.1   Summary of the global carbon cycle, showing sizes of main reservoirs (boxes of various shapes) and annual fluxes (arrows) in units of Gt (10^9 t or 10^{15} g) of carbon. (After several sources, including Bolin et al., 1979, 1983; Kempe, 1979; Mopper and Degens, 1979; De Vooys, 1979; NERC, 1989.)

Organisms and carbon cycling   In the atmosphere carbon exists mainly as carbon dioxide, which is taken up by growing plants during photosynthesis. The amount of solar energy captured by plants during photosynthesis is referred to as **gross primary production** and can be measured by the amount of carbon dioxide that has been fixed. Some of the gross primary production is used to provide the energy needed for the performance of normal biochemical processes, which are collectively termed **respiration**. Respiration can be viewed as the 'burning' of organic compounds to release stored energy and as such is effectively the opposite of photosynthesis, releasing carbon dioxide back into the atmosphere. The part of gross primary production that is not respired, but is available for growth and reproduction, is called **net primary production**.

From Fig. 1.1 it can be seen that the annual net primary production for land plants and marine plants is about the same (ca. 60 and 40 Gt, respectively), although the biomass of terrestrial plants is much greater than that of marine plants. This is an important demonstration of the fact that biomass is not necessarily a guide to productivity. In the ocean 95% of primary production is accounted for by the phytoplankton, which are free-floating, microscopic organisms, mainly unicellular algae. On a global basis macroscopic, multicellular algae (seaweeds) make only a minor contribution to marine primary production. Phytoplankton are short-lived compared with terrestrial plants, especially trees, and do not need to produce supportive structural tissue, which is largely photosynthetically inactive. The larger biomass of terrestrial plants is due to the storage of organic matter as woody material during greater life spans. Virtually the whole of the net primary production of phytoplankton is directed towards reproduction and growth, but much of this is grazed by small herbivorous animals, called zooplankton. As a consequence of this efficient grazing, phytoplankton biomass is low and the ratio of animal to plant biomass in the oceans is greater than on land.

All organisms rely on solar energy but only plants and some bacteria (e.g. cyanobacteria) are able to make direct use of it in the conversion of carbon dioxide into carbohydrates, the process of photosynthesis. Such organisms are termed **phototrophs**. Some bacteria are able to use chemical energy sources such as ammonia, nitrite and ferrous ions rather than light, and are therefore called **chemotrophs**. Generally, however, the chemosynthesis of organic matter is minor compared with photosynthesis. The phototrophs and chemotrophs are collectively known as **autotrophs** because they can manufacture their organic constituents directly from inorganic sources and do not rely on other organisms for energy supplies. All other organisms gain their energy supplies and organic substrates by feeding, directly or indirectly, upon autotrophs and are termed **heterotrophs**.

Energy is passed from the primary producers to the various heterotrophs along food chains and ultimately reaches the top carnivores. **Herbivores** represent the first link in the **grazing food chain**. Plant tissue that is not consumed by herbivorous animals eventually dies. In a steady state system losses of plant tissue by grazing and death are balanced by net primary production of new tissue. Dead plant matter together with the faecal material of animals and their remains upon death are collectively termed **detritus** and pass to organisms of the **detrital food chain** in soil, water and sediments. Invertebrate animals form one group of detrital-feeding organisms, the **detritivores**, but the most important group is the **decomposers**, which comprises bacteria and fungi.

Accumulation of sedimentary organic matter

In the marine environment as a whole most detrital organic matter (ca. 95%) is recycled by **pelagic** organisms (i.e. swimming or floating organisms) within the surface waters. The small amount that does reach the sediment is mostly recycled by detrital-feeding **benthic** organisms (which live on top of and within the sediment). Any detrital organic material that survives, under the right conditions (which will be discussed in Chapter 3), may be preserved and incorporated into the sediment and form organic-rich deposits. From Fig. 1.1 it can be seen that the total amount of carbon (organic and inorganic) accumulating in sediments is $<1\,Gt/a$. It is estimated that the accumulation rate of organic carbon in sediments is ca. $4\,Mt/a$ ($4 \times 10^6\,t/a$), which suggests that only ca. 0.01% of marine net primary production is preserved annually in sediments (Tissot and Welte, 1984). Even if the accumulation rate has been underestimated it seems probable that annual preservation of organic matter in sediments is $<0.1\%$ of marine net primary production.

Organic matter is virtually completely decomposed and recycled in soils. Preservation of organic matter in terrestrial environments is, therefore, confined mostly to peat formation in moorland bogs and low-lying swamps, but the size of this carbon reservoir (ca. 160 Gt) is not known with any accuracy. While the present annual accumulation rate of peat is low compared with that of organic-rich marine sediments, the presence of large coal deposits in the sedimentary record suggests that peat formation was of much greater importance in the past. Present-day net primary production in freshwater environments is $<1\,Gt/a$, although carbon-rich deposits are being formed under certain conditions in some lakes. However, lakes contain only a very small volume of water compared with the oceans and so it is in marine sediments that most of the global preservation of organic carbon is occurring.

Deviation from the steady state system

As expected for a steady state system, there appears to be a balance between uptake (by photosynthesis) and emission (by respiration) of carbon by terrestrial plants and animals. In contrast, there is an imbalance between uptake and release of carbon dioxide in the marine environment; the oceans seem to be a net sink for carbon dioxide, most of which may be taken up by phytoplankton during photosynthesis. This imbalance is mainly the result of fossil fuel burning and deforestation ('slash and burn'), which are responsible for a net annual surplus input of carbon dioxide to the atmosphere of ca. 3 Gt. The effects of this perturbation to the steady state system will be considered in more detail in Section 8.2.1. Carbon also enters the atmosphere in the form of trace gases such as methane and carbon monoxide. Methane is quantitatively the most important of these trace gases and is mainly produced during the microbial degradation of organic matter in the absence of oxygen. Carbon monoxide is released into the air by natural processes, such as atmospheric oxidation of methane, and by anthropogenic (man-made) processes, such as incomplete combustion of fossil fuels. These trace gases will be considered in more detail in Section 8.2.2.

Marine primary production and sedimentation

Terrestrial plants take up carbon dioxide directly from the atmosphere in gaseous form. However, aquatic plants utilise carbon dioxide dissolved in water, and it is for this reason that the photosynthesis and respiration flux arrows in Fig. 1.1 do not point directly to the marine biota. Molecules of carbon dioxide are constantly exchanging between the atmosphere and oceans, so that a dynamic equilibrium exists. In solution there are further dynamic equilibrium (reversible) reactions which favour the bicarbonate (HCO_3^-) ion:

$$CO_2 + H_2O \rightleftharpoons H_2CO_3 \rightleftharpoons HCO_3^- + H^+ \rightleftharpoons CO_3^{2-} + 2H^+ \qquad [1.1]$$

These equilibria mean that, at constant temperature, increasing the level of carbon dioxide in the atmosphere causes more to dissolve in the oceans, producing more bicarbonate and carbonate. In seawater most carbon exists in solution, mainly in inorganic forms and particularly bicarbonate. Carbon is chiefly assimilated as bicarbonate and is used in the production of organic tissue as well as, in some cases, the formation of carbonate skeletal material (**tests**). Carbon is also present in seawater as organic compounds, some being dissolved and some being present as suspended particles. The organic remains of organisms and faecal pellets represent a particulate organic carbon (POC) reservoir of ca. 30 Gt (Mopper and Degens, 1979). Dissolved organic carbon (DOC; ca. 1000 Gt, Fig. 1.1), therefore, accounts for some 97% of the total marine organic carbon. Fluvial particulates contribute little to the open ocean, as they are mainly deposited in estuaries and deltas as a result of slackening currents and flocculation caused by salinity differences.

Carbon-rich sediments may contain both inorganic and organic carbon. Inorganic carbon can be present as **biogenic** or **abiogenic** carbonate (i.e. of biological or non-biological origin, respectively). The former derives from the carbonate tests secreted by some planktonic organisms (e.g. coccolithophores), while the latter is formed by the chemical process of precipitation of carbonate out of seawater solution. Sedimentary organic carbon is mainly in the form of POC that has settled out from the overlying water column. Detritus is, therefore, generally important in the formation of carbon-rich sediments.

After the incorporation of detritus into marine sediments, recycling of carbon can continue in the upper sediment layers, the reactive sediments in Fig. 1.1. Inorganic carbonate and soluble organic compounds produced by decomposition processes can be leached out of these sediments into pore waters, which are able to exchange with water overlying the sediments. With increasing sedimentation this process of exchange ceases and long-term preservation can occur as sediments become consolidated and form sedimentary rocks. The amount of carbon in sedimentary rocks can be calculated from estimates of the total volume of different types of sedimentary material and their average carbon contents (Kempe, 1979). Such an approximation gives the value of ca. 55 Pt (55×10^{15} t) for the sedimentary rock carbon reservoir in Fig. 1.1. The carbon content of the entire Earth's crust has recently been estimated at 90 Pt (NERC, 1989). Of the organic carbon entering sediments ($< 0.1\%$ of marine net primary production), much is decomposed, ultimately to CO_2. However, a proportion is preserved in sedimentary rocks, but the amount of this transformed into fossil fuels on an annual basis is not known with any degree of certainty. The cycle for this organic material is completed by human exploitation of fossil fuels and by the action of natural processes, such as volcanism and the weathering of uplifted and exposed sedimentary rocks.

1.1.3 Biochemical and geochemical subcycles

In this book we are concerned with the organic components of the carbon cycle. It is convenient to consider the organic carbon cycle as comprising two subcycles. The larger (ca. 12 Pt of C, or 12×10^{15} t, calculated from organic content of sedimentary rocks in Fig. 1.1) involves sedimentary rocks and residence times of millions of years, and may be thought of as the **geochemical subcycle**. The smaller (ca. 3 Tt of C, or 3×10^{12} t, calculated from DOC, POC, biota biomass and soil organics in Fig. 1.1) involves biological recycling and residence times of up to a hundred years or so only, and may be thought of as the **biochemical subcycle**. These two subcycles are linked by a small two-way flux. That from biochemical to geochemical subcycle is the rate of incorporation of organic carbon into sedimentary rocks, estimated at presently < 1 Gt/a, although it has varied significantly in the past. In a steady state there would be an equal flux in the opposite direction corresponding to erosion of sedimentary rocks. However, human exploitation of fossil fuels has greatly augmented this flux from geochemical to biochemical subcycle.

While the larger, geochemical, subcycle is quantitatively the most important (ca. 99.95% of total organic carbon), all the organic matter in this cycle has originated from the biochemical subcycle (i.e. from plant and animal tissue). This demonstrates that the examination of biological processes in the production and modification of organic matter is as important as considering sedimentary processes in understanding the conditions under which organic-rich sediments are formed. The link between the biochemical and geochemical subcycles is also reflected by the relationship between the quantity of free oxygen in the atmosphere and the amount of reduced carbon compounds preserved in sediments and rocks. This relationship stems from the production of oxygen when carbon dioxide is 'fixed' during photosynthesis, but its removal from the atmosphere when reduced organic compounds (e.g. fossil fuels) are released from the geosphere and are oxidised to carbon dioxide (see Box 1.1).

Box 1.1

Oxidation and reduction

The most obvious definition of **oxidation** is the gain of oxygen by a chemical species, as in the burning of methane:

$$CH_4 + 2O_2 \rightarrow CO_2 + 2H_2O \qquad [1.2]$$

A further example is provided by the oxidation of ferrous ions (iron(II)) to ferric (iron(III)) during the sedimentary deposition of iron oxide:

$$4Fe^{2+} + O_2 + 4H_2O \rightarrow 2Fe_2O_3 + 8H^+ \qquad [1.3]$$

Oxidation can also be defined as the loss of hydrogen, as occurs with methane above (Eq. [1.2]). A further definition of oxidation is the loss of electrons. This is the net process undergone by iron in the above oxidation of iron(II) to iron(III) (Eq. [1.3]), and can be represented by:

$$Fe^{2+} \rightarrow Fe^{3+} + e^- \qquad [1.4]$$

All three definitions of oxidation are encountered in geochemistry.
   Reduction is the opposite of oxidation.
   Oxidising conditions in sedimentary environments are termed **aerobic** or **oxic** and are related to free oxygen being available for oxidative reactions to take place. In **anaerobic**, or **anoxic**, conditions there is no such available oxygen and conditions are described as reducing. In water (whether in water bodies or in sedimentary pore waters) dissolved oxygen levels of $>0.5\%_{oo}$ (parts per thousand, or per mil) correspond to oxic conditions, while those of $<0.1\%_{oo}$ correspond to anoxic conditions. Conditions related to intermediate values of oxygen concentration are generally described as **dysaerobic** (or **suboxic**).

Organic-rich sediments and primary production
The average carbon content of the Earth's crust is $<0.5\%$ (by weight), but in organic-rich sediments such as oil shales, coal seams and organic-rich limestones it can rise to $>5\%$. All the organic carbon in these sediments is biogenic. Formation of coal from compressed terrestrial plant remains (peat), particularly those which grew in Carboniferous swamps (a geological chronology is given in Fig. 1.6), is well known. Oil too has a biological origin, which became apparent when a ubiquitous red pigment, directly related to the compound (chlorophyll) responsible for photosynthesis in plants, was first isolated from oil in the 1930s. It was suggested that marine plants, the phytoplankton, were the chief source of the organic matter from which oil is generated. It is clear, then, that photosynthesising organisms have played a vital role in the formation of organic-rich sediments.

1.2   Photosynthesis and the evolution of life

1.2.1   Atmospheric oxygen, photosynthesis and the first organisms

Since the formation of the Earth the conditions for the synthesis, deposition and preservation of organic matter have changed considerably. The primordial atmosphere contained mainly hydrogen and helium, most of which escaped the gravitational field of the Earth to be replaced by juvenile volatiles from the interior of the planet. In view of the composition of volcanic emissions today these volatiles probably comprised mainly water vapour, nitrogen, carbon dioxide, carbon monoxide, sulphur dioxide and hydrogen chloride. Opinions vary over the importance of reducing gases like methane and ammonia, but it is clear that no free oxygen was present. Reducing conditions prevailed until the advent of oxygen-liberating (oxygenic) photosynthesis by unicellular organisms led to the large-scale production of organic matter and the development of oxidising conditions. Only with the availability of free oxygen was it possible for multicellular organisms to develop and diversify. The composition of the atmosphere has both affected and been affected by the development of life on the planet. Important stages in the evolution of the Earth's surface are presented in Fig. 1.2.

There is conjecture over whether the basic organic compounds required for the evolution of life on Earth originated primarily from photochemical reactions or from

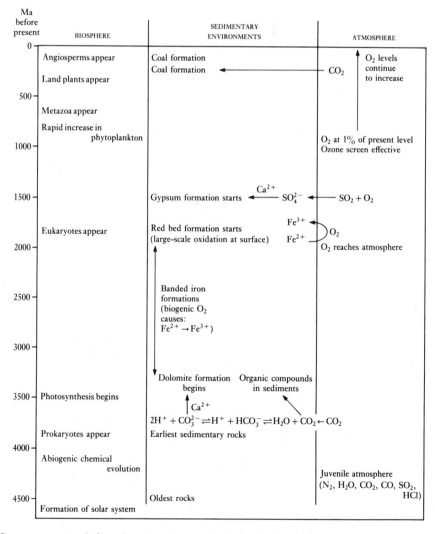

Figure 1.2   Important events influencing the carbon cycle during the Earth's evolution.

the carbonaceous comets and asteroids which bombarded the Earth between 4500 and 3800 Ma ago (Chyba et al., 1990). However these compounds originated (described as abiogenic chemical evolution in Fig. 1.2) it seems most likely that the early stages in the evolution of life were confined to aquatic environments, as water is an essential requirement for all life. The earliest organisms were anaerobic prokaryotic heterotrophs (see Box 1.2) that metabolised the simple abiogenic organic compounds at the Earth's surface. This food supply would have been limited and the proliferation of life required the evolution of autotrophs. The first photosynthesising organisms were also prokaryotic anaerobes, and they used hydrogen sulphide (H_2S) as a source of the hydrogen needed for carbohydrate synthesis, yielding sulphur as a by-product, not oxygen. Similar photosynthetic prokaryotes, such as the purple and green sulphur bacteria, exist today and are all anaerobes.

Box 1.2

Classification of organisms

Organisms (excluding viruses) can be broadly separated into **prokaryotes** and **eukaryotes** on the basis of cell structure. Prokaryotic cells contain no nuclear membranes and their DNA is not arranged in chromosomes. Eukaryotic cells always contain a nuclear membrane and their DNA is usually in chromosomal form. There are other differences between the two cell types but these are the most useful general distinguishing features. The prokaryotes comprise all bacteria (including cyanobacteria) and are therefore unicellular (sometimes termed the Monera). All other organisms are eukaryotes, either unicellular (the Protista) or multicellular.

The eukaryotes are often divided into two kingdoms: animals and plants. This distinction is blurred in the case of flagellates, which can fall into either kingdom and some can be classified in both. Further categorisation results in the main **taxonomic ranks** shown in Table 1.1, in which the modern anthropogenic classification is given as an example.

It is possible to use subcategories of the main ranks. The rank below kingdom is generally termed division for plants but phylum for animals. For the plant kingdom the following divisions are generally recognised: algae (Phycophyta), fungi (Mycophyta), lichens (Lichenes), bryophytes (Bryophyta), pteridophytes (Pteridophyta) and spermatophytes (Spermatophyta). The last two divisions comprise the vascular plants, while the bryophytes include mosses, liverworts and hornworts. Some organisms are difficult to classify, particularly unicellular organisms, and changes in classification can occur as our knowledge of these organisms grows.

There is flexibility in the application of classification ranks which can confuse the inexperienced reader (e.g. Holmes, 1983). For example some classes of algae are considered to be sufficiently distinct to warrant divisional status and so the algae as a whole become a subkingdom (even a kingdom in some classifications), and the cyanobacteria can be treated as a division of this subkingdom (the Cyanophyta). The prokaryotes can be treated as a division (Schizophyta) and the bacteria and cyanobacteria as classes (Schizomycetes and Schizophyceae, respectively). The fungi are sometimes considered to be a kingdom with two main divisions comprising the slime moulds (Myxomycota) and the true fungi (Eumycota).

A new arrangement for the tree of life has recently become possible, based on genetic relationships derived from ribosomal RNA sequencing (Woese et al.,

Box 1.2
(continued)

> 1990). A simplified version is shown in Fig. 1.3 in which there are three kingdoms: archaebacteria, eubacteria and eukaryotes. This classification demonstrates the importance of bacteria, which are divided into the **archaebacteria** (methanogens, halophiles and thermoacidophiles; Woese and Wolfe, 1985) and true bacteria, the **eubacteria**. Archaebacteria are considered the most ancient extant form of cellular life and occur in many environments that are hostile to other forms of life (e.g. hydrothermal vents, fumaroles and hypersaline lakes). The remaining eukaryotes occupy a much less significant position than in older classifications.

The first organisms able to perform photosynthesis with the liberation of oxygen were **cyanobacteria** (previously classified as blue–green algae) which are believed to have evolved from anaerobic, photosynthetic bacteria. Cyanobacteria also inhabited aquatic environments, which provided a ready supply of the water used as a source of hydrogen in carbohydrate synthesis. Evidence for oxygenic photosynthesis by cyanobacteria dating back 3500 Ma has been found in the form of **stromatolites**

Table 1.1   **Major taxonomic ranks and the corresponding anthropogenic classification**

Taxonomic rank	Classification for humans
Kingdom	Animalia
Division or Phylum	Chordata
Class	Mammalia
Order	Primates
Family	Hominidae
Genus	*Homo*
Species	*sapiens*

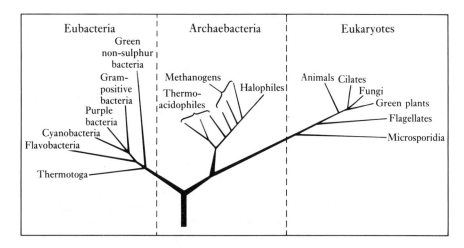

Figure 1.3   Evolutionary relationship of organisms based on ribosomal RNA sequence comparisons. This form of the 'tree of life' is divided into three fundamental kingdoms: eubacteria, archaebacteria and eukaryotes. (After Woese et al., 1990.)

(Rothschild and Mancinelli, 1990), the fossilised remains of microbial mats formed primarily by large colonies of this organism in shallow water. Around 1500 extant species of cyanobacteria have been identified, some of which are colonial mat formers while others are non–colonial and inhabit open water.

Box 1.3 summarises the complex series of processes that occur during oxygenic photosynthesis. Light energy in the visible part of the spectrum emitted by the sun is adsorbed by a green pigment, chlorophyll. This results in the transfer of hydrogen atoms from water to carbon dioxide molecules (i.e. reduction of CO_2) to build up carbohydrate units, while oxygen is liberated from the water molecules. The overall reaction for the formation of a carbohydrate such as glucose can be simplified to:

$$6CO_2 + 6H_2O \xrightarrow[\text{chlorophyll}]{\text{sunlight}} C_6H_{12}O_6 + 6O_2 \qquad [1.5]$$

Box 1.3	**Photosynthesis and chemosynthesis**

Chlorophyll-a is the primary pigment of **photosynthesis** and adsorbs photons of light energy in order to convert carbon dioxide into carbohydrates. Water is essential as it provides hydrogen, while oxygen is expelled into the environment as a by-product. Photosynthesis can be divided into a light (photochemical) stage and a dark (chemical) stage. The former requires light but is unaffected by temperature, while the latter does not require light and proceeds more rapidly with increasing temperature. These stages are presented in Fig. 1.4 and can be summarised as follows:

Light stage

(a) Generation of energy: an electron (e^-) is liberated from chlorophyll-a upon adsorption of light energy.

(b) Storage of energy: the high-energy electron can recombine with positively charged chlorophyll and the excess energy is used to convert ADP (adenosine diphosphate) into ATP (adenosine triphosphate) by the addition of inorganic phosphate.

(c) Storage of reducing power: an electron from chlorophyll may be captured by a hydrogen ion (H^+), produced from the self-ionisation of water, to yield a hydrogen atom, which is immediately taken up by NADP (nicotinamide adenine dinucleotide phosphate), storing the reducing power in the form of NADPH. The chlorophyll ion can regain an electron from a hydroxyl ion (OH^-), which is also formed during the self-ionisation of water, and the resulting hydroxyl radical ($OH\cdot$, which has no charge) combines with others to form oxygen and water.

Dark stage

Overall, the hydrogen stored in NADPH is used to reduce CO_2 to carbohydrate units (CH_2O). This is not a direct reaction because the CO_2 is first combined with a C_5 compound, ribulose diphosphate (RDP), which then spontaneously splits into two identical C_3 compounds, phosphoglyceric acid (PGA). Most of this PGA is used to synthesise further RDP but some is reduced by NADPH, using energy supplied by the ATP/ADP system, to give triose phosphate, which in turn is converted into the glucose phosphate from which various

Box 1.3
(continued)

carbohydrates are synthesised. This assimilatory path is known as the **Calvin cycle** and is involved in all autotrophic carbon fixation, whether photosynthetic or chemosynthetic.

Plants that use the Calvin cycle alone in carbon fixation are termed **C3-plants** because of the involvement of PGA, which contains three carbon atoms. Most plants and the cyanobacteria use the C3 path. However, under conditions of high temperature and low CO_2 levels much energy can be lost by C3-plants due to photorespiration. Two smaller groups of plants overcome this by using an additional biochemical pathway which fixes CO_2 at night. The carbon dioxide is released again within the plant tissue during the day for incorporation into the Calvin cycle. In this way stomatal pores can be closed during the day to reduce photorespiration without cutting off essential supplies of CO_2. These two groups are named after their additional mechanisms: C4-plants and CAM-plants.

Photosynthetic bacteria (a term that usually excludes the cyanobacteria) are strict (i.e. obligate) anaerobes and never use water as a source of hydrogen. They generally use H_2S and so liberate sulphur rather than oxygen. Some species make use of simple organic compounds (e.g. the purple non-sulphur bacteria *Rhodospirillum*).

Chemosynthesis is performed by some types of bacteria, most of which are obligate aerobes. The initial energy-generating process does not involve light or water, but utilises the energy stored in chemicals, mainly the reduced forms of simple inorganic species. An enzyme, dehydrogenase, is used to liberate the energy (in the form of electrons) and reducing power (in the form of protons) from a chemical species such as H_2S. While the necessary compounds for chemosynthesis are generally found throughout the water column they are in highest concentrations in anoxic waters beneath areas of high productivity, resulting from the breakdown of organic constituents in detritus. Chemosynthetic bacteria are, therefore, found at the oxic/anoxic boundary.

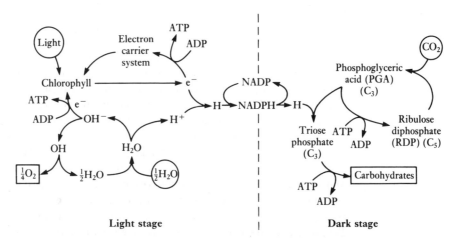

Figure 1.4   Summary of the chemical processes involved in oxygenic photosynthesis. Net inputs are shown in circles, products in rectangles. (ADP/ATP = adenosine di-/triphosphate; NADP/NADPH = nicotinamide adenine dinucleotide phosphate and its reduced form, respectively.)

Carbohydrates are used in the biosynthesis of other organic compounds and to provide an energy store for the performance of normal cell functions. In the absence of free oxygen, fermentation would have provided the energy-releasing step for the earliest organisms:

$$C_6H_{12}O_6 \longrightarrow 2C_3H_4O_3 + \quad 4H \qquad\qquad [1.6]$$

glucose $\qquad\qquad$ pyruvic $\quad$ combined with
$\qquad\qquad\qquad$ acid $\qquad$ other groups

Fermentation is one of a number of anaerobic respiration processes that will be discussed more fully in Section 3.2.2.

Oxygen liberated by early photosynthetic organisms would have been rapidly immobilised in the generally reducing environment, such as in the oxidation of iron from $Fe(II)$ to $Fe(III)$ (see Box 1.1). Banded iron formations (Fig. 1.2) are believed to represent localised oxidation involving the free oxygen liberated during photosynthesis. As the oxygen-producing photosynthetic organisms proliferated, large-scale oxidation of $Fe(II)$ occurred, generating widespread red beds (Fig. 1.2).

Eventually, free oxygen reached the atmosphere enabling the release of energy from carbohydrates by aerobic respiration, the oxidation of carbohydrates to carbon dioxide, which is the reverse of photosynthesis (Eq. [1.5]). Aerobic respiration is a much more efficient process than fermentation, releasing about 18 times more energy, and so mass production of organic material appears to have started about 2000 Ma ago with the development of an oxygenated atmosphere. With atmospheric oxygen came the development of an ozone layer (Fig. 1.2), which protected the delicate photosynthesising organisms in surface waters from the harmful effects of UV radiation. This would have permitted the evolution and proliferation of another important group of oxygenic photosynthesising organisms, the unicellular algae.

It is probable that aerobic respiration followed closely on the development of oxygenic photosynthesis and that the Archaean era (3800–2500 Ma ago) was not anoxic. Even taking low estimates for oxygenic photosynthetic production (e.g. 250–300 Mt/a, which is $< 1\%$ of the current level), sedimentation of an unrealistically large amount of iron(III) and oxidation to carbon dioxide of an unreasonable amount of biogenic methane would have been required to prevent oxygen build-up in the atmosphere. It has been suggested that aerobic respiration would have been the most likely mechanism for restricting oxygen levels (Towe, 1990), requiring a low but stable level of oxygen of ca. 0.2–0.4% (1–2% of current levels) throughout the Archaean. Such conditions would have resulted in the removal of most of the photosynthetic oxygen. Banded iron formations could then be accounted for by processes operating in a generally stratified ocean with oxygenated surface waters and deep sources of $Fe(II)$ (i.e. hydrothermal sources, from the interaction of hot water with igneous material associated with ocean floor spreading). Upwelling of nutrient-rich waters in certain nearshore areas, as occurs in the ocean today (see Section 3.2.2), would have supported high primary productivity, while compensatory downwelling in certain offshore areas would have brought oxygenated waters in contact with deep $Fe(II)$ sources, resulting in the deposition of banded iron formations. This process, together with oxidation of reducing gases from volcanic activity, would have provided a subordinate control on atmospheric oxygen levels. **Tectonic activity** (i.e. processes associated with movement of the Earth's crust) decreased in the early Proterozoic (from 2500 Ma ago), leading to reduced inputs of hydrothermal $Fe(II)$ and reducing gases, and coinciding with the cessation of red bed formation and increase in levels of atmospheric oxygen.

The presence of large amounts of oxygen in the atmosphere would have put severe ecological pressure on many species of bacteria that had evolved to live in reducing conditions. These anaerobes became confined to suitable habitats like anoxic waters and sediments. Some of these ancient bacteria are represented today by the methanogens. Aerobic heterotrophs, in contrast, became widespread and important in the recycling of organic material.

1.2.2 Evolution of marine life

Throughout the Precambrian, prokaryotes, in the form of photosynthetic bacteria and cyanobacteria, were the main producers of organic carbon. They were joined in this role during the Cambrian to Silurian by the unicellular algae, which are eukaryotes and are the photosynthesising plants that inhabit the surface waters of the oceans. These organisms, together with free-floating cyanobacteria in surface waters, comprise the phytoplankton (see Box 1.4). The early phytoplankton, which included the first members of the green algae, possessed organic cell walls. During the Mesozoic calcareous nanoplankton (i.e. secreting calcium carbonate tests), mainly **dino-flagellates** and **coccolithophores**, dominated primary production. Siliceous phytoplankton (i.e. secreting silica tests), especially **diatoms** and **silicoflagellates**, became important in the late Cretaceous and Cenozoic. The main primary producers within the phytoplankton today are the dinoflagellates, diatoms and, to a lesser extent, the coccolithophores.

Box 1.4 | **Plankton classification**

Plankton are organisms living primarily in the upper part of the water column and although often capable of some motion, particularly vertical migration in zooplankton, are unable to maintain their overall lateral position and drift with the ocean currents. Buoyancy aids such as oil bodies are sometimes present. The plankton can be divided into phytoplankton and zooplankton.
Phytoplankton are photosynthesising micro-organisms, usually dominated by unicellular algae. However, free-floating cyanobacteria and photosynthetic bacteria should also be strictly included by this definition. **Zooplankton** are animals ranging from unicellular micro-organisms (**protozoa**) to multicellular organisms (**metazoa**). Many of the smaller zooplankton are herbivores, feeding on phytoplankton, and are, in turn, food for larger carnivorous zooplankton.

Classification of the plankton as a whole is often made on the basis of size (Fig. 1.5). The **ultrananoplankton** is composed almost entirely of bacteria, the **nanoplankton** of algae (phytoplankton), and the macro- and megaplankton of animals (zooplankton), mainly invertebrates.

The abundance of the fossilised remains of herbivorous zooplankton (see Box 1.4) is relatively low for the early Palaeozoic but increases subsequently for orders such as the **radiolarians** and **foraminiferans** (both protozoa of the class Rhizopoda), which first appeared in the Precambrian. Grazing of phytoplankton by zooplankton greatly reduces the direct contribution of the former to sedimentary organic matter. However, there is still an indirect input via detrital material from zooplankton, comprising their faecal pellets and their remains upon death, and in

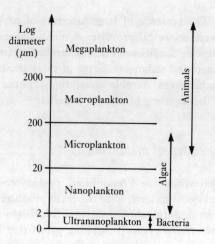

Figure 1.5   Size classification of plankton.

this respect the most important zooplankton today are **copepods** (small crustaceans) and foraminiferans.

Multicellular (macroscopic or macrophytic) algae appeared during the Cambrian and increased in numbers during the Ordovician and Silurian. These organisms were generally non-planktonic, being attached to the substrate, and so can be considered as benthic organisms. They can be classified according to colour: green (Chlorophyta), red (Rhodophyta) and brown (Phaeophyta). The Devonian saw significant changes in the evolution of the algae, with many forms dying out and others replacing them. Some forms were different from earlier and later types and may represent experimental land plants.

Fungi appear to have evolved alongside the algae. Their evolutionary record is, however, difficult to piece together due to the lack of preservable parts. Fossil remains have been found in Precambrian marine sediments and today fungi are highly successful in terrestrial environments. Fungi probably invaded the land at the same time as plants.

1.2.3   Evolution of terrestrial life

The last important group of photosynthesising organisms to appear were the terrestrial **vascular plants**, probably evolving from ancestral green algae. They are commonly termed the **higher plants** in order to differentiate them from the lower plants such as algae. The bryophytes (mosses, hornworts and liverworts) probably originated from ancestral algae at about the same time as the vascular plants; they can be important contributors to peat but not usually to sedimentary organic material. Colonisation of the land by vascular plants began in the Silurian, when the atmospheric oxygen level was ca. 2% and there was an ozone layer providing protection against harmful UV radiation.

A major requirement for land plants was the development of supportive structural tissue and protection against dehydration, both of which had been unnecessary for their marine plant predecessors. Areas with standing water or subject to flooding (e.g. tidal mud-flats) were ideal primary habitats as water was needed for the transport of spores and gametes to enable reproduction to take place. *Cooksonia* was the most successful of the first small vascular land plants, from which all other vascular plants evolved. Diversity and distribution of land plants increased greatly during the Devonian, the major vascular plant types belonging to the psilophytes, a class of the pteridophytes (see Box 1.2).

Dense vegetation stands grew during the late Devonian and Carboniferous, when the clubmosses and horsetails reached their peak of development, forming the great coal forests of the period. These plants grew to considerable size, with heights of 45 m or so for the largest clubmoss, *Lepidodendron*. When the climate changed and the swamps dried out at the end of the Carboniferous some of the larger of the pteridophytes died out, but survivors remain today among the ferns and clubmosses, and one horsetail, *Equisetum*.

The early vascular plants reproduced by spore formation. Seed-bearing plants (spermatophytes), in the form of the **gymnosperms**, emerged during the late Devonian from pteridophyte ancestors. Important orders of the gymnosperms were cycads, conifers and ginkgos, and they dominated the terrestrial flora until the emergence of another group of seed-bearing plants, the **angiosperms**, in the Cretaceous. Although of diminished importance, members of the gymnosperms exist today, most notably the conifers of the temperate regions. The angiosperms, or flowering plants, are highly versatile and successful. They have been dominant since the Tertiary and their productivity is reflected in the large coal deposits of the Cretaceous and Tertiary.

1.2.4 Regional variations in ecosystems

We have considered temporal variations in the biota, some of which have been the result of gradual evolutionary changes and others the result of relatively rapid and sometimes catastrophic events. Such events may occur on a global scale, such as the mass extinctions that have been attributed to meteorite/cometary impacts with the Earth at the Cretaceous/Tertiary boundary (Alvarez et al., 1980). Others may be more regional, such as the periodic advances and retreats of the polar ice-caps.

Regional variations in flora and fauna can be important in determining the main contributors to sedimentary organic material. While the spread of marine organisms throughout the oceans is largely unhindered by topographical obstacles, a uniform distribution does not always result even after sufficient time has elapsed for extensive colonisation to occur by a newly evolved species. For example, there are latitudinal variations in distributions of both abundance and species of phytoplankton which are related to climate and ocean circulation patterns. These variations and the factors that control micro-organism distributions in saline and freshwater environments are discussed in Chapter 3.

Geographical barriers have played an important role in the variations of terrestrial communities. In the Devonian and Carboniferous the land masses were united in two main regions, Laurasia in the northern hemisphere and Gondwanaland in the southern. Over large areas of Laurasia the climate was almost uniformly warm and moist all year (as in the present-day tropics). Under such conditions the *Lepidodendron* flora of the Laurasian coal swamps flourished, while a large area of Gondwanaland was covered by ice sheets. Widespread uniform conditions led to only minor variations in the composition of plant communities.

Regional differences in flora occurred towards the end of the Carboniferous, as the climate became more variable and plants adapted to different conditions and spread into more varied habitats. *Glossopteris* became the dominant gymnosperm in Gondwanaland in the late Carboniferous as the ice sheets receded. The *Glossopteris* flora lasted until the end of the Permian and produced significant coal deposits in the southern continents (Antarctica, Australia, South America and India). Meanwhile, in Laurasia other gymnosperms became established in the drier climate as the swamps dried out. The distinctions between the plant communities of various habitats became more pronounced during the Permian and Triassic as the land masses split up and drifted apart.

1.3 Major contributors to sedimentary organic matter

1.3.1 Major present-day contributors

Autotrophs, and particularly phototrophs, provide the energy needed by all other organisms. Energy is stored in the form of organic tissues and so energy transfer along a grazing food chain occurs by consumption of organisms at one link, or **trophic level**, in the food chain by those at the next higher trophic level. However, energy transfer between adjacent trophic levels is not very efficient. Of the organic matter consumed at a particular trophic level some is lost in excreted material, and of the energy that is assimilated only part is available for growth and reproduction (the elements that constitute net production). Heterotrophs at higher trophic levels (i.e. carnivores) often expend a considerable proportion of assimilated energy on movement in obtaining food and the resulting high respiration levels generally lead to low net production.

The efficiency of energy transfer between adjacent trophic levels can be measured from the ratio of the net production at the higher level to that at the lower level. Values of this ratio, known as the **transfer efficiency**, in terrestrial and marine ecosystems are typically only ca. 10%. Net production is, therefore, significantly higher for the main primary producers, phytoplankton and higher plants, than for organisms at higher trophic levels. Phytoplankton are obviously important contributors to marine and freshwater sedimentary organic matter, while higher plants are important in coal swamps. However, a considerable amount of higher plant material is also transported to sedimentary environments in coastal areas and lakes. Today, marine and terrestrial primary production are approximately equal and have been so since about the Cretaceous, but earlier in the Earth's history marine production was dominant.

The transfer efficiency from phytoplankton to zooplankton, the first step in the marine grazing chain, appears to be a little higher than average at ca. 20%. This is partly due to the greater digestibility of phytoplankton compared with terrestrial plants, particularly the more woody species. Zooplankton can, therefore, be a significant source of organic matter for sediments, whereas other animals do not appear to be important in this respect. As a result of the predator–prey link concentrations of zooplankton tend to be greatest where phytoplankton production is high. However, the transfer efficiency to herbivorous zooplankton is often lower in areas of highest primary productivity and there is a consequential increase in the proportion of phytoplankton remains reaching the sediment and passing to the detrital food chain or undergoing preservation.

There is one further major contributor to sedimentary organic matter, bacteria. A large proportion of the energy flow in ecosystems can pass through the detrital food chain, in which heterotrophic bacteria are prominent participants. Heterotrophic bacteria are important in all sedimentary environments, and although they consume organic detritus they supplement the organic matter with their own remains. In some environments (e.g. Black Sea) autotrophic bacteria may also be important.

While bacterial biomass may be relatively small, bacterial productivity can be very high in aquatic environments. For example, in the Caspian Sea the biomass and productivity of heterotrophic bacteria have been estimated to be about half the corresponding values of the phytoplankton (Bordovskiy, 1965), while in the Black Sea bacterial production (autotrophic and heterotrophic) may be an order of magnitude greater than that of the phytoplanktonic algae. Consequently, the organic-rich remains of bacteria may make significant contributions to most, if not all, sedimentary organic matter.

There are, therefore, four major sources of sedimentary organic matter, in general order of importance: phytoplankton, bacteria, higher plants and zooplankton. Fungi do not appear to make significant contributions to sedimentary organic matter. Although they are important organisms in terrestrial environments they are much less so in marine environments, where degradation is primarily carried out by bacteria.

1.3.2   The fossil record of major contributors

The fossil record can be used to gain information on the relative importance of organisms throughout geological time, as represented in the bar diagram in Fig. 1.6. However, among the limitations of this technique is the inherent tendency to give undue weight to those organisms that have recognisable preservable parts, while organisms comprised of only soft tissue are unlikely to be represented. This means that organism numbers and productivity may not be accurately reflected. For example, remains of the early unicellular organisms that cannot be unambiguously identified in the fossil record are often referred to as **acritarchs** and may be algal or bacterial in origin. In addition, the predominant phytoplankton of the Palaeozoic (acritarchs, green algae and cyanobacteria) had cell walls composed of organic material and so were less likely to leave evidence of their existence than the siliceous and calcareous phytoplankton that became dominant in more recent times.

Formation of dark, organic-rich, marine shales appeared to be widespread in the early Palaeozoic but was generally less common after the Silurian. This change coincides with an increase in abundance of herbivorous zooplankton remains in the fossil record and so could be interpreted as reflecting an increasing importance of the grazing food chain at that time, resulting in less detritus reaching marine sediments in general. However, as will be seen in Chapter 3, other factors are more important in determining whether organic matter undergoes long-term preservation, and it is possible that herbivores were no less abundant during the early Palaeozoic than later but may have lacked identifiable preservable parts. Indeed, a large range of soft-bodied invertebrate phyla appeared explosively at the beginning of the Cambrian (Gould, 1991), and the grazing food chain probably became important then. Stromatolites are far less widely distributed after the Precambrian. This is consistent

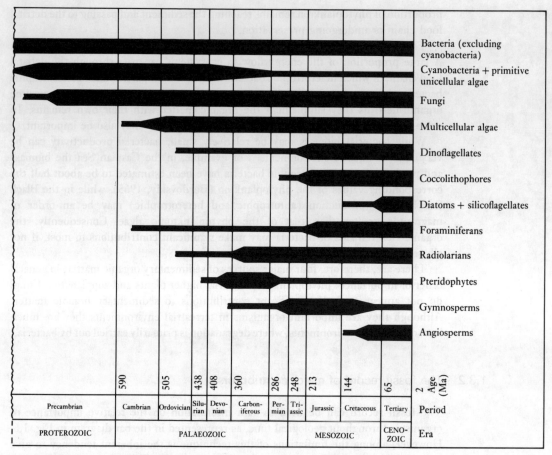

Figure 1.6   Evolution of some important contributors to sedimentary organic matter. Bar widths reflect relative abundance trends within each group of organisms.

with grazing pressure permitting the emerging groups of benthic algae to gain a foothold in favourable shallow-water environments; resulting competition would restrict cyanobacterial mat communities to the harsher environments in which they are presently found.

Because of the problems associated with differing preservation rates in interpreting the fossil record, the width of bars in Fig. 1.6 is only intended as a guide to trends in the relative abundance within each group of organisms, and even this limited assessment is extremely difficult for bacteria and fungi. In the absence of preserved hard parts, chemical evidence in the form of specific compounds may provide information on contributing organisms, but this approach is limited with respect to fungi.

1.4   Photosynthesis and stable isotopes of carbon

It is possible to assess the importance of photosynthesis throughout geological time by examining the isotopic distribution of carbon in carbonates and organic-rich sediments of various ages (see Box 1.5). The mean isotopic signature for the Earth as a whole can be assumed to be that inherited from the parent solar nebula, as represented by the isotopic signature of the mantle ($\delta^{13}C$ ca. $-5‰$). As can be seen from Fig. 1.7, all autotrophic processes favour incorporation of the lighter isotope of carbon into live organic tissue, the isotopic signature of which is preserved in sedimentary organic matter ($\delta^{13}C$ ca. $-26‰$). The two surface reservoirs of carbon dioxide (atmosphere and ocean) are correspondingly depleted in the lighter isotope. The reservoir of CO_2 used by terrestrial plants is the atmosphere ($\delta^{13}C - 7‰$). For phytoplankton the reservoir is dissolved CO_2, which rapidly equilibrates with carbonate and bicarbonate ions, predominantly favouring the bicarbonate ion species (Eq. [1.1]). The marine reservoir of CO_2 can, therefore, be considered as bicarbonate ion ($\delta^{13}C - 1‰$), the isotopic signature of which is broadly preserved in sedimentary carbonate ($\delta^{13}C$ ca. $0‰$). We have seen that the surficial compartment of the carbon cycle is only ca. 0.1% of the carbon incorporated into sediments. It can further be seen (Fig. 1.7) that the total standing biomass (marine and terrestial) comprises ca. 20% of the surface compartment of carbon, the rest being mainly in the form of bicarbonate.

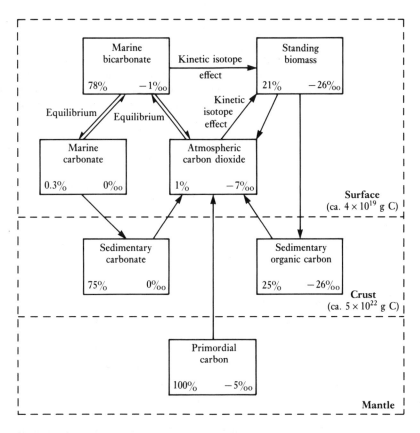

Figure 1.7   Carbon reservoirs and their exchange pathways within each of the three main compartments: surface, crust and mantle. Approximate relative sizes of reservoirs within each compartment are shown together with average $\delta^{13}C$ values. (Data after Schidlowski, 1988.)

Box 1.5   **Carbon isotopes**

Apart from a short-lived radionuclide, ^{14}C, carbon is a mixture of two stable isotopes, ^{12}C and ^{13}C. **Isotopes** are atoms of the same element which contain the same numbers of protons and electrons, so are chemically identical, but contain different numbers of neutrons, so their masses are different. Each element has an individual atomic number, equal to the number of electrons (or protons) in an atom (six for carbon). Electrons carry a unit negative charge but very little mass. The negative charge of the electrons in an atom is offset by an equal number of positively charged particles, protons, which have masses considerably greater than the electron. The protons exist in a nucleus, around which the electrons orbit. Also in the nucleus are uncharged particles called neutrons, with similar masses to the protons. Isotopes of an element differ in the number of neutrons in their nuclei and, therefore, in their atomic mass, which is the sum of the protons and neutrons (12 and 13 for the stable isotopes of carbon). In the Earth as a whole the relative abundances of ^{12}C and ^{13}C are 98.894% and 1.106%, respectively. Carbon compounds of biological origin are relatively enriched in the lighter isotope, while the heavier isotope is retained in the main forms of inorganic carbon (e.g. carbonate, bicarbonate and carbon dioxide). The lightness of biogenic substances results from isotope fractionation processes in the main assimilatory pathways and, to a lesser extent, in ensuing metabolic reactions. **Isotopic fractionation** occurs when the molecules of a particular compound react at different rates depending on their isotopic compositions, and so the isotopic composition of the reaction product is controlled by what can be termed a **kinetic isotope effect**.

The ratio of ^{13}C to ^{12}C in a geological sample is measured by mass spectrometry after converting the carbon to CO_2. However, it is difficult to measure the absolute amounts of $^{12}CO_2$ and $^{13}CO_2$ accurately and so the ratio in a sample is compared with that of a standard analysed at the same time. The isotopic composition of organic matter in a sample is normally expressed by δ values (with units of per mil, or $^o/_{oo}$) relative to the standard, which is usually PDB (a belemnite from the Cretaceous Peedee Formation):

$$\delta^{13}C^o/_{oo} = \left(\frac{(^{13}C/^{12}C \text{ in sample})}{(^{13}C/^{12}C \text{ in standard})} - 1 \right) \times 1000 \qquad [1.7]$$

For PDB $^{12}C/^{13}C = 88.99$ and its $\delta^{13}C$ value $= 0^o/_{oo}$ by definition, so negative values for a sample indicate depletion in the heavier isotope compared with PDB.

The bulk of autotrophic production of organic matter proceeds by the photosynthetic mechanism shown in Box 1.3, the C3 pathway. The contemporary biomass reflects the isotopic signature of photosynthesis by the C3 pathway of between ca. $-19^o/_{oo}$ and $-33^o/_{oo}$, with a mean of $-26^o/_{oo}$ ($-26^o/_{oo} \pm 7^o/_{oo}$). This isotopic lightness is a consequence of $^{12}CO_2$ being taken up at a faster rate than $^{13}CO_2$ during the first carboxylation reaction with ribulose diphosphate (RDP, Fig. 1.4). The range of $\delta^{13}C$ values about a mean of $-26^o/_{oo}$ reflects the effect of other equilibration and dissimilatory processes, which vary from species to species. All common photosynthetic routes discriminate against ^{13}C, mainly as a result of the kinetic isotope effect (see Box 1.5) in the first carboxylation reaction with RDP and also from the faster diffusion rate of carbon dioxide containing the lighter isotope into plant tissue. The excess ^{13}C is retained mainly as dissolved bicarbonate in seawater (Fig. 1.7).

The carbon isotope record in sediments of both carbonate ($+0.5\%_{oo} \pm 2.5\%_{oo}$) and organic matter (mainly $-26\%_{oo} \pm 7\%_{oo}$) appears fairly constant throughout the sedimentary record and is similar to that of today for marine bicarbonate ($-1\%_{oo}$) and the mean isotopic signature for the standing biomass ($-26\%_{oo}$). These values appear to reflect the remarkably consistent isotopic signal of autotrophic carbon fixation over at least the last 3500 Ma, suggesting that photosynthesis has remained quantitatively the most important process of CO_2 assimilation and has undergone little evolutionary change.

The residence time of bicarbonate ion in seawater is presently ca. 10^5 years. This means that any changes in the amount of carbon fixed by photosynthesis would be reflected in a change in the isotopic signature of carbon in bicarbonate in a short period of time in geological terms, which would in turn be preserved in sedimentary carbonate. Consequently, the constant isotopic signatures throughout the sedimentological record can be argued to imply that a steady state in the balance between organic and carbonate carbon was reached at an early stage in the evolution of life (Schidlowski, 1988). A corollary is that productivity rapidly reached a limiting level and then remained more or less constant. For the ancient Earth the limiting agent for primary production by the marine photosynthetic bacteria may have been the supply of essential mineral nutrients (see Section 3.2.2). The occurrence of stromatolites in the earliest sediments demonstrates the high levels of productivity achieved by cyanobacterial mats, which today can reach $8-12\,g\,C_{org}/m^2\,d$. It is possible, therefore, that photosynthesis improved little in effectiveness and quantitative importance since the first microbial communities became established.

It should not be thought that the carbon isotopic record has been entirely uniform throughout geological time. Some excursions have occurred, which can be attributed to natural perturbations and which were followed by a relatively rapid return to the mean isotopic levels described above. For example, unusually heavy carbonates (having $\delta^{13}C$ values $> +5\%_{oo}$) have been recorded for two periods (2700 and 1800–1900 Ma ago) in the Precambrian, coinciding with the two most powerful stages of magmatism and granitisation (Rhodesian and Belomorian) in the Earth's history (Galimov, 1976). The high tectonic activity caused the injection into the atmosphere and ocean of huge quantities of metamorphogenic carbon dioxide, relatively rich in ^{13}C, which was incorporated in biogenic carbonates until levels once again returned to their equilibrium values.

Chapter 2

Chemical composition of biogenic matter

2.1 Structure of natural products

2.1.1 Introduction

Having established the major contributors to sedimentary organic matter, it is important to consider their chemical composition in order to understand the processes that lead to the fossil forms of carbon preserved in ancient sediments. All organisms are composed of the same basic chemical classes, the most important of which, geochemically, are: carbohydrates, proteins and lipids. These three classes are discussed in Sections 2.2, 2.3 and 2.4, respectively. In addition, higher plants contain significant quantities of lignin, a major component of their supportive tissue, which is described in Section 2.5. Lipids are believed to play a dominant role in petroleum formation and so organisms relatively rich in lipids, such as the plankton (ca. 10% lipids, dry weight), are important contributors to oil source rocks.

In the following sections we shall review the composition of the main chemical classes and their biochemical functions. These functions are mainly controlled by the reactive (functional) groups (see Box 2.1) attached to the basic carbon skeleton and by the shape of the molecule, in which the way atoms bond together is an important factor.

Box 2.1

Geochemically important functional groups

An organic compound contains a basic skeleton of carbon atoms, which can be arranged as a simple straight chain, a branched chain, one or more rings, or even a combination of these. Many natural products involve cyclic structures, with rings of six carbon atoms being particularly common. The simplest organic compounds contain only hydrogen atoms bonded to the basic carbon skeleton and are called hydrocarbons. Atoms of other elements (**heteroatoms**) can be incorporated in the basic hydrocarbon structure, often in the form of peripheral **functional groups**. The most common heteroatoms in natural products are oxygen, nitrogen and sulphur. Each functional group confers specific properties on a compound, so that compounds are often considered collectively under the name of the functional group they contain, e.g. carboxylic acids for the compounds bearing a carboxyl group. The main types of functional groups encountered in natural products are listed in Table 2.1. More than one type of functional group may be present in a molecule.

Table 2.1   **Geochemically important functional groups**

Symbol*	Group name	Resulting compound name
R—OH	Hydroxyl	Alcohol (R = aliphatic group), Phenol (R = aromatic group)
—C=O | R	Carbonyl	Aldehyde (R = H), Ketone (R = aliphatic or aromatic group), Quinone (where C of CO group incorporated in aromatic ring)
—C=O | OH	Carboxyl	Carboxylic acid
—O—	Oxo	Ether
—NH$_2$	Amino	Amine
—C=O | NH$_2$	Amido	Amide
—SH	Thio	Thiol
(ring structure)	Y = CH$_2$ Indenyl Y = O Furanyl Y = NH Pyrryl Y = S Thiophenyl	Indene Furan Pyrrole Thiophene
(ring structure)	Y = CH Phenyl Y = N Pyridinyl	Benzene Pyridine
(ring structure)	Pyranyl	Pyran

*Where R groups are shown they are not part of the functional group, but are included to indicate the different types of compounds that can be formed.

2.1.2 Bonding in organic compounds

Atoms in organic molecules are held together by **covalent bonds**, which are formed by adjacent atoms sharing pairs of electrons. Single, double and even triple bonds can be formed, in which one, two and three electron pairs are shared, respectively, although triple bonds are rare among natural products. Compounds where all the carbon atoms are joined together by single bonds are called **aliphatic** (or **saturated**). Aliphatic hydrocarbons are termed **alkanes** and can be acyclic or cyclic. The simplest acyclic alkanes are straight–chain compounds, called the normal alkanes (*n*-alkanes).

Compounds that contain one or more pairs of adjacent C atoms joined together by a double bond (C=C) are termed **unsaturated**. An unsaturated aliphatic hydrocarbon is called an **alkene**. Particularly stable arrangements are formed where several C=C bonds are present and alternate with C—C bonds producing the pattern: $\sim$C—C=C—C=C—C=C$\sim$. This arrangement of the double bonds is described as **conjugated** and is frequently encountered in polyunsaturated natural products. Conjugation is possible in a ring of six carbon atoms, formally involving three double and three single bonds. Any compound containing a structure of this kind is termed **aromatic**. An aromatic compound is, therefore, a particular type of unsaturated species and possesses enhanced stability. The simplest aromatic compound is the hydrocarbon benzene (Table 2.1), but it is possible for a number of aromatic rings to be fused into polycyclic structures. The presence of an aliphatic or aromatic hydrocarbon group in a molecule can be represented by R (as in Table 2.1), where its actual identity is not important.

Another type of bond that can be important in the chemistry of natural products is the **hydrogen bond**. It is an electrostatic attraction and is weaker than a normal single covalent bond. It involves a slightly positively charged H atom, the charge resulting from partial withdrawal of electron density by an electronegative atom, usually O or N, covalently bonded directly to the H atom. This H atom is attracted to a second, slightly negatively charged, electronegative atom, which is usually in another molecule but sometimes in another part of the same molecule. Hydrogen bonding, therefore, can affect the shape adopted by molecules and their interactions with other molecules. Throughout this book hydrogen bonds are represented by dotted lines.

It is important to remember that, although compounds are usually represented as flat in diagrams for clarity, natural products are generally complex 3-D structures. The geometry of bonds at each carbon atom can be tetrahedral, trigonal or linear, depending on whether the atom is bonded to four, three or two substituents, respectively. This variation in bond geometry for carbon arises from the different spatial arrangements possible for the four bonds generally formed by carbon. If there are four single bonds a tetrahedral arrangement results; if there are two single and one double bond a trigonal geometry results; and if there are two double bonds or one triple and one single, a linear geometry arises. Conventions for representing the 3-D nature of structures are given in Box 2.2.

Box 2.2   **Structural representation of organic compounds**

Bonds between atoms are represented by lines, with the number of lines indicating the number of bonds. While it is possible to label each carbon and hydrogen atom by C and H, the overall structure of complex molecules is often clearer if H atoms are omitted and the carbon skeleton is shown as lines representing bonds between C atoms. For example, alternative representations of n-hexane and phenol are given in Fig. 2.1a. The presence of a carbon atom is inferred at each change in angle of the line drawing and at each end of the chain in n-hexane. Heteroatoms are represented by the normal chemical symbol, as in phenol (Fig. 2.1a). Sometimes the conjugated bond system of aromatic rings is represented by a circle, again as shown for phenol (Fig. 2.1a). This reflects the delocalisation of bonding that occurs in conjugated systems, such that bonds between all the C atoms involved are approximately identical and of intermediate strength between single and double bonds.

Box 2.2
(continued)

> Where the relative 3-D spatial arrangement of atoms in a molecule is important, symbols exist to represent bonds projecting above and below the plane of the paper. Thickened lines show bonds projecting above the paper, dotted lines show bonds projecting below it. It is sometimes more convenient in cyclic structures to show the relative orientation of a hydrogen atom rather than that of the adjacent C—C bonds. In the convention adopted in this book an open ring around the C atom to which the H atom is bonded indicates that the H atom lies below the plane of the paper, while a filled-in circle denotes that the H atom is above the paper plane. When representing a compound as a formula the corresponding configuration of H atoms can be expressed as α and β, respectively. An example is shown in Fig. 2.1b and the conventions are summarised in Table 2.2. It should be noted in the structure on the right side of Fig. 2.1b that it is not necessary to show the configuration for the hydrogen atom bonded to the bottom C atom in the left-hand ring. It must lie below the plane of the paper because the methyl group is shown projecting above the plane.

2.1.3   Stereoisomerism

The possibility for different spatial arrangements, or **configurations**, of groups on a C atom gives rise to the phenomenon of **stereoisomerism**. There are two forms of stereoisomerism, optical and diastereomeric.

Optical isomerism occurs where otherwise identical structures are mirror images of each other and are not superimposable. The simplest example is a carbon atom bonded to four different atoms or groups. It is possible to draw two versions of the tetrahedrally arranged groups, which are non–superimposable mirror images, e.g. ($+$)- and ($-$)-glyceraldehyde in Fig. 2.2a. Such stereoisomers are called **enantiomers** and most of their physical properties are identical but they rotate the plane of polarised light in opposite directions. The carbon atom around which the four different groups are arranged is termed **chiral**. Chiral carbons can exist as part of a cyclic system, as in carvone (Fig. 2.2a). At first sight it may appear that the

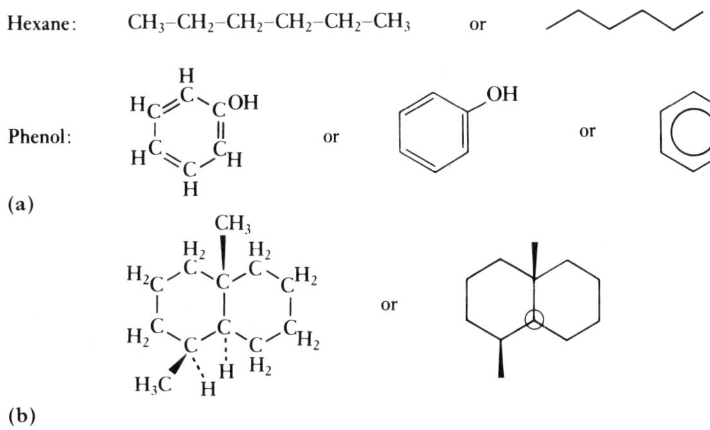

Figure 2.1   Representation of (a) carbon skeletons in molecules and (b) spatial arrangement of atoms and groups.

Table 2.2   **Symbols representing the spatial arrangement of bonds**

❚	C—C bond above plane of page	●	C—H bond above plane of page (β(H) configuration)
⋮	C—C bond below plane of page	○	C—H bond below plane of page (α(H) configuration)
│	C—C bond in plane of page	}	Unspecified C—C bond configuration (either possibility exists)

indicated chiral carbon is not surrounded by four different groups because it is bonded to two identical CH_2 units in the cyclic system. However, proceeding around this cycle in the clockwise direction we find a —CH_2— unit followed by a —CH≡, while in the anticlockwise direction a —CH_2— unit is followed by —CO—. The cyclic system, then, does not appear the same when viewed from each direction of attachment to the chiral carbon and, effectively, this atom is surrounded by four different groupings.

Natural products can have several chiral carbons. If there are n chiral carbons, there are 2^n possible stereoisomers, comprising 2^{n-1} pairs of enantiomers. In any particular pair of enantiomers the configuration at each chiral atom in one isomer is opposite to that of the corresponding atom in the other isomer. An enantiomer that rotates the plane of polarised light in a clockwise direction is termed dextrorotatory and is labelled (+). The other enantiomer is laevorotatory, is labelled (−) and rotates plane-polarised light by the same amount in an anticlockwise direction. This property provides a means of identifying relative configuration at a chiral atom. The true spatial orientation of groups at each chiral carbon in a molecule, the **absolute configuration**, can be represented by the labels R and S. An older system with D and L labels is commonly used to distinguish between enantiomers of amino acids and sugars (see Box 2.3). A 50:50 mixture of the enantiometers of a compound is optically inactive and is termed **racemic**.

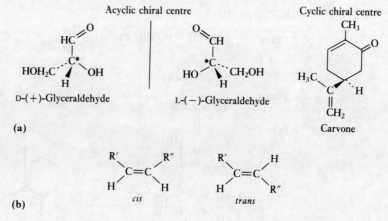

(a)

(b)

Figure 2.2   Examples of stereoisomerism: (a) optical isomerism involving acyclic and cyclic carbon atoms (∗ indicates chiral centre); (b) *cis* and *trans* diastereomers (formerly termed geometric isomers) in unsaturated compounds (R′ and R″ = aliphatic or aromatic groups).

Box 2.3

Systems for denoting absolute stereochemical configuration

Prior to the advent of X-ray crystallography the absolute configuration of chiral molecules was not known. However, it was still possible to compare the absolute configuration of different molecules by relating them to a reference chiral compound, glyceraldehyde (see Fig. 2.2a), which is biosynthetically related to a variety of other natural products. The absolute configuration of the (+) enantiomer of glyceraldehyde was arbitrarily assigned the D label, and that of the (−) enantiomer the L label (see Fig. 2.2b). A compound that can be converted into D-glyceraldehyde by reactions that conserve the configuration of the chiral carbon in the compound is also given the D label, while one that can be converted to L-glyceraldehyde is given the L label. It is important to remember that the D and L labels relate to the absolute configuration and not the direction of rotation of plane-polarised light. Although the D enantiomer of glyceraldehyde is dextrorotatory, for another compound the D enantiomer may be laevorotatory. For monosaccharides a single D/L label is used, denoting configuration at the highest-numbered chiral carbon (e.g. C-5 in aldohexoses).

A more recent convention for representing absolute configuration is the R and S system. It is based on a series of rules effectively concerning atomic mass priority among the substituents on a chiral carbon, which can be found in any basic organic chemistry text (e.g. Finar, 1975; Vollhardt, 1987; Solomons, 1988). In determining which assignment is given, the chiral carbon is viewed with the bond to the lowest priority group, usually a hydrogen atom in the compounds we are concerned with, pointing away from the viewer. The other three bonds then appear to form a trigonal arrangement. If the mass priority of these groups decreases in a clockwise direction the R configuration is assigned. The S configuration is given to an anticlockwise decrease in mass priority. An example is given in Fig. 2.3.

The R and S convention can be used for all chiral carbons, but sometimes the $\alpha\beta$ convention is used, particularly for atoms that are part of a cyclic system. In the latter convention α indicates that the labelled atom or group projects below the plane of the paper when the structure is drawn and β indicates that it projects above (as noted for H atoms in Box 2.2). The principles of the RS system can also be applied to describing the configuration about a C=C bond (formerly termed geometric isomerism), but two new symbols, E and Z, are substituted. Where the groups with highest atomic mass priority lie on the same side of the C=C bond (e.g. the *cis* isomer in Fig. 2.2a) the Z label is used. This system is most useful when either or both of the C atoms involved in the C=C bond have two alkyl substituents because the *cis* and *trans* system then becomes difficult to apply.

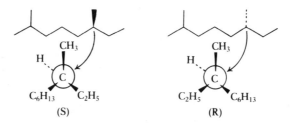

Figure 2.3   R and S configurational assignment in 2,6–dimethyloctane.

All stereoisomers that are not enantiomers are termed **diastereomers**. For example, in a molecule containing two or more chiral centres any pair of stereoisomers that are not enantiomers are diastereomers. In other words, a pair of diastereomers share the same configuration at at least one, but not all, of their chiral atoms and, unlike enantiomers, they have different physical properties. Two diastereomers that differ in configuration at only one of a number of chiral atoms are commonly called **epimers**.

Diastereomers are also encountered in unsaturated acyclic compounds. All the bonds (one double and four single) to the two C atoms joined together by a double bond lie in the same plane as the C=C bond. If each of these two carbon atoms is bonded to a H atom and a hydrocarbon (alkyl) chain, the alkyl chains can either be on the same side of the C=C bond as each other or on opposite sides and the resulting diastereomers (which used to be known as **geometric isomers**), shown in Fig. 2.2b, are termed *cis* and *trans*, respectively. Again, these diastereomers have different physical properties (see also Box 2.3). Optical isomerism is not possible for a C=C bond (configurations that are mirror images are also superimposable and are, therefore, identical).

While complex natural products may have a number of chiral centres, they tend to exhibit only a limited number of stereoisomeric variations (e.g. see Finar, 1975). This is because biosynthesis results in the selective formation of one configuration at many of the stereoisomeric centres, as will be seen in subsequent sections. The overall shape of a molecule can be very important in controlling its biochemical functions.

2.2 Carbohydrates

2.2.1 Composition

The term **carbohydrate** derives from the fact that many members of this group of compounds have the general formula $C_n(H_2O)_n$, i.e. they contain only carbon, hydrogen and oxygen, with H and O being in the same ratio as in water. They are polyhydroxy-substituted carbonyl (i.e. aldehyde or ketone) compounds. The simplest molecules are **monosaccharides**, which are named according to the number of carbon atoms present; e.g. tetroses, pentoses, hexoses and heptoses contain four, five, six and seven carbons, respectively. Both aldehyde and ketone derivatives of these units exist, called respectively aldoses and ketoses. Hence a C_6 monosaccharide may be an aldohexose (e.g. glucose) or a ketohexose (e.g. fructose). Most naturally occurring monosaccharides are hexoses or pentoses, some examples of which are given in Fig. 2.4a. They mainly exist as cyclic systems, with those forming five-membered rings being called furanoses and those forming six-membered rings pyranoses after the simplest parent compounds furan and pyran (Table 2.1). Another way of representing these cyclic systems is by the Howarth structure, which is shown for glucose in Fig. 2.5.

Some modified monosaccharides exist, such as D-glucosamine and D-galactos-amine, in which the hydroxy group at C-2 has been replaced by an amino group (Fig. 2.4a). Other modifications include deoxygenation, e.g. the absence of oxygen

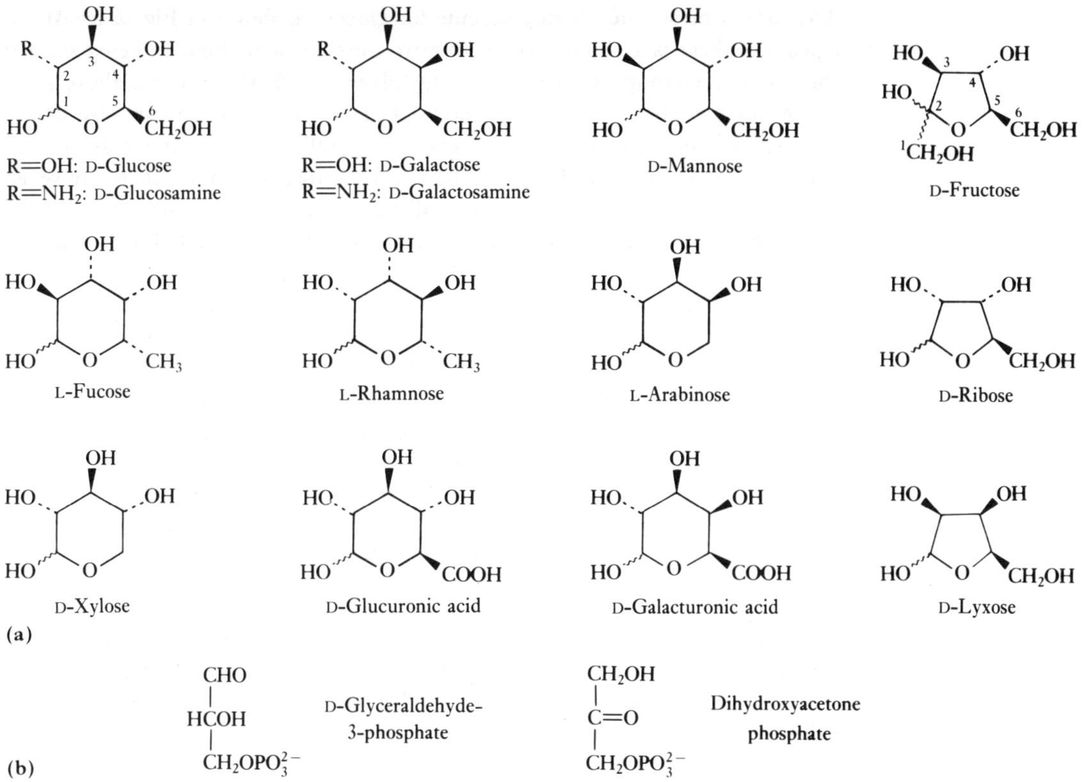

Figure 2.4   (a) Geochemically important monosaccharides (carbon numbering convention shown for D-glucose and D-fructose) and (b) their biological precursors.

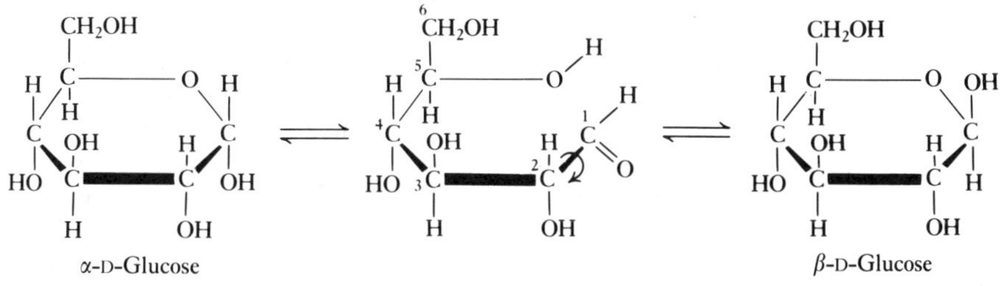

Figure 2.5   Anomerism of glucose (arrow in intermediate structure denotes bond rotation).

at C-6, as in L-rhamnose and L-fucose (Fig. 2.4a; which can also be called 6-deoxy-L-mannose and 6-deoxy-L-galactose, respectively), and the presence of carboxylic acid groups, giving rise to the **uronic acids** (e.g. glucuronic and galacturonic acids, Fig. 2.4a).

We have already considered the photosynthetic formation of carbohydrates (Box 1.2) and have seen that the basic building block is a C_3 compound, a triose phosphate. This unit is actually either D-glyceraldehyde-3-phosphate or dihydroxyacetone phosphate (Fig. 2.4b), the latter being formed from the former by enzymatic isomerisation.

The formal carbon numbering scheme for glucose is shown in Fig. 2.4a. At the C-1 position there is a chiral carbon, the two possible stereoisomers being termed α where the OH group lies below the ring plane and β where it lies above. The other four carbon atoms in the ring system of glucose (C-2 to C-5) are also chiral, giving rise to 2^4 or 16 possible stereoisomers. Naturally occurring monosaccharides have a fixed configuration at C-5 (usually D). Therefore, in reality, there are seven diastereomers of glucose, each having a different name and possessing α and β stereoisomers at C-1. Isomerism at C-1 is discussed in further detail in Box 2.4.

Box 2.4

Ring formation in monosaccharides

In solution the α and β isomers of monosaccharides interconvert to give equilibrium mixtures of both isomers. This results from ring opening by cleavage of the bond between the O and C-1 atoms in the ring, yielding an aldehyde group for the aldoses. In the open-chain form C-1 of the free aldehyde group is no longer chiral. Rotation about the bond joining the aldehyde group to the rest of the aldose allows either isomer, referred to as **anomers**, to be formed upon ring closure. This process of anomerism (or mutarotation) is shown for glucose in Fig. 2.5. Anomerism occurs when polysaccharides are broken down by hydrolysis into monosaccharide units (e.g. during compositional analysis). Configurational specificity is unavoidably lost at C-1, the carbon linking monosaccharide units together. In polysaccharides only the cyclic (pyranose and furanose) forms are present.

The fact that glucose is an aldehyde derivative is more clearly seen in the intermediate open-chain form than in the cyclic structures (Fig. 2.5). The reaction of the aldehyde group with the alcohol group at C-5 is a typical reaction of a carbonyl with an alcohol. Reaction with the C-5 OH group is favoured as a six-membered ring is formed, imposing minimal strain on the bonds forming the ring. In aqueous solution <1% of D-glucose exists in the intermediate open-chain form, while the most abundant form is the β anomer (63%), which has the more stable orientation for the C-1 OH group.

The forms of glucose in aqueous solution are quite complex and small proportions exist as furanose structures. This arises from the reaction between the OH group at C-4 and the aldehyde group. The five-membered ring system is quite stable but pyranose forms are favoured over furanose forms by most monosaccharides; notable exceptions are the ketohexose fructose and the aldopentose ribose. Too much strain would be involved in reducing bond angles to produce smaller rings, so trioses and tetroses do not form intermolecular cyclic systems.

A final consideration in describing the structure of a monosaccharide is the spatial arrangement of atoms in the ring system, the **conformation**. For minimum bond strain the tetrahedral arrangement of bonds about the carbon atoms in the ring must be maintained, which dictates a non-planar structure. Most aldopyranoses adopt a **chair** conformation, the most stable form being that where as many as possible of the bulkier OH and CH_2OH groups are **equatorial** (i.e. bonds to them are parallel to the plane of the main part of the ring system, as shown for glucose in Fig. 2.6) rather than **axial** (bonds perpendicular to the main ring plane). A corresponding 'envelope' conformation is the most stable for furanoses (as for fructose in Fig. 2.6).

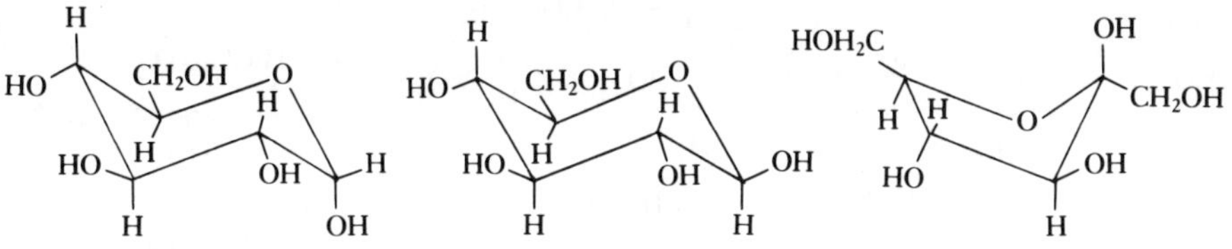

| α-D-Glucopyranose | β-D-Glucopyranose | β-D-Fructofuranose |

Figure 2.6   Stable conformations adopted by cyclic monosaccharides: chair form for six-membered rings (e.g. glucose) and envelope form for five-membered rings (e.g. fructose). Each carbon atom in the cyclic system possesses one axial and one equatorial substituent; the axial bonds are represented by vertical lines. All OH groups and the CH_2OH group are equatorial in β-D-glucose.

Monosaccharide units can be linked together by **condensation** (a reaction that joins together two molecules with the elimination of a simple molecule, water in this instance). The resulting linkage is called the **glycosidic bond**, and condensation usually involves OH groups on C-1 and C-4. Condensation of two monosaccharides yields a **disaccharide**, such as sucrose, which contains a glucose and a fructose unit (Fig. 2.7). Further units can be linked to give tri- and tetrasaccharides, etc. Those formed from two to ten monosaccharide units are generally termed **oligosaccharides**, while those with more units, such as cellulose (Fig. 2.7), are called **polysaccharides**. The mono- and disaccharides are commonly termed **sugars**.

The number of possible monosaccharide units and their order and orientation in polysaccharides leads to an immense variety of possible structures. For example, amylose and cellulose are both glucose polymers but differ in the configuration of the bridging C-1 atom (Fig. 2.7). Branching of the polysaccharide chain may also occur, as in amylopectin (Fig. 2.7), leading to further structural variety. Polysaccharides formed from only one type of monosaccharide, like cellulose and amylose, are termed **homopolysaccharides**, while those formed from different types of monosaccharides are **heteropolysaccharides**.

2.2.2   Occurrence and function

Carbohydrates function as food reserves, structural material and antidesiccants. Polysaccharides are major components in most **cell walls**, which provide a rigid, reinforcing layer around the cell membranes in plants, bacteria and fungi. D-glucose is by far the most abundant monosaccharide. It is important as an energy source and it is the basic unit of the polysaccharide **cellulose**, the main structural building material of plants, a molecule of which contains ca. 10 000 glucose units. Cellulose is the most abundant natural organic compound, with higher plants containing the largest amounts while some algae appear to have none.

As an energy reserve D-glucose is stored in the form of polysaccharides: starch in plants and glycogen in animals. Utilisation of these reserves initially involves breaking down the polysaccharides into glucose units which then undergo **glycolysis**, a respiration process that yields pyruvic acid and some energy. Glycolysis is common to all organisms, aerobes and anaerobes alike. In aerobes the pyruvic acid is converted into acetyl coenzyme A and then enters the **Krebs cycle**, in which oxidation produces

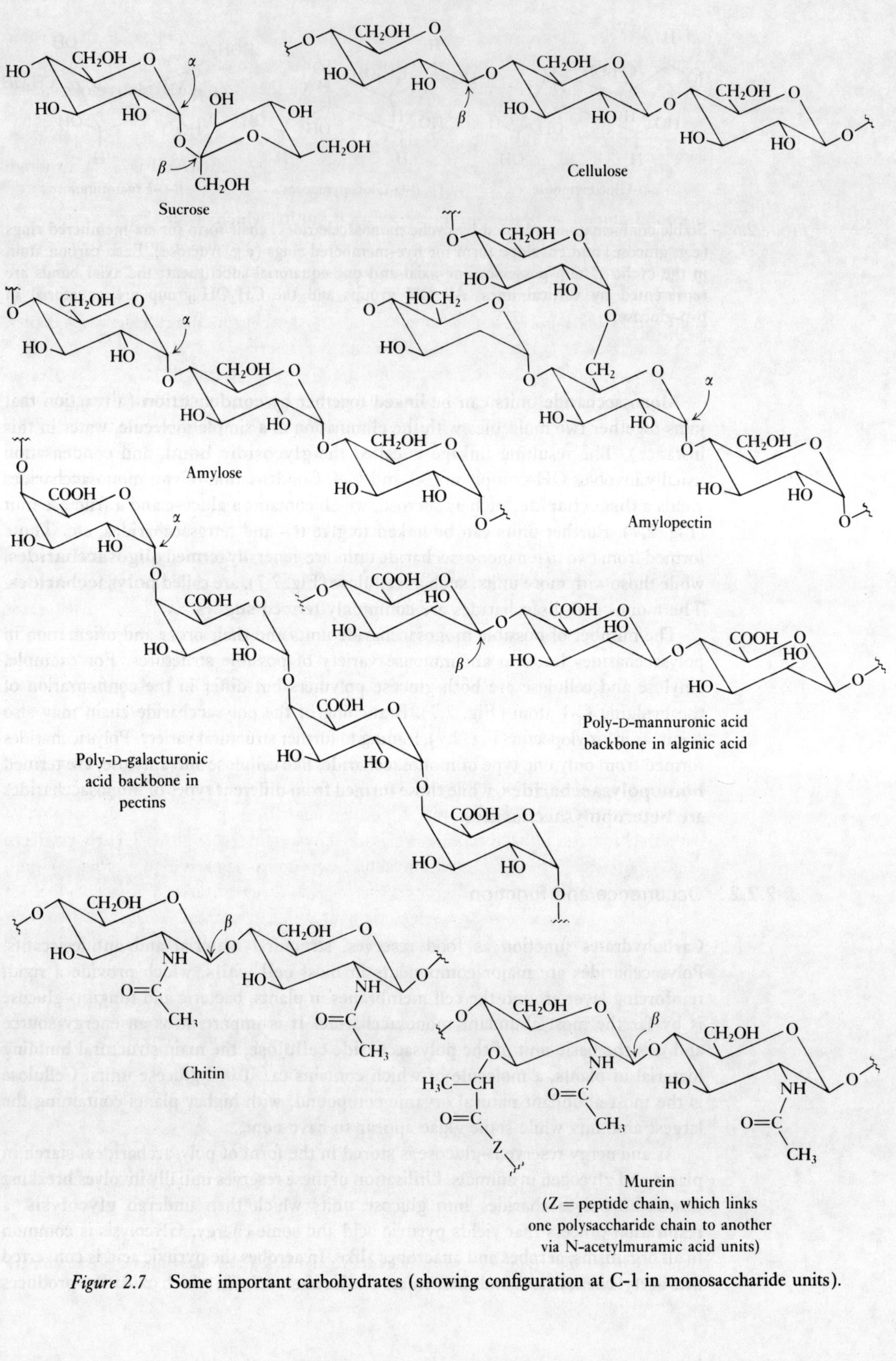

CH₂OH

Sucrose

Cellulose

Amylose

Amylopectin

Poly-D-galacturonic
acid backbone in
pectins

Poly-D-mannuronic acid
backbone in alginic acid

Chitin

Murein
(Z = peptide chain, which links
one polysaccharide chain to another
via N-acetylmuramic acid units)

Figure 2.7   **Some important carbohydrates (showing configuration at C-1 in monosaccharide units).**

CO_2 with the release of energy (in the form of ATP). A summary of the aerobic respiration of carbohydrates is given in Fig. 2.8. Although glycolysis contributes only ca. 5% of the energy available to aerobes from the complete oxidation of glucose, it is an important source of energy for some anaerobes (anaerobic respiration is considered in more detail in Section 3.3.2).

Starch normally comprises 80% amylopectin and 20% amylose (Fig. 2.7), although compositional variations exist. Its structure, therefore, differs from that of cellulose in containing a C-6 branch in the amylopectin units, which occur about every 20–25 glucose units, and in the configuration at C-1. Glycogen has a similar structure to amylopectin but with more frequent branching.

D-fructose is the most important ketose and is found in all living plant tissue. In the form of homopolysaccharides it serves as a short-term energy reserve in plants, and similar homopolysaccharides are found in bacteria.

After cellulose the next most abundant group of carbohydrates is the **hemicelluloses**, compounds that form a matrix surrounding the cellulose fibres in plant cell walls. They are a complex mixture of polysaccharides, mainly containing 50–2000 monosaccharide units, the most abundant of which are L-arabinose, D-galactose and D-xylose, with lesser amounts of others (e.g. D-mannose). Hemicelluloses contain some homopolysaccharides but heteropolysaccharides predominate. Also present in higher plant primary (i.e. non-woody) cell walls and in intercellular layers are pectins. They are present in only minor amounts in woody (i.e. secondary cell wall) tissues but are more abundant in fruits. **Pectins** are complex mixtures of mainly heteropolysaccharides: polygalacturonic acid structures predominate (Fig. 2.7) with lesser amounts of various other monosaccharides.

Cellulose is replaced as a structural material in most fungi, some algae, arthropods (e.g. insects and crustaceans) and molluscs by **chitin** (Fig. 2.7), a homopolysaccharide of N-acetyl-D-glucosamine. All eubacterial cell walls contain **murein**, which comprises polysaccharide chains of alternating N-acetyl-D-glucosamine and N-acetylmuramic acid units that are cross-linked by chains of amino acids (Fig. 2.7). Because of its formation from peptide and carbohydrate units this material is often referred to as a peptidoglycan, and it can account for up to 75% (dry weight) of bacterial biomass. Eubacteria can be classified by a stain test as either **Gram-positive** (e.g. *Clostridium, Bacillus*, actinomycetes) or **Gram-negative** (e.g. *Pseudomonas, Methylomonas*), reflecting differences in cell wall architecture. A major difference is the presence in Gram-negative bacteria of an outer membrane, covering the murein layer, in which **lipopolysaccharides** (compounds in which lipids are bound to

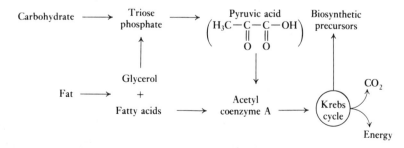

Figure 2.8   Summary of aerobic respiration of carbohydrates and fats. (The biosynthetic precursors are used in the synthesis of fatty acids, amino acids, terpenoids, etc.)

polysaccharides, see Section 2.4.1) are major constituents. Again, N-acetyl-D-glucosamine is an important unit in these polysaccharides. Gram-positive bacteria have a capsule of polysaccharides outside their cell walls in which D-glucose and D-galactose are major components.

Archaebacterial cell wall composition can vary markedly. The cell walls of some species contain only protein, while others comprise glycoprotein, polysaccharide or a type of peptidoglycan, but murein is never present. The Gram stain test can also be applied to archaebacteria (e.g. the *Halobacterium* are Gram-negative).

Fungal polysaccharides are mainly homopolysaccharides of D-glucose, D-galactose and D-mannose. D-glucose and D-galactose are also important constituents of algal polysaccharides, such as agar. The algae provide examples of how certain carbohydrates can be characteristic of, although not exclusive to, various groups of organisms. Among marine algae the Chlorophyta have relatively large amounts of L-rhamnose, the Rhodophyta contain abundant D-galactose, while the Phaeophyta are rich in D-ribose and some species contain up to 40% (dry weight) of alginic acid, an important component of which is D-mannuronic acid (Fig. 2.7). In contrast, freshwater algae and higher aquatic plants generally contain significant amounts of L-arabinose and D-xylose.

Carbohydrates are important in the production of fats and proteins. As well as generating energy the Krebs cycle (Fig. 2.8) provides precursors for the biosynthesis of fatty acids, amino acids and terpenoids, which are discussed in the following sections.

2.3 Proteins

2.3.1 Composition

Proteins account for most of the nitrogen present in organisms. They are polymers (polypeptides) of **α-amino acids** (i.e. the amino (NH_2) and carboxylic acid (COOH) groups are attached to the same carbon atom). In the general structure in Fig. 2.9a the α-carbon atom can be seen to be chiral for all amino acids but glycine (in which the C atom is bonded to two H atoms and so is not chiral). This reflects the stereospecificity of enzymes involved in the formation and utilisation of these compounds. In fact, proteins only contain L-amino acids, while D-amino acids (e.g. D-alanine and D-glutamic acid) are found in the peptidoglycans of bacterial cell walls. The existence of acid and amine groups in the same molecule leads to zwitterion formation, i.e. the transfer of the acidic proton to the amine group (Fig. 2.9a). The zwitterion ammonium carboxylate form is naturally adopted by amino acids, although in solution both forms exist in equilibrium.

The different types of amino acids can be broadly classified as neutral, acidic or basic (see Box 2.5), depending on whether the number of carboxylic acid groups in a molecule equals, is greater than, or is less than the number of amine groups respectively. Proteins are made up from some 20 different amino acids, shown in Fig. 2.9b. Sulphur is an important component in some amino acids, e.g. cysteine. In plants amino acids are generally synthesised from glutamic acid (Fig. 2.9b) by transfer of the amino group to other carbon skeletons (**transamination**). Animals cannot synthesise all the amino acids they need for protein formation and so must obtain these essential amino acids (there are nine for humans) directly or indirectly from plants.

Box 2.5

Acids and bases

In the context of organic geochemistry the most appropriate definition of an **acid** is a proton donor, i.e. it contains a hydrogen atom that can be transferred to another chemical species as H^+ (a proton, which is a hydrogen atom less an electron). In addition to traditional inorganic acids, e.g. H_2SO_4, certain organic compounds contain groups that possess acidic hydrogens, such as the —COOH group of carboxylic acids. **Bases** are compounds that can accept a proton, and include amines. This behaviour of acids and bases is demonstrated in Eqs. [2.1] and [2.2]:

$$CH_3COOH \longrightarrow CH_3COO^- + H^+ \qquad [2.1]$$
$$\text{acid}$$

$$C_2H_5NH_2 + H^+ \longrightarrow C_2H_5NH_3^+ \qquad [2.2]$$
$$\text{base}$$

A carboxylic acid can react with an amine to form an amide, and water is liberated:

$$CH_3COOH + C_2H_5NH_2 \longrightarrow CH_3CONHC_2H_5 + H_2O \qquad [2.3]$$
$$\text{acetic acid} \qquad \text{ethylamine} \qquad \text{N-ethylacetamide} \qquad \text{water}$$

An important reaction in natural product chemistry is the reaction of organic acids with alcohols to form esters, like ethyl acetate, with the liberation of water:

$$CH_3COOH + C_2H_5OH \longrightarrow CH_3COOC_2H_5 + H_2O \qquad [2.4]$$
$$\text{acetic acid} \qquad \text{ethanol} \qquad \text{ethyl acetate} \qquad \text{water}$$

The reverse of reactions [2.3] and [2.4] is achieved by **hydrolysis**, so named because it involves the addition of water. Hydrolysis reactions can be speeded up (catalysed, see Box 2.6) by the presence of an inorganic acid or water-soluble inorganic base (i.e. the alkalis NaOH and KOH). The alkaline hydrolysis of fatty esters is often termed **saponification**.

An acid group on one amino acid molecule can undergo a condensation reaction with an amine group on another, with the elimination of water. The resulting amide group (see Box 2.5) that joins the two amino acids together to form a dipeptide is usually called the **peptide linkage**:

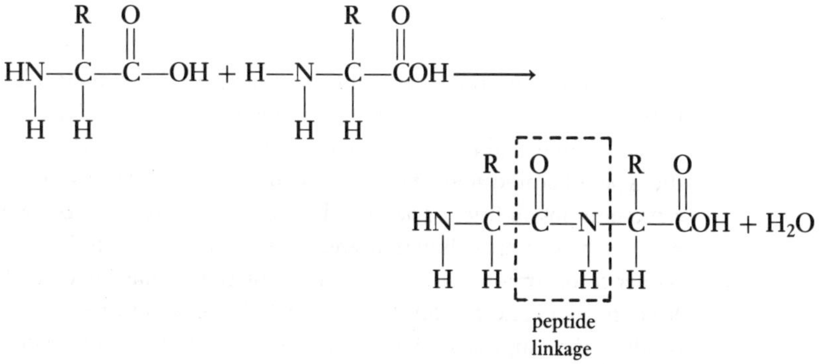

peptide
linkage

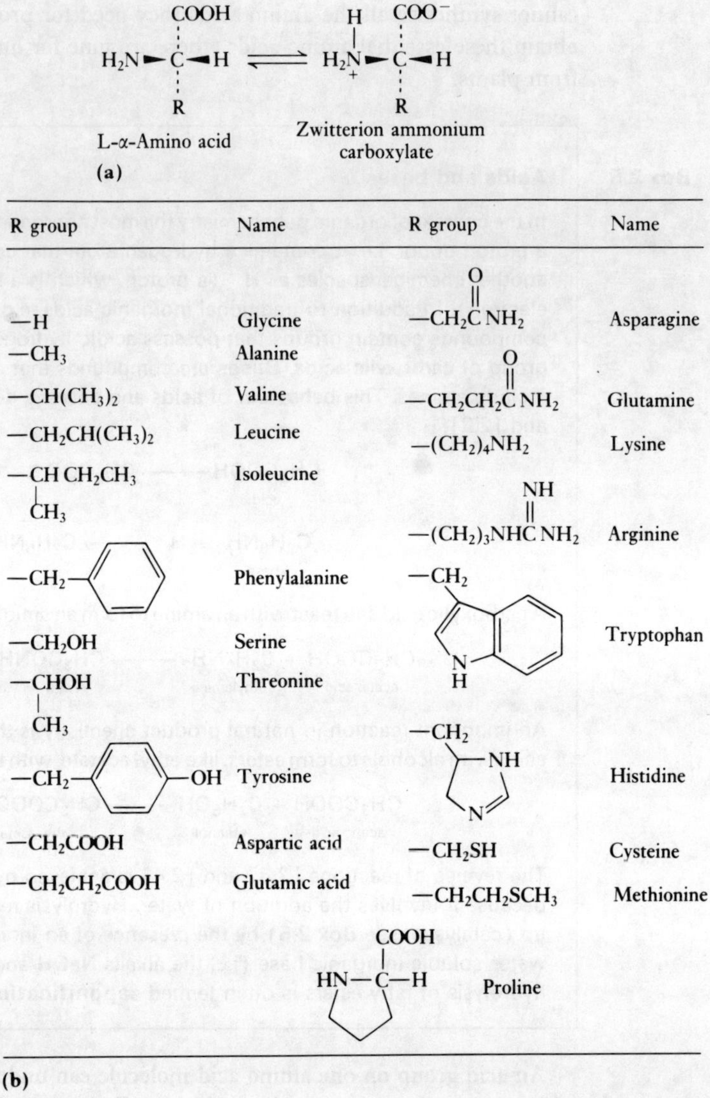

Figure 2.9   (a) General structure of L-α-amino acids and (b) important members of this compound class (R refers to group in general structure, but for proline the complete structure is shown).

Condensation reactions can continue until very large molecules, **polypeptides** (or polyamides), have been built up. Proteins are large polypeptides and can contain > 8000 amino acid units and have molecular weights $> 10^6$. In such large molecules the types of amino acids incorporated and their order leads to a multitude of different structural possibilities. The overall shape of a protein molecule is inextricably linked with its biochemical function and is governed by a number of factors, such as the sequence of amino acids, the rigidity of the amide bond, hydrogen bonding (see Section 2.1.2) and the formation of S—S bonds between the SH groups of cysteine residues that approach close enough for this bonding to occur.

2.3.2   Occurrence and function

Proteins can be a sizable fraction of the bulk organic material in an organism. The polypeptide chains can be folded into regularly repeating structures allowing fibres to be formed. These fibrous proteins serve as supportive tissues in animals, e.g. in skin and bone (collagen), in hooves and claws (keratin), in silk and in sponge. This is in contrast to plants, where cellulose and lignin (see Section 2.5) perform the structural role, although they can be associated with a collagen-like protein. Many fibrous proteins adopt an α-helix structure, stabilised by hydrogen bonding, which allows them to be packed together in bundles. There are also globular proteins, which perform many important functions and include enzymes (the biochemical catalysts (see Box 2.6)), hormones (which regulate metabolism, e.g. insulin), antibodies (which are glycoproteins, i.e. they contain carbohydrates) and transport and storage units (e.g. haemoglobin for oxygen transfer and cytochrome-c for electron transfer).

Box 2.6

Catalysis

A **catalyst** is a substance that increases the rate at which a reaction occurs but is not itself consumed by the reaction. We can represent the energy changes taking place during the course of a simple reaction by Fig. 2.10. Two substances that are to react must be brought close together and certain bonds need to be weakened to allow the transfer of atoms and formation of new bonds that lead to the generation of products. This requires energy to be supplied. As we move along the reaction path in Fig. 2.10 we reach a maximum in the energy curve; thereafter energy decreases as products are formed and move apart. Most reactions result in the overall evolution of excess energy (ΔH) as heat, as in Fig. 2.10. The initial amount of energy needed for the reaction to proceed, the **activation energy** (E_{act}), can be reduced by a catalyst. The catalyst partially bonds to the reactants, weakening one or more of the bonds that have to be broken, which means less external energy is required. When the products are formed the catalyst reverts to its original state. The overall energy change in the reaction (ΔH) is not affected by the action of the catalyst. Although there is no net energy gain, the benefits of a catalyst are that the reaction can proceed faster and less initial energy (in the form of ATP in cells) is required to supply the necessary activation energy.

Catalysts are used throughout the chemical industry and **enzymes** are the cell's catalysts. In addition to reducing the energy needed to break bonds, enzymes also act as templates to hold the reactants in the correct orientation for the desired reaction. This is particularly important for complex polymers where a specific reaction at one point only is involved. Mineral surfaces can also behave as catalysts in geochemical reactions.

While proteins comprise the bulk of nitrogen-containing organic material in organisms, free amino acids and peptides are found, but in lesser amounts. There are other nitrogen compounds, present in low amounts relative to proteins, which perform essential cell functions like reproduction. Biosynthesis of proteins is controlled by nucleic acids (RNA and DNA), which comprise **nucleotide** units made up of a phosphate, a pentose sugar and a nitrogen-containing organic base.

Non-catalysed reaction

Energy

E_{act}

Reactants

ΔH

Products

Reaction path

Catalysed reaction

Energy

Catalyst surface

E_{act}

Reactants

ΔH

Products

Reaction path

Figure 2.10   Effects of a catalyst on the energetics of a reaction.

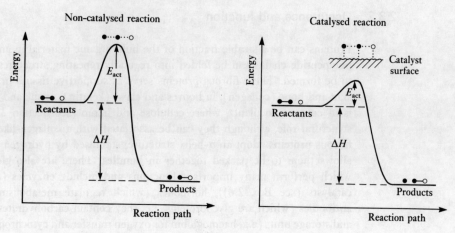

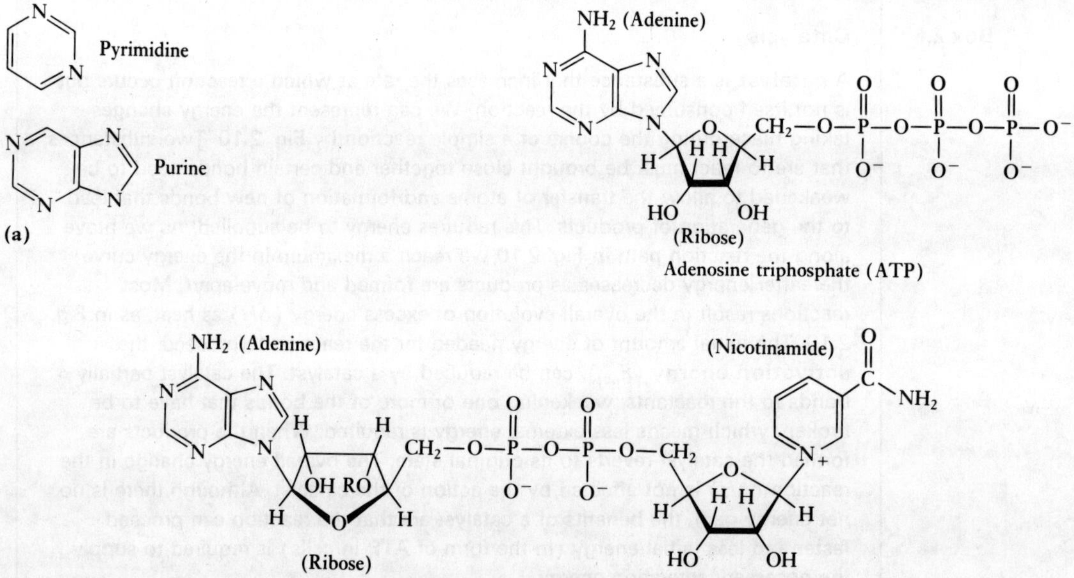

Pyrimidine

Purine

(a)

NH_2 (Adenine)

Adenosine triphosphate (ATP)

NH_2 (Adenine)

(Ribose)

(Nicotinamide)

(Ribose)

R=H: Nicotinamide adenine dinucleotide (NAD)
R=PO_3^{2-} : Nicotinamide adenine dinucleotide phosphate (NADP)

NH_2

(b)

Coenzyme A

Figure 2.11   (**a**) Parent nitrogen-containing bases for nucleotides and (**b**) structures of ATP, NAD, NADP and coenzyme A.

The base is either a purine or pyrimidine derivative (Fig. 2.11a), generally containing additional amino or carbonyl groups. We have already met two important nucleotides, ATP and NADP (Fig. 2.11b), which contain the purine derivative adenine and are involved in many biochemical reactions in addition to photosynthesis. Another important nucleotide, closely related to NADP and used by all organisms in oxidation and reduction processes, is NAD (Fig. 2.11b). Adenine forms part of coenzyme A, another important compound in many biochemical reactions (Fig. 2.11b).

2.4 Lipids

Lipids can be defined as all the substances produced by organisms that are effectively insoluble in water but extractable by solvents that dissolve fats (e.g. chloroform, hexane, toluene and acetone). This broad definition is suitable for our purposes and encompasses a wide variety of compound classes, including photosynthetic pigments. However, application of the term lipid can vary, sometimes being restricted to fats, waxes, steroids and phospholipids, and sometimes to fats alone. Simple organic compounds like aliphatic carboxylic acids and alcohols can be found among the lipids, but most lipids exist as combinations of these simple molecules with one another (e.g. wax esters, triglycerides, steryl esters and phospholipids) or with other compound classes such as carbohydrates (glycolipids) and proteins (lipoproteins). We shall consider the quantitatively and geochemically most important classes of these lipids.

2.4.1 Glycerides

Glycerides are esters of the alcohol glycerol. A glycerol molecule contains three hydroxyl groups and so it can react with one, two or three carboxylic acid molecules, forming mono-, di- and triglycerides. Among the important types of glycerides are the fats and phospholipids.

Fats   **Fats** are triglycerides, formed from straight-chain, aliphatic carboxylic acids, called **fatty acids**. Each of the fatty acids in a triglyceride molecule can be different:

$$
\begin{array}{l}
\mathrm{H_2C{-}OH} \quad \mathrm{RCOOH} \\
\,\mid \\
\mathrm{HC{-}OH} + \mathrm{R'COOH} \\
\,\mid \\
\mathrm{H_2C{-}OH} \quad \mathrm{R''COOH} \\
\text{glycerol} \qquad \text{fatty acids}
\end{array}
\longrightarrow
\begin{array}{l}
\mathrm{H_2C{-}OOCR} \\
\,\mid \\
\mathrm{HC{-}OOCR'} + 3\mathrm{H_2O} \\
\,\mid \\
\mathrm{H_2C{-}OOCR''} \\
\text{triglyceride} \qquad \text{water}
\end{array}
\qquad [2.6]
$$

The fatty acids are typically of C_{12} to C_{36} chain length and in animals they are saturated (called alkanoic acids), while in plants unsaturated (alkenoic acids) and often polyunsaturated acids predominate. For the same chain length an unsaturated acid has a lower melting point than a saturated fatty acid and so, at ambient temperatures, plant-derived fats are often oils while animal fats are solids. In animals C_{16} and C_{18} saturated fatty acids predominate, while the major fatty acids in plants

Table 2.3 **Common fatty acids**

Common name	Systematic name	Structure
Lauric acid	*n*-Dodecanoic acid	$CH_3(CH_2)_{10}COOH$
Myristic acid	*n*-Tetradecanoic acid	$CH_3(CH_2)_{12}COOH$
Palmitic acid	*n*-Hexadecanoic acid	$CH_3(CH_2)_{14}COOH$
Stearic acid	*n*-Octadecanoic acid	$CH_3(CH_2)_{16}COOH$
Arachidic acid	*n*-Eicosanoic acid	$CH_3(CH_2)_{18}COOH$
Palmitoleic acid	*n*-Hexadec-9-enoic acid	$CH_3(CH_2)_5CH{=}CH(CH_2)_7COOH$
Oleic acid	*n*-Octadec-9-enoic acid	$CH_3(CH_2)_7CH{=}CH(CH_2)_7COOH$
Linoleic acid	*n*-Octadec-9,12-dienoic acid	$CH_3(CH_2)_4CH{=}CHCH_2CH{=}CH(CH_2)_7COOH$
Linolenic acid	*n*-Octadec-9,12,15-trienoic acid	$CH_3CH_2CH{=}CHCH_2CH{=}CHCH_2CH{=}CH(CH_2)_7COOH$
Arachidonic acid	*n*-Eicos-5,8,11,14-tetraenoic acid	$CH_3(CH_2)_4CH{=}CHCH_2CH{=}CHCH_2CH{=}CHCH_2CH{=}CH(CH_2)_3COOH$

are the C_{18} mono-, di- and triunsaturated forms. Polyunsaturated fatty acids are more common in algae than higher plants. The number of double bonds and their geometric configuration are important factors in the function of these compounds. Most unsaturated fatty acids adopt the *cis* configuration. The common and systematic names and the structures of some fatty acids are given in Table 2.3.

As can be seen from Table 2.3 systematic names and formulae can become cumbersome, while trivial names require memorising. These problems can be overcome by the use of shorthand notations for fatty acids and two of the more common are described in Box 2.7.

Box 2.7 | **Simple notation schemes for fatty acids**

The important attributes of a fatty acid are its carbon chain length, the number of double bonds present and their position, which can be represented in this order by a simple notation scheme. This scheme is best illustrated by an example: oleic acid can be represented by *cis*-18:1ω9, where the number following ω is the position of the double bond from the opposite end to the acid group. As double bonds in polyunsaturated acids are usually conjugated (see Section 2.1.2) it is only necessary to give the position of the first double bond, as all others follow on alternate carbon atoms. Hence arachidonic acid is 20:4ω6, in which the first C=C bond occurs between C-6 and C-7, numbering from the opposite end to the acid group, and the other three C=C bonds are between C-8 and C-9, C-10 and C-11, and C-12 and C-13.

An alternative convention is also often used in which the position of the first double bond is numbered from the end of the molecule bearing the functional group, the ω symbol being replaced by Δ. Thus vaccenic acid, which is most commonly found as the *trans* geometric isomer, can be represented as *trans*-18:1ω7 or *trans*-18:1Δ11. The alternative E/Z system for labelling geometric isomers (see Box 2.3) is not commonly applied to fatty acids.

Fatty acids have predominantly even numbers of carbon atoms because they are effectively formed from acetyl (C_2) units, which are derived from glucose in the presence of various enzymes, coenzymes and carrier proteins. An overall scheme for saturated fatty acid biosynthesis is presented in Fig. 2.12, in which it can be seen that the first step is the formation of acetyl coenzyme A. One molecule of acetyl coenzyme A undergoes addition of CO_2 to form malonyl coenzyme A, while the acetyl group on another molecule is transferred to an enzyme (fatty acid synthase). The malonyl unit (C_3) is added to the enzyme-bound acetyl unit, which produces a butyryl group following loss of CO_2, dehydration and reduction. Six further steps of combined malonyl addition, decarboxylation, dehydration and reduction occur to yield palmitate (C_{16}). Higher acids are built from palmitate in a similar way but using different enzymes. Enzymatic desaturation of these acids (i.e. dehydrogenation) can occur, resulting in conjugated double bonds in the unsaturated products.

Fats are used as energy stores by animals and plants. On a weight-for-weight basis during complete oxidation fats liberate just over twice as much energy as carbohydrates because they contain less oxygen to start with. Fats, therefore, are particularly useful where a compact energy source is required (e.g. in seeds and fruits). Aerobic respiration of fats proceeds by hydrolysis of triglycerides to release fatty acids, which then undergo successive loss of C_2 units, which in turn, in the form of acetyl coenzyme A, are oxidised to CO_2 in the Krebs cycle (Fig. 2.8) with the release of energy as ATP. The glycerol part of fats also contributes to energy generation in the Krebs cycle after conversion to pyruvic acid (Fig. 2.8). This mechanism of fatty acid oxidation is termed β-oxidation and occurs in all aerobes.

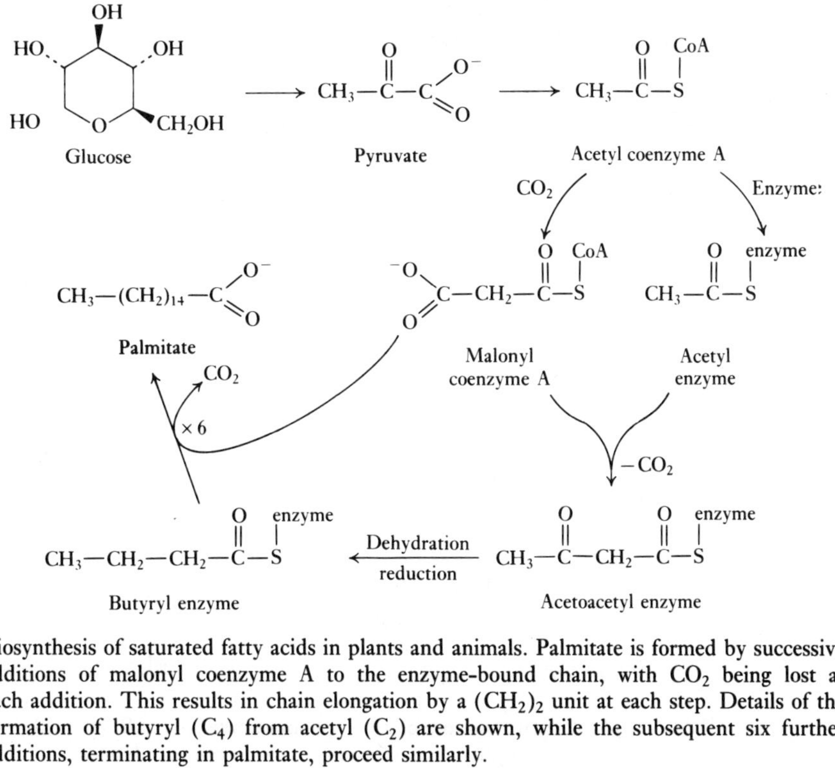

Figure 2.12 Biosynthesis of saturated fatty acids in plants and animals. Palmitate is formed by successive additions of malonyl coenzyme A to the enzyme-bound chain, with CO_2 being lost at each addition. This results in chain elongation by a $(CH_2)_2$ unit at each step. Details of the formation of butyryl (C_4) from acetyl (C_2) are shown, while the subsequent six further additions, terminating in palmitate, proceed similarly.

However, aerobic oxidation can also occur via enzymatic removal of single carbon atoms from fatty acids, termed α-oxidation. It operates in bacteria, and in plants it is often the most important of the two oxidation pathways.

Phospholipids, glycolipids and ether lipids

These lipids are all important constituents of the membranes that isolate the contents of cells from the surrounding environment in various organisms. These cell **membranes** (or plasma membranes) chiefly comprise lipids and proteins, and phospholipids are the main type of membrane lipid (up to 65% in plants).

Phospholipids (or phosphatides) are triglycerides containing one phosphoric acid and two fatty acid units. The phosphate group is often linked to a nitrogen base, such as choline in the phospholipid lecithin, which is found in all animals and plants (Fig. 2.13a). The phospholipids are arranged in a bilayer, with the non-polar (hydrophobic) alkyl chains of the fatty acids directed towards the interior of the bilayer and the polar (hydrophilic) phosphate ends lying on each surface of the

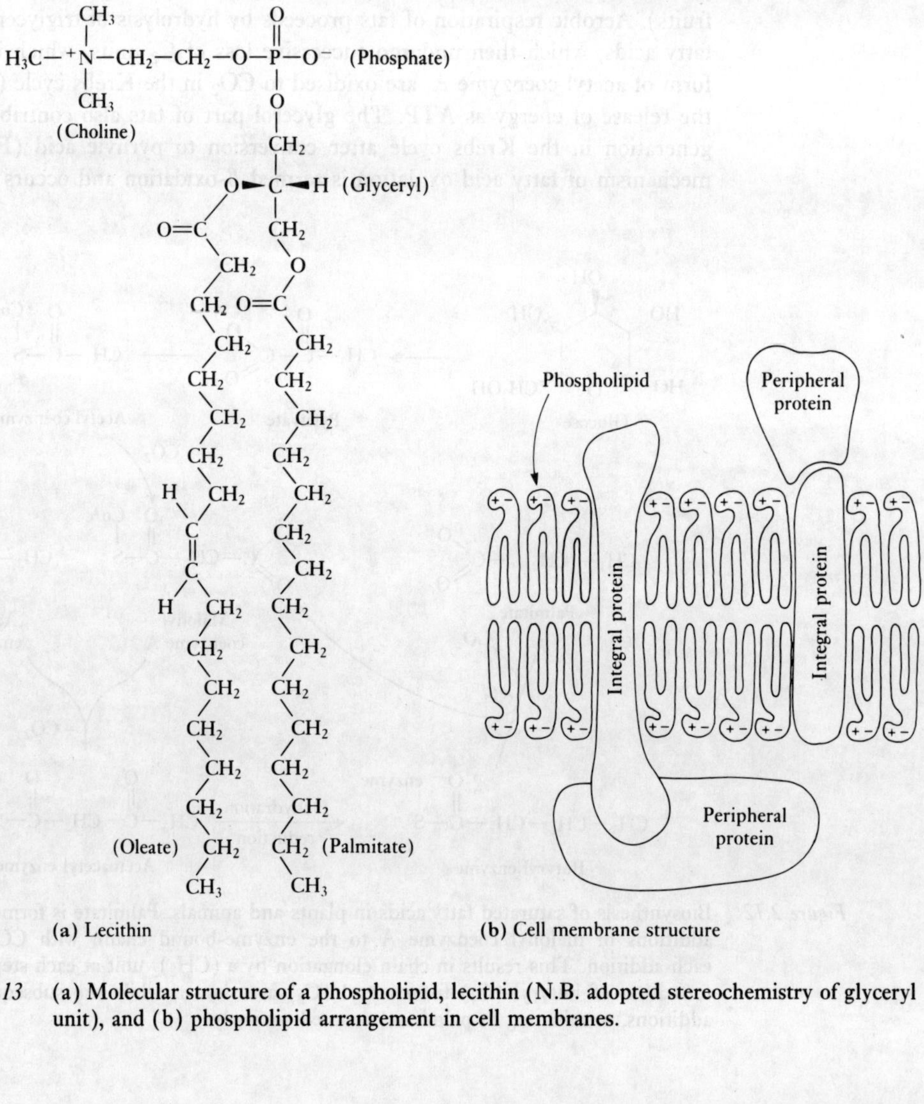

(a) Lecithin (b) Cell membrane structure

Figure 2.13 (a) Molecular structure of a phospholipid, lecithin (N.B. adopted stereochemistry of glyceryl unit), and (b) phospholipid arrangement in cell membranes.

membrane (Fig. 2.13b). Cell membrane proteins are of two types: integral, which bridge the membrane and are involved in transfer processes across it; and peripheral, which are located on the membrane surface and are bonded to integral proteins (Fig. 2.13b).

The term **glycolipid** can be applied to any compound in which lipids and carbohydrates are combined. Glycolipids in which the phosphate group of a phospholipid has been replaced by a sugar are found in plant cell membranes (accounting for up to 20% of membrane lipids) and are important components in the cell membranes of Gram-positive bacteria. They are also major components of the membranes surrounding chloroplasts (the photosynthesising organelles) in higher plants, algae and cyanobacteria (e.g. the diacylgalactosylglycerol in Fig. 2.14a). Teichoic acids are another type of glycolipid and are major components in the cell membranes and walls of Gram-positive bacteria. Glycerol teichoic acids, polymers of glycerol phosphate in which the OH groups of the glycerol units are substituted by D-glucose, D-alanine and fatty acids (Fig. 2.14a), are present in both cell walls and membranes, while ribitol teichoic acids (in which the open-chain form of ribose replaces glycerol, Fig. 2.14a) are found in cell walls only, where they are bonded to murein (see Section 2.2.2). In contrast, the outer membrane of Gram-negative bacteria is composed of lipopolysaccharides, comprising polysaccharide chains linked to glycolipids containing sugar, fatty acid and phosphate units.

Ether lipids are glycerides, but instead of being formed from fatty acids they are formed from fatty alcohols, leading to ether rather than ester linkages (see Table 2.1). These lipids can contain phosphate and sulphate units and also sugar residues. The cell membranes of anaerobic bacteria contain large amounts of plasmalogens, which have both ester and ether linkages (Fig. 2.14a). Archaebacterial cell membranes differ from those of other organisms in being formed from ether lipids containing only phytanyl chains (i.e. formed from the saturated counterpart of the alcohol phytol, Fig. 2.16c). These glycerol ether lipids comprise diphytanyl diethers and biphytanyl tetraethers, in which the biphytanyl groups (formed from linking two phytyl units tail-to-tail) link two glyceride units (Fig. 2.14b). Halophile and methanogen cell membranes contain mixed di- and tetraethers, while thermoacidophiles contain mainly tetraethers. One species of methanogen has been found to synthesise a cyclic biphytanyl diether (Fig. 2.14b). Archaebacterial cell walls lack the murein that provides rigidity and strength in eubacterial cell walls. However, these functions may be performed in archaebacteria by the biphytanyl tetraethers that span the cell membrane (Fig. 2.14b), locking the two halves of the lipid bilayer together, providing enhanced rigidity and strength to the membrane.

2.4.2   Waxes and related compounds

Waxes   **Waxes** mainly function as protective coatings, such as leaf cuticular waxes. They are mixtures of many constituents with high melting points, important members being esters of fatty acids with straight-chain saturated alcohols (*n*-alkanols, sometimes called fatty alcohols). Both fatty acids and alcohols in these **wax esters** have similar chain lengths, mainly in the range C_{24} to C_{28}. They have predominantly an even number of carbon atoms because the alcohols are biosynthesised from fatty acids by enzymatic reduction (Eq. [2.7]). Lesser amounts of ketones, branched

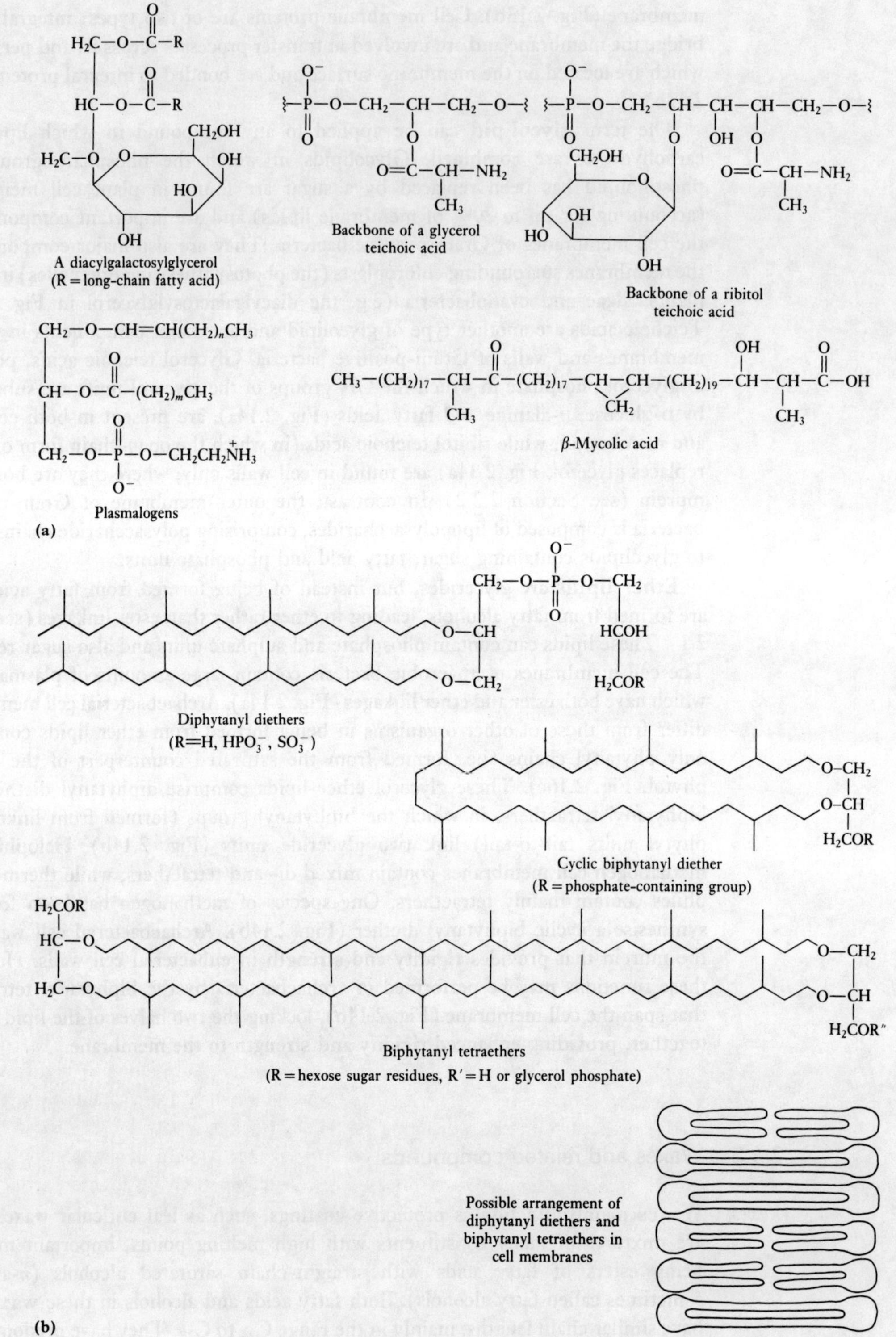

A diacylgalactosylglycerol
(R = long-chain fatty acid)

Backbone of a glycerol
teichoic acid

Backbone of a ribitol
teichoic acid

β-Mycolic acid

Plasmalogens

(a)

Diphytanyl diethers
(R = H, HPO₃⁻, SO₃⁻)

Cyclic biphytanyl diether
(R = phosphate-containing group)

Biphytanyl tetraethers
(R = hexose sugar residues, R' = H or glycerol phosphate)

Possible arrangement of
diphytanyl diethers and
biphytanyl tetraethers in
cell membranes

(b)

Figure 2.14 (a) Examples of lipids in the membranes and cell walls of eubacteria. (b) Phytanyl ether lipids in archaebacterial cell membranes.

alkanes and aldehydes are present.

$$CH_3(CH_2)_nCH_2COOH \longrightarrow CH_3(CH_2)_nCH_2CHO \longrightarrow$$
$$\text{acid} \qquad\qquad\qquad \text{aldehyde}$$
$$\text{(where } n = \text{odd number)}$$

$$CH_3(CH_2)_nCH_2CH_2OH \qquad [2.7]$$
$$\text{alcohol}$$

Plant waxes also contain hydrocarbons, mainly long-chain n-alkanes. In contrast to the fatty acids and alcohols, the n-alkanes contain chiefly odd numbers of C atoms, generally in the range C_{23} to C_{33}, with C_{27}, C_{29} and C_{31} predominating. This results from biosynthesis of the alkanes from acids by enzymatic decarboxylation:

$$CH_3(CH_2)_nCH_2COOH \longrightarrow CH_3(CH_2)_nCH_3 + CO_2 \qquad [2.8]$$
$$\text{acid} \qquad\qquad\qquad \text{alkane}$$

$$\text{(where } n = \text{odd number)}$$

Fungal cell walls contain n-alkanes similar to those in the higher plants and possibly fulfilling a similar role. In contrast, waxes are absent in most bacteria. However, the mycobacteria and the related nocardiae and corynebacteria, which contain greater amounts of lipids than other types of bacteria, contain high-molecular-weight waxy molecules in their cell walls. These molecules comprise various mycolic acids, such as β-mycolic acid (Fig. 2.14a), which are bonded to polysaccharides and, via phosphate groups, to murein.

A particular type of wax ester, a **steryl ester**, is produced where the alcohol unit joined to a fatty acid is a sterol. These are also waxy solids, an example being lanosteryl palmitate in lanolin, which is found in sheep's wool.

Cutin and suberin   Related to the plant waxes are **cutin** and **suberin**, which form protective coatings for plant tissue. They are polymerised and cross-linked structures of hydroxy fatty acids, are resistant to oxidation and to microbial and enzymatic attack, and help prevent excessive evaporative loss of water. Cutin is mainly a polymer of hydroxy (often dihydroxy) fatty acids and forms a layer on the outer surface of exposed plant tissue (i.e. the cuticle). Suberin is mainly associated with underground parts of plants and with protecting wounds. Its constituents are mainly α, ω-diacids (i.e. fatty acids with a COOH group at each end) and ω-hydroxy fatty acids (i.e. OH group at opposite end of molecule to COOH group). The units from which cutin and suberin are composed chiefly contain an even number of carbons in the C_{16} to C_{26} range, with C_{16} and C_{18} dominating (Holloway, 1982).

In addition to the polyester cutin, plant cuticles can contain a highly aliphatic polymer, lacking ester linkages, formed predominantly from chains of CH_2 units. This polymethylenic material appears to be bonded to a polysaccharide and has been termed **cutan**. While the cuticles of some plants contain a mixture of cutin and cutan, others contain only one or the other. A polymethylenic counterpart of suberin also exists, termed **suberan**.

2.4.3   Terpenoids

Terpenoids are a class of lipids displaying a great diversity of structures and functions, ranging from volatile sex pheromones to natural rubber. There is, however, a unifying theme in that they are all constructed from C_5 isoprene units. Classically, the terpenoids are grouped on the basis of the number of constituent isoprene units: **monoterpenoids** contain two; **sesquiterpenoids** three; **diterpenoids** four; sesterterpenoids five; **triterpenoids** six; **tetraterpenoids** eight; and polyterpenoids contain a greater number of units (an example being natural rubber). Most naturally occurring terpenoids contain oxygen, commonly in alcohol, aldehyde, ketone and carboxylic acid groups. The suffix '-ene' is sometimes used to denote a whole class of compounds or just the alkenes within a class. In order to avoid such confusion we shall apply the general suffix '-oid' to all classes of compounds within a group (e.g. diterpenoids), '-ane' to denote alkanes within a group (e.g. diterpanes) and '-ene' to specify alkenes (e.g. diterpenes). As we will see, many terpenoids form cyclic systems but there are also non-cyclic terpenoids which are often referred to as **acyclic isoprenoids**.

Terpenoids do not necessarily contain exact multiples of five carbons and allowance has to be made for the loss or addition of one or more fragments and possible molecular rearrangements during biosynthesis. In reality the terpenoids are biosynthesised from acetate units derived from the primary metabolism of fatty acids, carbohydrates and some amino acids (see Fig. 2.8). Acetate has been shown to be the sole primary precursor of the terpenoid cholesterol. The biosynthesis of terpenoids is summarised in Fig. 2.15. Acetyl coenzyme A is involved in the generation of the C_6 mevalonate unit, a process that involves reduction by NADPH. Subsequent decarboxylation during phosphorylation (i.e. addition of phosphate) in the presence of ATP yields the fundamental isoprenoid unit, isopentenyl pyrophosphate, from

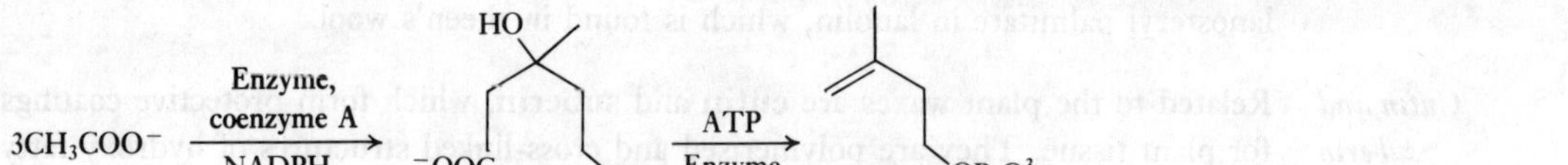

Figure 2.15   Biosynthesis of terpenoids.

which the terpenoids are synthesised by enzymatic condensation reactions. We shall consider some of the more common examples of the main classes of terpenoids.

Monoterpenoids   Monoterpenoids are abundant compounds in higher plants and algae. Because of their volatility they serve as attractants (e.g. insect pheromones) but are probably best known as components of the essential oils of plants (e.g. menthol in peppermint oil). Esters of chrysanthemic acid, found in pyrethrum flower heads, are natural insecticides. Some examples are shown in Fig. 2.16a.

Sesquiterpenoids   Some sesquiterpenoids function as essential oils in plants while others act as fungal antibiotics. The acyclic compound farnesol is widely distributed in nature, being found in many plants and in the chlorophyll of some bacteria (see Section 2.4.4). Mono- and dicyclic sesquiterpenoids are common in plants. Some examples are given in Fig. 2.16b.

Diterpenoids   The most important acyclic diterpenoid is **phytol**. It forms part of the chlorophyll-a molecule and is present in many other chlorophylls (see Section 2.4.4). The saturated analogue of phytol, dihydrophytol (or phytanol), is present in a variety of bacterial glyceride ether lipids (Section 2.4.1).

Most diterpenoids are di- and tricyclic compounds, and are especially common in higher plants. The resin acids are particularly abundant in conifers and include agathic, abietic, communic and pimaric acids. Resins also contain alkenes, such as kaurene and hibaene. Most of these resin diterpenoids exist as isomers differing in configuration at a chiral centre or in $C{=}C$ bond position, which gives rise to a large number of components. They act as general protective agents, sealing wounds and discouraging insect and animal attack.

Other diterpenoids include compounds that give the bitter taste in plants (bitter principles, e.g. columbin). Important, but quantitatively minor, constituents of plants are the growth-regulating gibberellins (e.g. gibberellic acid). Representative structures are shown in Fig. 2.16c.

Triterpenoids   All triterpenoids appear to be derived from the acyclic isoprenoid squalene ($C_{30}H_{50}$, see Fig. 2.19), which is a ubiquitous component in organisms (e.g. shark oil, vegetable oils, fungi). Most triterpenoids are either pentacyclic or tetracyclic. The latter belong primarily to the important class of compounds, the **steroids**, which will be considered separately. Pentacyclic triterpenoids are commonly found in higher plants, chiefly as resin constituents. Three major series can be distinguished among these higher plant triterpenoids: the oleanane (e.g. β-amyrin), ursane (e.g. α-amyrin) and lupane (e.g. lupeol) series (Fig. 2.17). Another important series of pentacyclic triterpenoids is found in bacteria, the hopane series (e.g. diploptene, Fig. 2.17).

Some prominent triterpenoids in organisms are considered to be direct precursors of hydrocarbons in fossil sediments and petroleum, examples of which are given in Fig. 2.17. Most C_{30} pentacyclic triterpenoids with a six-membered E-ring (see Box 2.8) are of higher plant origin. In contrast, the **hopanoids** have a five-membered E-ring and are often called the bacteriohopanoids because of their derivation from compounds like bacteriohopanetetrol (Fig. 2.17), which control fluidity in bacterial cell membranes (Ourisson et al., 1979). However, as can be seen from Fig. 2.17, this distinction between sources based on E-ring size is not absolute.

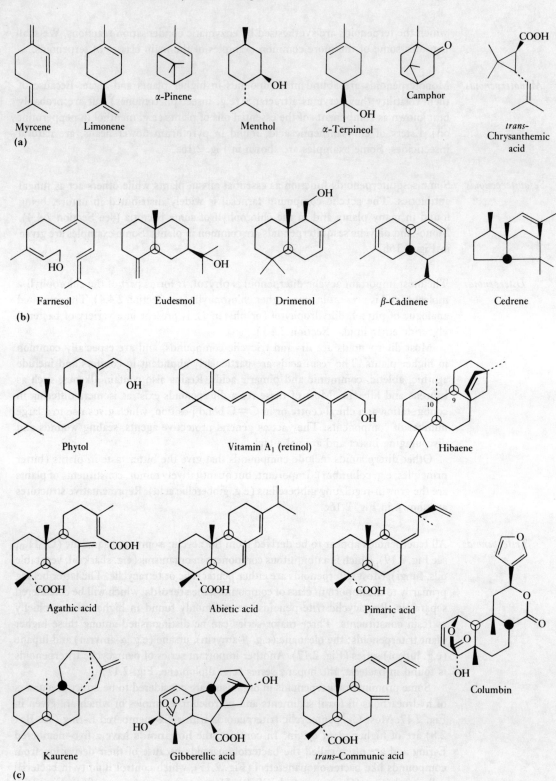

Figure 2.16   Examples of (a) monoterpenoids, (b) sesquiterpenoids and (c) diterpenoids.

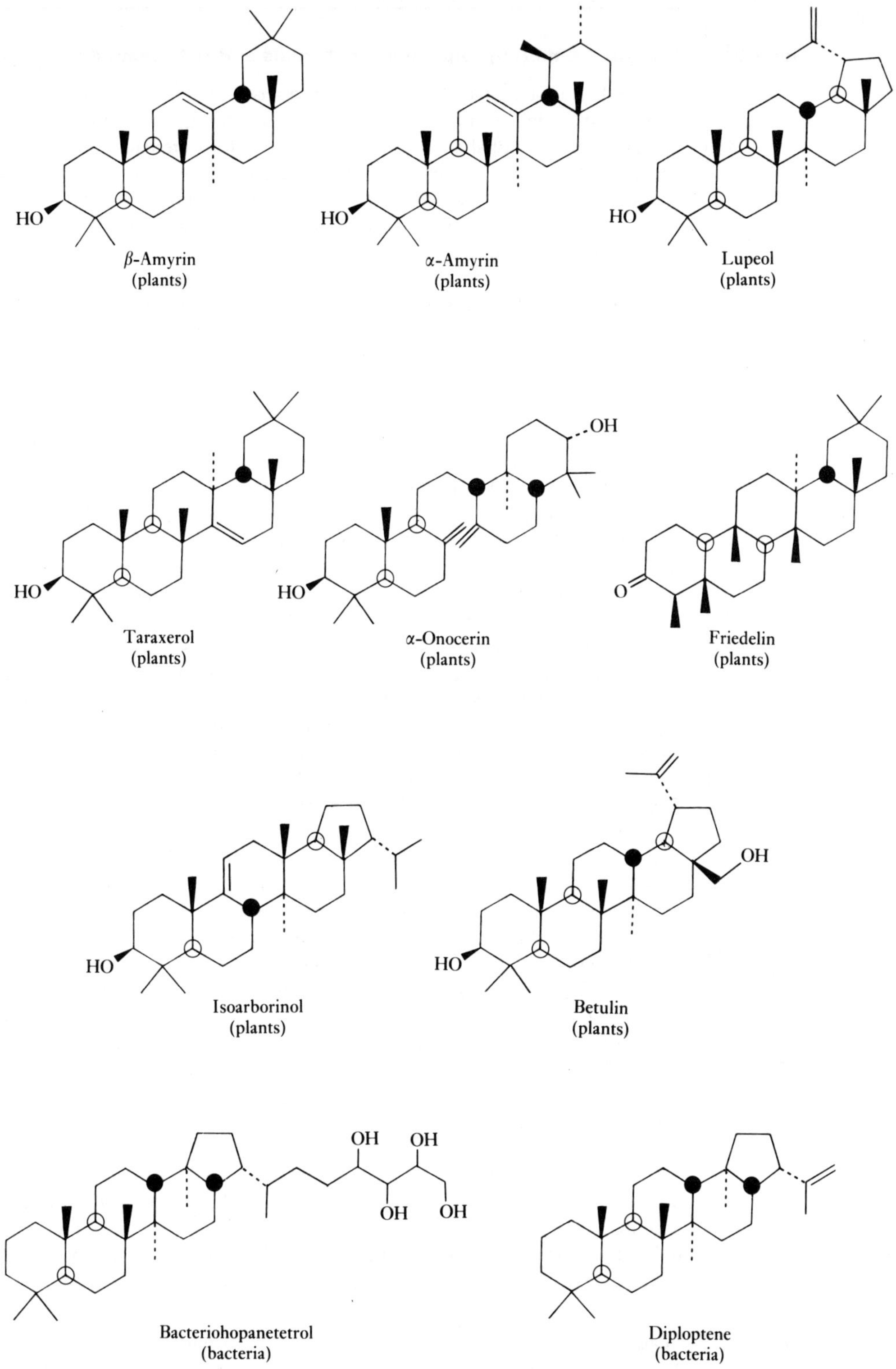

β-Amyrin
(plants)

α-Amyrin
(plants)

Lupeol
(plants)

Taraxerol
(plants)

α-Onocerin
(plants)

Friedelin
(plants)

Isoarborinol
(plants)

Betulin
(plants)

Bacteriohopanetetrol
(bacteria)

Diploptene
(bacteria)

Figure 2.17   Some geochemically important polycyclic triterpenoids.

Box 2.8   **Carbon numbering sequence for steroids and triterpenoids**

The method of identifying particular carbon atoms in these polycyclic compounds is achieved by a systematic numbering sequence. Individual rings are also denoted alphabetically. The conventions for steroids and hopanoids are shown in Fig. 2.18a. Numbering for higher plant triterpenoids follows a similar sequence to that of hopanoids, with C-22 being incorporated in the six-membered E-ring, and C-29 and C-30 being the substituents on the E-ring at C-19 and/or C-20.

As in carbohydrates the six-membered rings are stable but the tetrahedral bond angle of the saturated carbon atoms requires a puckered arrangement. Ring junctions are generally *trans*, allowing the least strained 'all-chair' conformation to be adopted, in which methyl and hydrogen atoms at ring junctions occupy axial positions as shown in Fig. 2.18b for $5\alpha(H)$-cholestan-3β-ol and hydroxyhopanone.

Steroids can be formally named using the basic C_{27} cholestane structure (i.e. the sterane formed by reduction of cholesterol), e.g. cholesterol can be called cholest-5-en-3β-ol. The position of the hydroxy group and the first (lowest numbered) carbon atom involved in the double bond (denoted by 'en') are indicated, and the second (higher numbered) carbon atom of the C=C bond is usually only given (in parentheses after the first number) where more than one possibility exists. Where alkyl groups additional to the cholestane skeleton are present the position of the carbons to which they are attached is noted, and the configuration at various positions (where the stereochemistry is known) can be given, e.g. stigmasterol can be called 24α-ethylcholesta-5,22E-dien-3β-ol. The absence of a methyl group from the basic cholestane structure is indicated by the position number of the missing carbon atom followed by 'nor'. Diasteroids share the same numbering scheme as the regular steroids, with the 'rearranged' methyl groups C-18 and C-19 being attached to the cyclic system at C-14 and C-5, respectively, rather than at C-13 and C-10.

In hopanoids nomenclature is based on the C_{30} compound hopane. The presence of additional carbon atoms in the alkyl chain at C-21 is indicated by their position number followed by 'homo', e.g. the C_{35} hopanoidal alkane in Fig. 2.18a is called 31, 32, 33, 34, 35-pentakishomohopane. However, additional methyl groups elsewhere are often denoted by their attachment position (see steroids described above, e.g. 3-methylhopane). As with steroids, the absence of carbon atoms is indicated by the prefix 'nor' with the relevant position numbers. For example, the absence of the alkyl chain attached to the E-ring of hopane results in the C_{27} compound 22, 29, 30-trisnorhopane (Tm, Fig. 2.18c). There are also variations on the regular hopanoidal structure which give rise to other series of hopanoids, such as the neohopanoids (e.g. Ts, Fig. 2.18c), in which the C-28 methyl is attached to C-17 rather than C-18 at the D,E-ring junction.

Sometimes the position of C=C bonds in steroids and triterpenoids is denoted by a Δ followed in superscript by the numbers of the carbon atoms involved, as described above. The absence of a C-C bond in the ring system is indicated by the carbon numbers involved followed by 'seco', e.g. 8,14-secohopanoids.

Steroids   Formation of the steroids results from enzymatic oxidation of squalene followed by cyclisation. This produces either cycloartenol, the precursor of most plant steroids, or lanosterol, the precursor of animal and fungal steroids, and also some plant steroids

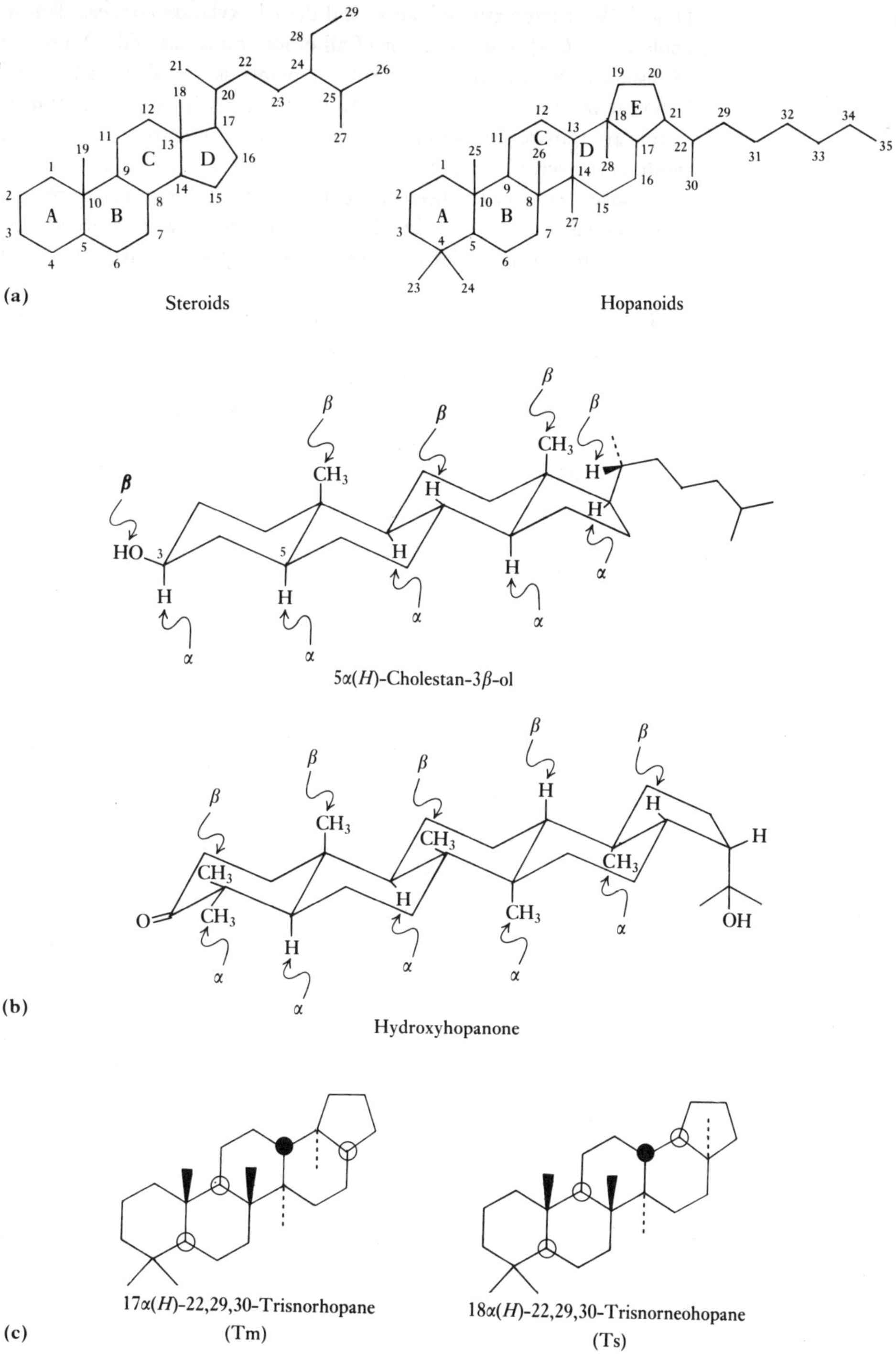

(a)   Steroids   Hopanoids

$5\alpha(H)$-Cholestan-3β-ol

(b)   Hydroxyhopanone

17α(H)-22,29,30-Trisnorhopane   18α(H)-22,29,30-Trisnorneohopane
(c)   (Tm)   (Ts)

Figure 2.18 (a) Ring numbering conventions for steroids and hopanoids; (b) examples of 'all-chair' conformations (with *trans* ring junctions) for steroids and hopanoids; (c) hopanoidal nomenclature system applied to two C_{27} alkanes. (T_s may also be called $18\alpha(H),21\beta(H)$, 17α-methyl-18,22,29,30-tetrakisnorhopane.)

(Fig. 2.19). Enzymatic oxidation and decarboxylation converts lanosterol (C_{30}) into cholesterol (C_{27}), the precursor of all other animal steroids. Most of the cholesterol in animals and the related sterols in plants is found in cell membranes and in lipoproteins. **Lipoproteins** are the main means by which hydrophobic lipids like sterols are transported within organisms. In cell membranes sterols appear to act as rigidifiers, inserted between the adjacent fatty acid chains on phospholipids to give the precise geometry required of the membrane. The equivalent role in many bacteria is performed by the hopanoids. Steroids are rare in bacteria, their limited occurrence probably reflecting ingestion (i.e. heterotrophic uptake) rather than bacterial biosynthesis.

Sterols of geochemical significance are mainly C_{27} to C_{30} compounds with a β-hydroxy group at C-3, a C=C bond within the ring system (usually at the 5, 6 position, i.e. Δ^5) and a side chain on the D-ring (at C-17) with some branching and often some unsaturation. Some examples are shown in Fig. 2.20. The main higher plant sterols are β-sitosterol and stigmasterol. The term **sterol** is commonly used to denote steroidal alcohols, which may or may not be unsaturated. More specific

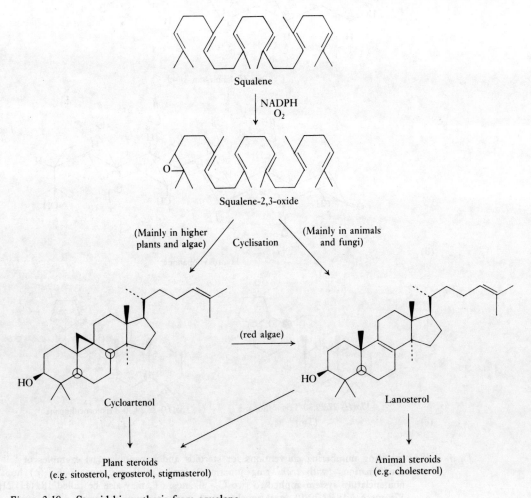

Figure 2.19   Steroid biosynthesis from squalene.

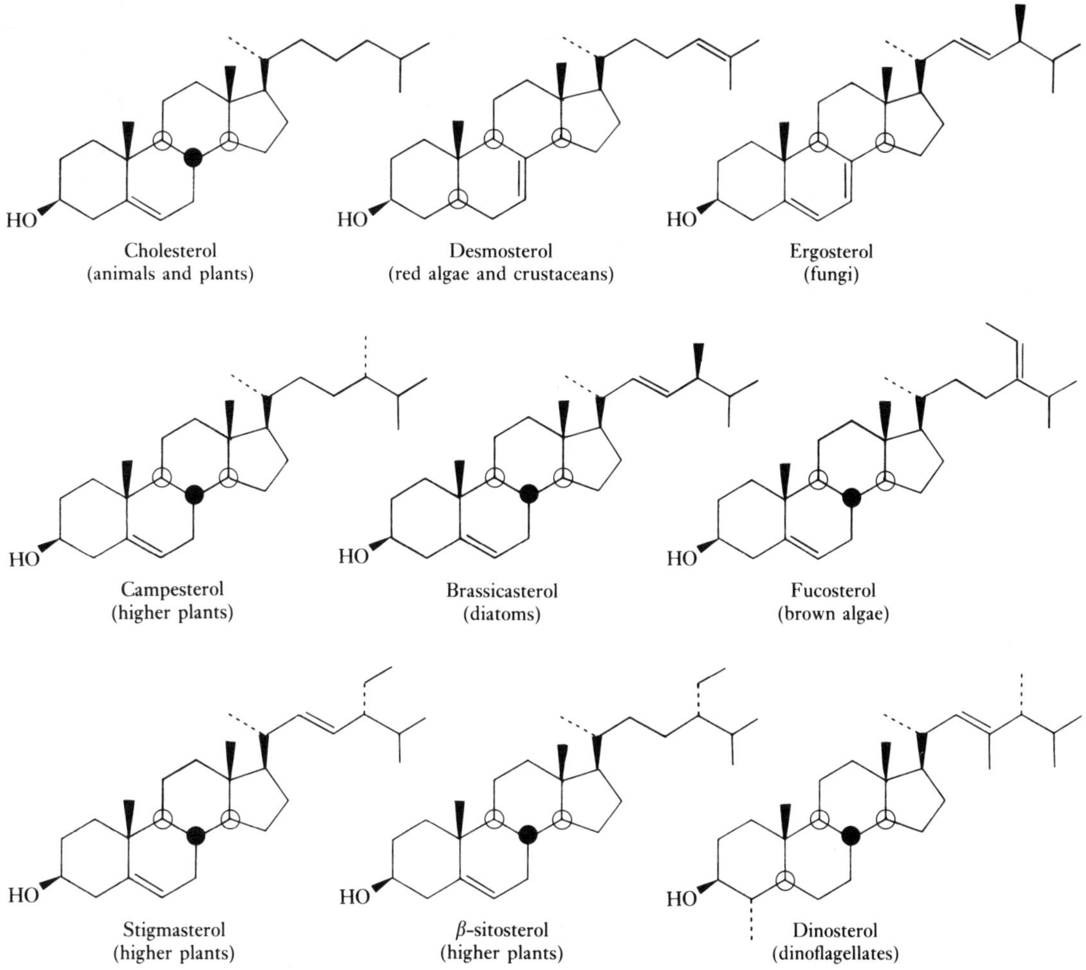

Figure 2.20   Some geochemically important sterols.

nomenclature can be used to differentiate the saturated alcohols, **stanols**, and unsaturated alcohols, **stenols**. The main sterols in photosynthetic organisms are sometimes referred to as **phytosterols** and include the main higher plant sterol, stigmasterol, and brassicasterol, an abundant sterol in diatoms. Of less importance, geochemically, are the animal bile acids (e.g. cholic acid) and hormone steroids (e.g. testosterone, cortisone, aldosterone). There are also plant steroids that can have medicinal uses (e.g. digitonin, a cardiac stimulant, and conessine, which is active against amoebic dysentery).

Tetraterpenoids   The most important members of this group are the **carotenoid** pigments, which are widely distributed compounds with a variety of functional groups and a broad range of reactivities. They are subdivided into hydrocarbons, the **carotenes**, and oxygen-containing compounds, the **xanthophylls**. They are highly unsaturated and the conjugation of C=C bonds (see Section 2.2.1) enables carotenoids to adsorb light of a particular wavelength, endowing them with a yellow/red coloration. Some representative structures of carotenoids are presented in Fig. 2.21.

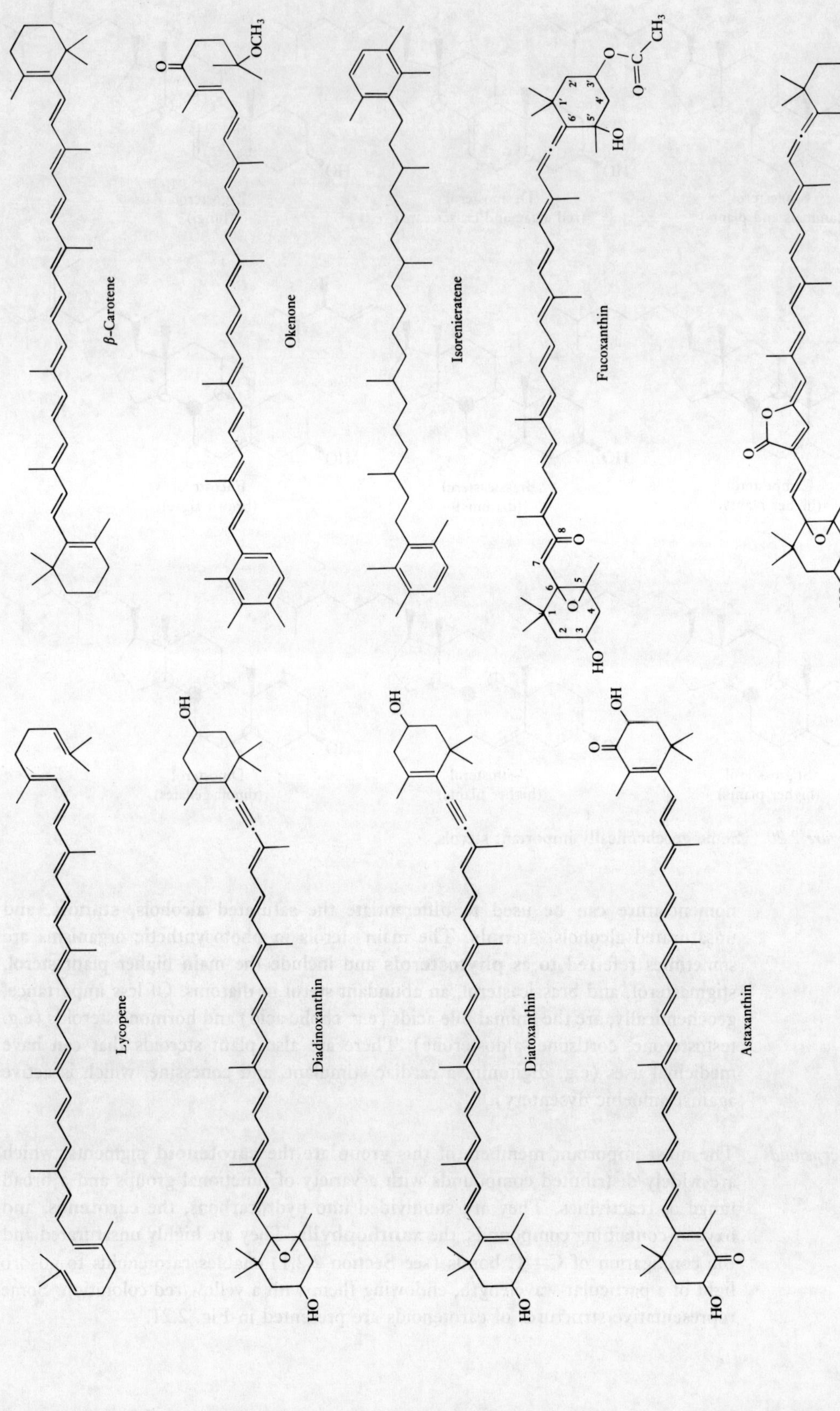

Figure 2.21   Some geochemically important carotenoids (ring numbering scheme shown for fucoxanthin).

Carotenoids are found in most organisms, including non-photosynthetic bacteria, fungi and mammals (in which they are essential for vitamin A production). They are abundant in higher plants and algae, sometimes constituting $>5\%$ of the organic carbon of algae, and are also important in photosynthetic bacteria. Carotenoids are present in all marine phytoplankton and are responsible for the colour of the 'red tides' encountered during periods of high dinoflagellate productivity (blooms). The major role of carotenoids is as accessory pigments to chlorophyll-a in photosynthesis (see Box 2.9). They also provide protection against **free radicals**, which are highly reactive chemical species containing an odd electron and which undergo damaging reactions with cellular chemicals. Although most of the 300 or so naturally occurring carotenoids are tetraterpenoids, there are also some C_{30} and C_{50} compounds unique to non-photosynthetic bacteria.

Box 2.9

Accessory pigments

Chlorophyll-a does not adsorb light uniformly over all the visible region of the electromagnetic spectrum (400–700 nm wavelength). Maximum adsorbance occurs at the red and blue ends of the spectrum (Fig. 2.22), resulting in the green colour of most photosynthesising organisms (particularly terrestrial plants). **Accessory pigments** include carotenoids and a variety of chlorophylls and related tetrapyrroles (see Section 2.4.4), all of which enable more light of shorter wavelengths to be adsorbed (Fig. 2.22), increasing the ability of organisms to photosynthesise at lower light levels. This is especially useful where daylight hours may be restricted (e.g. winter at high latitudes) or light intensity may be reduced (e.g. a low angle of incidence, again at high latitudes). Light intensity falls off quite rapidly with increasing water depth, due to adsorption processes, and so there is a limit to the depth at which photosynthesis can occur. This limit is about 200 m in clear water, which is much greater than would be possible with chlorophyll-a alone and is due to light 'scavenging' by accessory pigments. Light at the blue end of the spectrum penetrates to greatest depths and is adsorbed by accessory pigments, particularly carotenoids (which consequently appear red-brown). The utilisation of light energy during photosynthesis is represented by **action spectra**, which show the amount of light adsorbed at different wavelengths by the pigments present. The action spectrum in Fig. 2.22 demonstrates the effectiveness of accessory pigments.

Accessory pigments do not perform any function other than capturing light energy in photosynthesis and must pass on the adsorbed light energy to a special form of chlorophyll-a. The colours of some accessory pigments tend to mask the green of chlorophyll-a and often give rise to the names of certain classes of multicellular, generally benthic algae, the Rhodophyta (red algae) and the Phaeophyta (brown algae, e.g. fucoids and kelps). In contrast, chlorophylls-a and b dominate in the Chlorophyta (green algae), which form a large group, including both unicellular phytoplankton and multicellular macrophytes (e.g. *Enteromorpha* and *Ulva*), which do not tend to live at such great depths as the red and brown algae.

Distributions of carotenoids can be characteristic for various groups of photosynthetic organisms. Fucoxanthin is characteristic of diatoms (Bacillariophyceae) and peridinin is found in many dinoflagellates (Dinophyceae), while diatoxanthin and

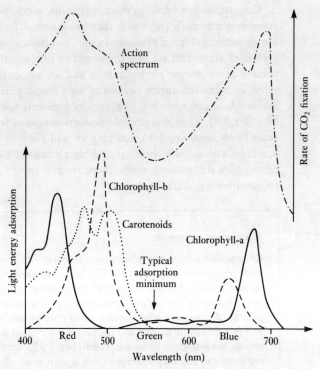

Figure 2.22   Light adsorption characteristics of some photosynthetic pigments and their relationship with utilisation of light energy during photosynthesis (action spectrum).

diadinoxanthin occur in both these phytoplankton classes. Although it is a less specific marker compound, β-carotene is abundant in cyanobacteria. Photosynthetic bacteria produce acyclic and aromatic carotenoids (e.g. lycopene and okenone, respectively). Astaxanthin and its esters are major constituents of marine zooplankton. All these carotenoids are shown in Fig. 2.21.

Sporopollenin, the polymeric material forming the protective coatings of spores and pollens, may contain carotenoid and/or carotenoid ester structures like astaxanthin dipalmitate. Recent research has suggested, however, that the major part of sporopollenin is synthesised from unbranched aliphatic chains, probably derived from fatty acids. Polyhydroxybenzene units similar to lignin precursors (see Section 2.5) have also been identified. Sporopollenin is found in the pollen grains of angiosperms and gymnosperms, in the spores of algae, fungi and lycopods and in fossil spores (e.g. tasmanites).

2.4.4   Tetrapyrrole pigments

These compounds contain four pyrrole units (see Table 2.1) linked together by =CH— groups, either in the form of an open chain or as a large ring (macrocyclic) system. They are all involved in photosynthesis, as primary or accessory pigments (see Box 2.9). Chlorophyll-a is just one of a variety of **chlorophylls** comprising the macrocyclic group of **tetrapyrrole pigments**, some of which are shown in Fig. 2.23a. The general name **porphyrin** is often applied to this cyclic tetrapyrrole structure. Phytol is incorporated in many chlorophylls.

Chlorophyll	Structure	R^1	R^2	R^3	R^4
Chlorophyll-a	I	$-CH{=}CH_2$	$-CH_3$	$-CH_2-CH_3$	Phytyl
Chlorophyll-b	I	$-CH{=}CH_2$	$-C(=O)H$	$-CH_2-CH_3$	Phytyl
Chlorophyll-c_1	II	$-CH{=}CH_2$	$-CH_3$	$-CH_2-CH_3$	H
Chlorophyll-c_2	II	$-CH{=}CH_2$	$-CH_3$	$-CH_2{=}CH_2$	H
Chlorophyll-d	I	$-C(=O)H$	$-CH_3$	$-CH_2-CH_3$	Phytyl
Bacteriochlorophyll-a	III	$-C(=O)CH_3$	$-CH_3$	$-CH_2-CH_3$	Phytyl, farnesyl or geranylgeranyl
Bacteriochlorophyll-b	III	$-C(=O)CH_3$	$-CH_3$	$-CH_2-CH_3$	Phytyl, farnesyl or geranylgeranyl
Bacteriochlorophyll-e	IV	$-C(H)(OH)CH_3$	$-C(=O)H$	$-CH_2-CH_3$, $-CH_2-CH_2-CH_3$ or $-CH_2-CH(CH_3)-CH_3$	Farnesyl

(a)

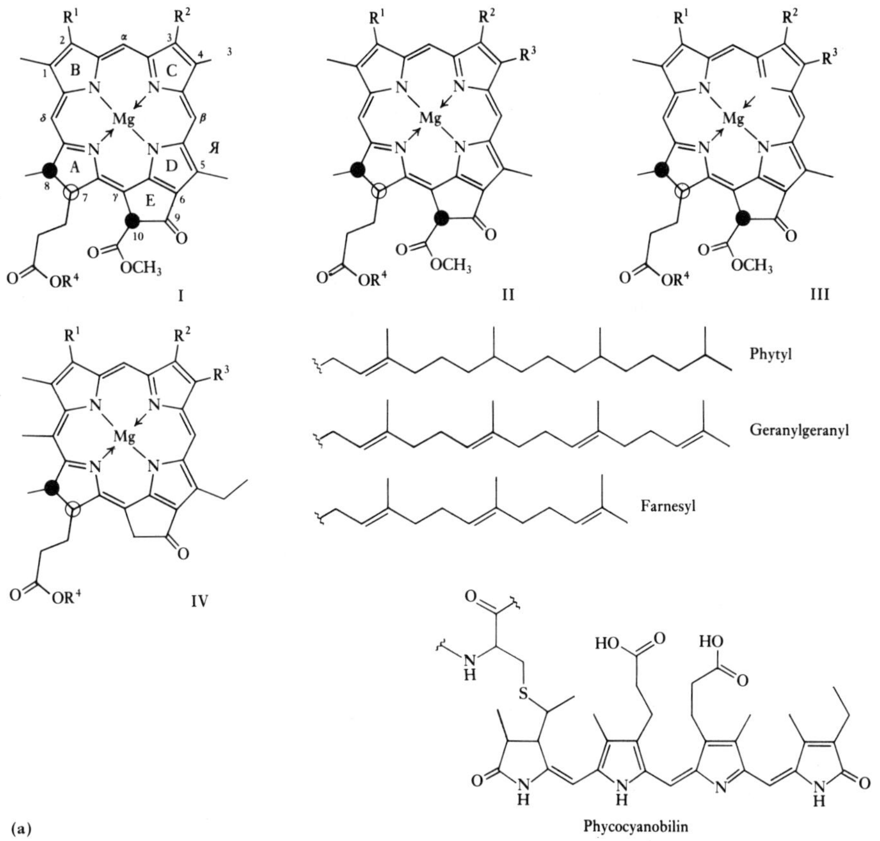

(a)

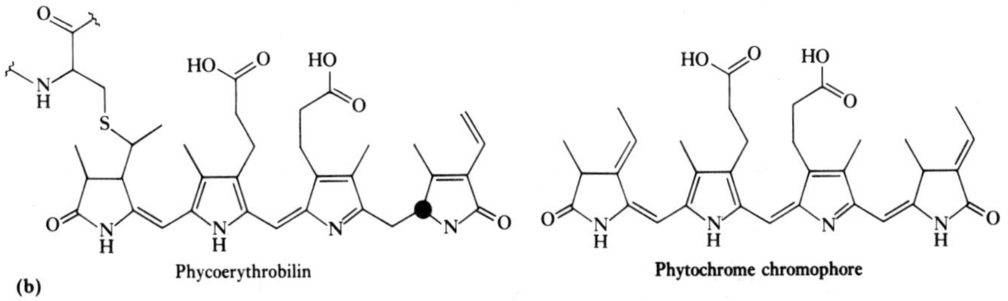

(b)

Figure 2.23 Some important tetrapyrrole pigments: (**a**) chlorophylls (ring numbering scheme shown); (**b**) non-macrocyclic structures.

Chlorophyll-a is quantitatively by far the most important chlorophyll and is present in all algae and higher plants and the cyanobacteria. In addition, all higher plants contain chlorophyll-b. Photosynthetic bacteria contain **bacteriochlorophylls** but, as in plants and cyanobacteria (see Box 2.9), the 'a' form is the primary photosynthetic pigment and the other forms act as accessory pigments, passing captured light energy on to the 'a' form. Bacteriochlorophyll-b is present together with the 'a' form in most species of photosynthetic bacteria. Species of the green photosynthetic bacteria *Chlorobium* also contain some unique chlorophylls, including bacteriochlorophyll-e (Fig. 2.23a), which is characterised by a methyl group in the δ position.

Oxygen transport in animals is performed by related pigments such as **haem** (which is associated with a protein, globin, in haemoglobin). The metal atom in haems (iron) and chlorophylls (magnesium) is important in electron transfer processes that occur during respiration and photosynthesis. Haems are also found in cytochromes, which perform hydrogen transfer in energy-releasing processes in most organisms (but not, for example, in methanogens).

Among the accessory pigments are non-cyclised, non-metallated tetrapyrroles. Phycobilins are such compounds, the most important examples of which are phycocyanobilin, found in cyanobacteria, and phycoerythrobilin, found in certain red algae (Fig. 2.23b). In these pigments the tetrapyrrole system is bonded to a protein (molecular weight ca. 20 000) through a cysteine residue. A related blue-green photochromic pigment, phytochrome, is found in all higher plants and it is also bound to a protein (molecular weight ca. 120 000). It controls various growth-related functions, such as germination, and a wide range of developmental and metabolic processes.

2.5 Lignin, tannins and related compounds

These higher plant components are characterised by phenolic (hydroxy-aromatic) structures. Such structures, which derive originally from monosaccharides, are common in plant but not animal tissue.

2.5.1 Lignin

Lignin forms a network around cellulose fibres in maturing xylem, the channelled woody core of terrestrial plants, which fulfils an important supportive function as plants grow. Cellulose is still an important component of wood (40–60%) but lignin comprises most of the remaining material. Lignin is a high-molecular-weight, polyphenolic compound, formed by condensation reactions (involving dehydrogenation and dehydration) between three main building blocks: coumaryl, coniferyl and sinapyl alcohols. These compounds are biosynthesised from glucose under enzymatic control and some of the intermediates are shown in the scheme in Fig. 2.24a. One intermediate, shikimic acid, occurs widely in plants and it is transformed into phenylalanine by a series of reactions which include NH_3 transfer from an amino acid (transamination).

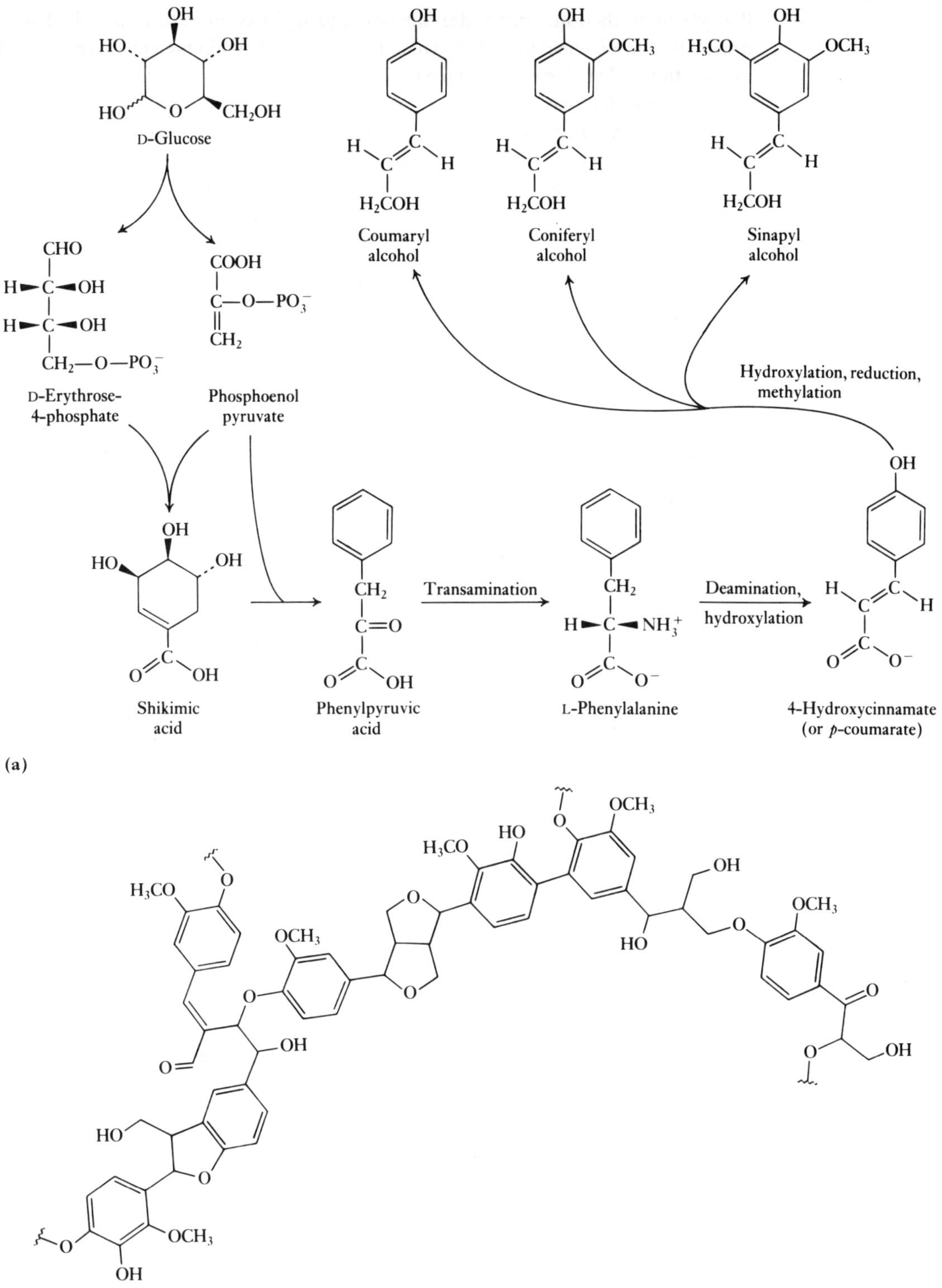

(a)

(b)

Figure 2.24 (a) Biosynthesis of major lignin precursors and (b) a partial structure of lignin illustrating the types of condensation products probably involved.

Phenylalanine then undergoes deamination and hydroxylation to form 4-hydroxy-cinnamate (or *p*-coumarate), from which the three main lignin precursors are formed by reactions that include carboxyl reduction, ring hydroxylation and partial methylation of hydroxyl groups.

The types of structural elements in lignin that are likely to be formed from the three precursor phenolic compounds are shown in Fig. 2.24b. Lignin composition varies according to plant type, for example coniferyl alcohol units dominate in lignins from conifers, while sinapyl alcohol units are most important in deciduous lignins.

2.5.2 Tannins and other hydroxy-aromatic pigments

Tannins are widespread in nature but are quantitatively and geochemically less important than lignin. They are important in making plants less palatable to herbivores and when extracted by steeping plant material, particularly bark and leaves, in water, they can be used for tanning leather. The structural units of tannins are polyhydroxy aromatic acids, such as gallic acid and ellagic acid (Fig. 2.25), which are produced by similar biosynthetic routes to lignin precursors. The condensed, polymeric structures of tannins generally incorporate glucose esters on residual carboxylic acid groups and have molecular weights of 500–3000.

Polyhydroxyflavonol units have also been recognised in some tannins. These units are directly related to the **flavonoid** higher plant pigments, which include anthocyanins, flavones and flavonols. All flavonoids share the same basic carbon skeleton, as shown for cyanidin chloride in Fig. 2.25, but differ in the number and position of hydroxy

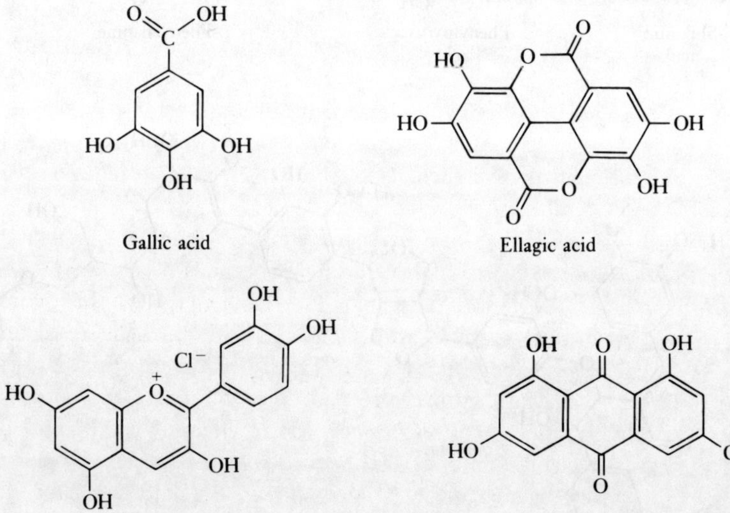

Figure 2.25   Some hydroxy-aromatic plant pigments and quinones.

and methoxy substituents. They are often found as **glycosides** (i.e. they are complexed with sugars in a similar way to tannins) and appear to offer protection against UV radiation as well as providing the colour in petals and autumn leaves.

Anthraquinones, another group of hydroxy-aromatic pigments, are found as glycosides in higher plant tissues, particularly bark, heartwood and roots. Although not generally associated with tannin formation these compounds occur in a range of organisms, in fungi, lichens and insects, as well as vascular plants. Emoldin is perhaps the most widely distributed (Fig. 2.25). Other quinones based on benzene and naphthalene units also occur widely in nature, such as the ubiquinones (also called coenzymes Q) involved in electron transport during respiration (Fig. 2.25).

2.6   Geochemical implications of compositional variation

2.6.1   Compositional variation of organisms

Despite the fact that all organisms contain broadly the same groups of chemical classes performing generally the same biochemical functions, there can be considerable variation in the relative amounts of each class of compound between different groups of organisms, reflecting varying physiological and metabolic requirements. An obvious contrast is the composition of higher plants, in which cellulose and lignin can account for up to 75% of the organic material, and phytoplankton, which do not contain these structural components. Diatoms and dinoflagellates contain ca. 25–50% protein, ca. 5–25% lipids and up ca. 40% carbohydrates (dry weight). Bacteria are quite variable in their chemical constituents, but as a guide their composition can be considered as similar to that of planktonic algae. Higher plants contain ca. 5% protein, ca. 30–50% carbohydrate (mainly cellulose) and ca. 15–25% lignin. The lipid content of higher plants is relatively low and chiefly concentrated in fruiting bodies and in cuticular leaf waxes.

We have also seen examples of how compound distributions within a given chemical class vary between different groups of organisms. We shall examine the use of individual compounds in sedimentary organic matter as indicators of contributing organisms in Chapter 6. However, there are some complicating factors that require consideration. Small organisms are particularly sensitive to changes in factors such as temperature and salinity because they have high surface area to volume ratios. As a result, variation in environmental conditions may bring about greater changes in lipid distributions within a species than exists between species of the same genera or even between more distantly related groups of planktonic organisms. For example, in copepods the abundance of lipids in general and wax esters in particular, which are used as energy stores, depends on factors such as activity, nutritional status and apparently also water temperature. It has been found that copepod species that live in cold waters (i.e. permanently below 250 m depth or in high-latitude surface waters) generally have more lipids and wax esters than those from warmer waters. An indication of the abundance range of total lipids, wax esters and triglycerides observed for copepods is given in Table 2.4.

It can now be appreciated that the nature of the organic material deposited in sediments will depend on the types of organisms contributing to the sediments. For

Table 2.4   **Abundance range of copepod lipids** (after Lee et al., 1971)

Total lipids (% of dry weight)	3–73
Triglycerides (% of lipids)	1–42
Wax esters (% of lipids)	0–92

example, higher plant lignins are the main source of aromatic compounds in contemporary sediments, while plankton and bacteria contribute primarily aliphatic material. Inputs of sedimentary organic matter are classified as **autochthonous** if they originate at or close to the site of deposition, or **allochthonous** if transported from another environment. Autochthonous inputs to most aquatic environments include the remains of phytoplankton and of organisms that feed directly or indirectly on phytoplankton (e.g. zooplankton and bacteria) and which live within the water column and upper layers of sediment. Allochthonous organic material mostly derives from higher plants, usually transported by water from adjacent areas of land to the deposition site. However, in peat swamps higher plants make a large autochthonous contribution. The composition of sedimentary organic matter depends on the relative contributions from the various autochthonous and allochthonous inputs and their chemical compositions.

2.6.2   Variations throughout geological time

The evolution of organisms throughout geological time means that we do not necessarily see the same types of sedimentary organic matter in what are otherwise similar environments. For example, stromatolites are more frequently found in the older sediments deposited before the cyanobacterial communities had to cope with predation by herbivores and competition for habitats with other photosynthetic organisms. In addition, higher plant inputs are not seen prior to the Silurian. The 'fingerprint' of organic compounds preserved in any particular environmental setting will reflect the stage of evolution of life-forms and, especially since the Devonian, the increasing diversity of species.

The geographical distribution of organisms is also an important consideration for terrestrial ecosystems. In the previous chapter we noted how regional variations in flora occurred in the Carboniferous and became more pronounced with the splitting up of the supercontinents in the Triassic and the development of more varied climates. The marine environment is more uniform than the terrestrial and there is less restriction to the spread of organisms. However, marine habitats have been significantly affected by variation in tectonic activity and the related changes in land mass distributions over geological time, as we shall see in the following chapter when examining the conditions necessary for the production and preservation of organic-rich sediments.

Chapter 3

Conditions for the accumulation of organic-rich sediments

3.1 Introduction

Having considered the main biological sources of sedimentary organic matter and its general chemical composition, it is time to examine the physical, chemical and geological processes that control the production of organic matter and its accumulation and preservation within sediments. The organic carbon content of sediments can range from zero up to ca. 20%, the latter value being recorded for the Holocene (i.e. $\leq$ 10 Ka old) sapropel underlying the Black Sea. In present-day sediments organic carbon content rarely exceeds 2% and over most of the ocean floor it is <0.25%. As was noted in Chapter 1, organic-rich sediments have not been laid down continuously throughout geological time over the whole face of the Earth. To understand the formation of ancient organic-rich sedimentary rocks it is necessary to examine the production and fate of organic material in present-day environments.

There are some general conditions that favour the formation of organic-rich sediments. Firstly, a sufficiently large amount of organic material is needed which, as we have seen in Section 1.3, is predominantly derived (directly or indirectly) from the main primary producers: higher plants on land and phytoplankton in aquatic environments. High primary productivity is, therefore, an important factor, permitting relatively large amounts of organic matter to reach and be incorporated into the sediment, rather than being recycled within the water column. Secondly, water velocities in depositional environments need to be low enough for fine-grained, organic-rich material to accumulate instead of being washed away, and inputs of inorganic material should be small to prevent dilution of organic matter. Finally, preservation of organic matter within the sediment must occur, rather than degradation by detritivores and decomposers, and this appears generally to be favoured by the development of anoxicity. The following sections are concerned with the major factors controlling the production and preservation of organic matter in contemporary sediments and their effects in various depositional environments.

3.2 Factors affecting primary production

Photosynthesis is the most important process in primary production and involves the conversion of carbon dioxide to cellular organic components by utilising light energy. Light, then, is an important factor in primary production. Water is vital for all life and provides a ready source of the hydrogen needed for aerobic photosynthesis. Most photosynthesising organisms are aerobes: vascular plants, macroscopic algae (seaweeds), unicellular algae (phytoplankton) and cyanobacteria. Water availability is obviously not a problem for aquatic organisms but it can be an important factor controlling terrestrial primary production. Some of the most productive areas, such as the tropical rain forests (net annual primary production ca. 15.3 Gt C; biomass ca. 340Gt C), have a more than ample water supply and it is probable that the great coal-forming forests of the past grew in swampy areas maintained by high rainfall.

For a long period of the year at low latitudes water is present as ice and is, therefore, unavailable for plant uptake. This is one reason for high terrestrial productivity being confined to the middle and lower latitudes. As well as this indirect effect, temperature directly affects photosynthesis because the dark reactions (see Box 1.3) are restricted to a certain temperature range. Within this range the rate of dark reactions tends to increase with increasing temperature. However, high temperatures tend to result in lower net production in C3 vascular plants, due to elevated photorespiration levels (see Box 1.3), and excessively high temperatures inhibit photosynthesis by destroying or altering enzymes and cell components.

Extremes of temperature probably affect species diversity but not necessarily productivity, as organisms adapted to high or low temperatures are likely to have less competition from other species and may consequently be more productive; for example, dinoflagellates are more productive at higher water temperatures than diatoms. Although the solubility of carbon dioxide in water decreases with increasing temperature, the concentration of dissolved CO_2 is usually high enough not to be a limiting factor in primary production.

The concentration of ions in water (**salinity**) has an effect on primary producer communities. In seawater, the major ions are chloride and sodium (accounting for 55.04% and 30.61% by weight, respectively). Fresh water (salinity $<0.5\%_{oo}$) and waters with salinity levels typical of open marine environments (ca. $35\%_{oo}$) contain the greatest numbers (diversity) of species. However, relatively few organisms can tolerate large fluctuations in salinity (e.g. where fresh water meets seawater in estuaries) and very saline (hypersaline) conditions. In **hypersaline** conditions (salinity $>40\%_{oo}$) phytoplankton diversity is much reduced but the species adapted to these environments can be very prolific and produce large amounts of organic material. In addition, herbivore abundance may be low and so much of the net primary production may be available for incorporation into sediments. Cyanobacterial mat communities tend to be successful in such environments, growing around the high water mark. In hypersaline lakes other anions, such as sulphate, may be particularly abundant.

Table 3.1 presents comparisons of the total levels of net primary production and production per unit area in various aquatic environments. It can be seen that freshwater primary production amounts to a little over 1% of total, autochthonous, aquatic primary production. Phytoplankton account for ca. 95% of marine primary production (which totals ca. 40 Gt C/a). Much of the remaining marine production (ca. 3%) is contributed by coastal ecosystems, such as the saltmarsh and mangrove swamp communities of the intertidal zone, in which macrophytes are important (e.g. *Rhizophora* in mangrove swamps, and eelgrass (*Zostera*) and cordgrass (*Spartina*)

Table 3.1   **Estimated primary productivity and biomass values for aquatic environments** (After several sources, including Whittaker and Likens, 1975; De Vooys, 1979)

Ecosystem type	Area (A) (10^6 km²)	Plant biomass (Gt C)	Annual net primary production (P) (Gt C/a)	P/A (g C/m² a)
Open ocean	332.0	0.46	32.1	100
Upwelling zone	0.4	0.004	0.2	500
Continental shelf	26.6	0.13	7.3	270
Algal bed and reef	0.6	0.54	0.5	830
Estuary (open water)	1.4	0.63	0.1	70
Swamp and marsh	2.0	10.4	1.3	650
Lake and stream	2.0	0.02	0.6	300

in saltmarshes. Mangrove swamps are characteristic of warm, wet climates, while saltmarshes can be considered their counterparts in cooler and/or drier coastal areas. Coral reefs, although extremely productive on a unit area basis, account for <1% of total aquatic, primary production.

While the contributions of photosynthetic and chemosynthetic bacteria to global primary production are relatively small they may be important on a local basis. In addition, these bacteria play an important role in the cycling of the essential mineral nutrients (see Section 3.2.3) required by all other organisms in both terrestrial and marine environments. The availability of these nutrients to the main photosynthesising organisms is another major factor affecting primary production.

Light and nutrient availability are probably the main factors influencing primary production in aquatic environments and they will be considered in more detail in Sections 3.2.2 and 3.2.3. However, the operation of these factors is affected by stratification of the water column and so we shall examine this phenomenon first.

3.2.1   Stratification of the water column

In water bodies of sufficient depth, stratification can occur as a result of density differences related to temperature. In such circumstances a stable condition arises in which water warmed by solar radiation (**insolation**) overlies colder, denser water. At the boundary between the two there is a layer of water in which temperature changes rapidly, termed a **thermocline**, which effectively isolates the warm layer of water from the cold body beneath.

As a consequence of the general deep circulation of the ocean (see Box 3.1), a permanent thermocline is present in temperate and tropical oceans (at depths of ca. 300 m and 100 m, respectively). The permanent thermocline persists throughout the year at middle and low latitudes but is absent at latitudes above ca. 60° because the cooler climate and reduced insolation do not cause sufficient heating of the surface water to produce an adequately steep, thermal gradient.

Box 3.1

General oceanic circulation

Deep currents and the permanent thermocline

The water in the oceans behaves as though it is made up of discrete bodies which lie at depths dictated by their densities. Cold, dense, oxygen-rich water sinks in polar regions and flows over the bottom of the ocean basins. This water gradually rises as it is displaced by further dense cold water and it offsets the downward transfer of heat from surface waters at middle and low latitudes that are warmed by insolation. The result is a layer of water, termed the main or **permanent thermocline**, in which temperature rapidly decreases with depth (often exceeding 5°C/100 m over a range of ca. 15°C to 5°C). As the name implies, the permanent thermocline is a constant feature at latitudes below ca. 60° (Fig. 3.1) and beneath it water temperature decreases steadily down to 2°C or less.

Surface currents

Surface currents in the oceans mostly result from the frictional stress imposed by wind action and, because of the inertia of water, there is a relatively constant circulation pattern broadly reflecting the main prevailing winds. These winds include the Trade Winds, predominantly north-easterlies in the northern hemisphere and south-easterlies in the southern hemisphere. They blow from the subtropical high-pressure belts at ca. 30° in both hemispheres towards the equatorial low-pressure system (Doldrums). The westerlies are another important set of winds, related to air movement from the subtropical highs towards the subpolar lows at ca. 60° in both hemispheres.

The different relative movements of water bodies cause various convergences and divergences at the surface. At a **divergence** surface waters in adjacent bodies move away from each other and are replaced by water welling up from depth. The converse occurs at a **convergence**. The permanent thermocline is affected in these areas, being nearer the surface at divergences and deeper (e.g. 300–400 m) at convergences (see Fig. 3.1).

Ocean gyres and coastal upwelling

The movement of water and air masses over the Earth are subjected to the Coriolis effect, which results from the fact that the velocity of the Earth's surface in the direction of spin (eastwards) is greatest at the equator and decreases to zero at the poles. Consequently, as water moves away from the equator at the western margins of the main oceans (where it is piled up by the action of the Trade Winds), it moves progressively faster than the Earth's surface in an easterly direction. In the northern hemisphere its motion is turned progressively clockwise, while in the southern hemisphere the imposed direction of motion is anticlockwise. With continental masses acting as barriers to the east and west the result is the formation of two large, roughly circular, surface circulation cells, or **gyres**, in both the Atlantic and Pacific Oceans between ca. 10° and 50° latitude. Only a single gyre is present in the Indian Ocean because it lies entirely within the southern hemisphere. The northern subtropical gyres have a clockwise circulation and the southern subtropical gyres an anticlockwise rotation, reflecting the movement of air around centres of high pressure (anticyclones) that generally overlie these regions. The Coriolis effect is also

Box 3.1
(continued)

responsible for cyclonic and anticyclonic air circulation and the direction of the westerlies and Trade Winds.

Where the subtropical gyres impinge on the western sides of continental masses the prevailing winds in both hemispheres cause a net transport of warm surface water away from land (a divergence), which is replaced by upwelling of cooler, deeper water (from up to ca. 500 m depth).

Above the permanent thermocline at middle and low latitudes smaller diurnal (i.e. daily) and seasonal thermoclines can develop, resulting from warming of surface waters by insolation on a daily or seasonal basis. For example, a **seasonal thermocline** can be established at depths of ca. 50–100 m in temperature latitudes during spring and summer, when winds are light, but it is destroyed by general cooling of surface waters and the turbulent effects of strong winds (the effects of which can extend to 200 m depth) as winter approaches, resulting in mixing of the water column down to the permanent thermocline. There is insufficient seasonal contrast in insolation in the tropics for a seasonal thermocline to form, and consequently the permanent thermocline is nearer the surface (see Fig. 3.1). However, a shallow diurnal thermocline can develop in calm waters.

Stratification can also occur because of large differences in salinity, and the resulting boundary layer of high salinity gradient is termed the **halocline**. Because temperature and salinity affect density, a rapid change in density is associated with thermoclines and haloclines and is termed a pycnocline. Solid material suspended in the water can also create density gradients, leading to pycnocline formation, and is important in glacial lakes and lakes fed by rivers with large suspended sediment loads.

Thermal stratification can also occurs in lakes. Where it persists for short periods of one day to a few weeks the lakes are termed **polymictic**. Stable stratification resulting in a seasonal thermocline lasting through the warm season only appears to be possible under present climatic conditions for lakes of at least 40 m depth at latitudes <70°. These lakes are termed either **dimictic** if they have ice cover

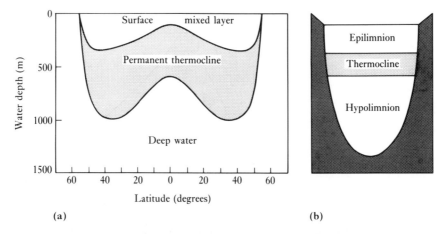

Figure 3.1   (**a**) Generalised thermal stratification of the oceans and (**b**) corresponding nomenclature in lakes.

regularly each winter, or **warm monomictic** if this freezing is sporadic. Lakes that do not stratify tend to be found at high latitude and are termed amictic if they are permanently ice-covered or cold monomictic if only covered with ice during winter.

In lakes, the warm, oxygenated and less dense layer above the thermocline is called the **epilimnion**. The lower layer is termed the **hypolimnion** and contains colder denser water which may become depleted in oxygen. Circulation within each layer is independent; it is likely to be more sluggish in the hypolimnion as it is not affected by wind action. Because thermocline formation is seasonal in dimictic and warm monomictic lakes, stratification in these lakes breaks down as ambient atmospheric temperatures decrease during the autumn and winter, causing cooling, and strengthening winds produce turbulence within the surface layers.

The density of fresh water reaches a maximum at 4°C, while it is still a liquid, allowing ice to float. In addition, density does not change uniformly with temperature; it decreases progressively more rapidly as temperature rises. For a given temperature difference between epilimnion and hypolimnion, more energy is required to mix the water masses in a stratified freshwater lake at higher temperatures because the density difference is greater. Consequently, stratification is more readily achieved in tropical than in temperate lakes. In contrast to fresh water, normal seawater exhibits a continuous increase in density with decreasing temperature.

3.2.2   Light

The quantity and quality of light affects the rate of photosynthesis and, therefore, primary production. The rate of photosynthetic production varies with the intensity of light and this is demonstrated in Fig. 3.2, which shows three distinct phases: a linear phase, a saturation plateau and a photo–inhibition phase. Each phase has different limiting factors: the linear phase is probably limited by insufficient light intensity; the saturation plateau is probably limited by the rate of dark reactions

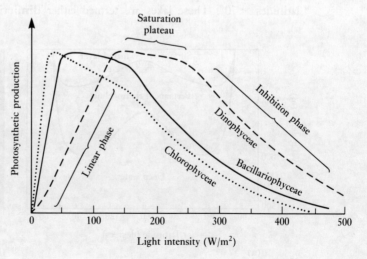

Figure 3.2   Variation in relative photosynthetic production with light intensity for three classes of phytoplankton: green algae (dotted line), diatoms (solid line) and dinoflagellates (broken line).

(see Box 1.3); and during the inhibition stage the high levels of light intensity may limit photosynthesis by damaging enzymes and possibly also cell structure. Most phytoplankton live just beneath the water surface to avoid cell damage and destruction of chlorophyll by excessive exposure to UV irradiation. The various classes of phytoplankton exhibit different abilities to cope with varying light levels (see Fig. 3.2).

A certain amount of photosynthetic production is needed to offset phytoplankton respiration before there is any surplus energy that can be channelled towards growth and reproduction. The point at which the photosynthetic rate is balanced by the rate of respiration is called the compensation light intensity, while the depth at which the oxygen consumed by respiration equals that produced by photosynthesis is known as the compensation depth.

In terrestrial ecosystems the amount of light available for photosynthesis is limited by factors such as day length and the angle of incidence of sunlight (both functions of latitude and season), and also by shading of individual leaves by the rest of the canopy. Low availability of light results in a short growing season and relatively low productivity at high latitudes. There are similar restrictions on aquatic primary production (including self-shading when phytoplankton numbers are very high), but there is the additional effect of water depth and clarity on light penetration. The surface layer of water in which light intensity is sufficient for photosynthesis to be possible is termed the **photic zone** (or euphotic zone). Water clarity is dependent on the amount of suspended or dissolved material in the water column, which can adsorb or scatter light, so reducing the depth of the photic zone. In exceptionally clear oceanic water the photic zone can extend down to ca. 200 m, while in some estuaries light may only penetrate a few centimetres because of a large suspended material load. Typical average depths for the photic zone in the open ocean and in coastal areas are 150 m and 30 m, respectively. In general, then, the photic zone lies above the permanent and seasonal thermoclines.

The maximum depth of the photic zone would be significantly shallower if only chlorophyll-a was involved in light adsorption. Accessory pigments, such as carotenoids, enable organisms to gain maximum use of the available light at greater depths (see Box 2.9). At depths of ca. 3700 m much of the ocean floor (the abyssal plains, Fig. 3.3) is below the photic zone but in coastal regions, when light penetrates

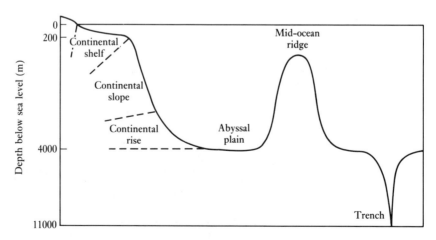

Figure 3.3   Major regions of ocean basin floors (non-linear depth scale with exaggerated gradients).

down to the sediment surface, benthic plant communities can be significant primary producers on a localised basis. For example, benthic diatoms can be prolific producers in estuaries, while productive macrophyte communities include seagrass (*Thalassia*) meadows in shallow coastal waters (down to ca. 15 m) and forests of kelp (e.g. *Laminaria, Macrocystis, Ascophyllum*) in deeper coastal waters (ca. 20–40 m depth).

3.2.3   Nutrients

The rate of growth and reproduction of organisms depends not only upon the availability of carbon but also upon a variety of essential mineral **nutrients**. In Chapter 2 a number of elements can be seen to be important, such as nitrogen in amino acids, phosphorus in ATP and magnesium in chlorophylls. Some of these essential elements are more abundant than others and can be categorised on this basis (Table 3.2).

Nitrogen, phosphorus or silicon availability usually limits primary production in the ocean and the situation in stratified lakes is similar. These three elements are, therefore, termed **biolimiting** (Table 3.2). Silicon is used by organisms secreting siliceous tests, such as the diatoms and silicoflagellates. Terrestrial primary production is mostly controlled by the availability of nitrogen and phosphorus. Phosphorus is utilised by phototrophs as phosphate, silicon as silicate and nitrogen mainly as nitrate but also sometimes as ammonium (by some terrestrial plants). In terrestrial ecosystems these and other mineral nutrients are obtained from the soil in solution via root systems, while in aquatic ecosystems they are absorbed directly from solution into the bodies of phytoplankton. The availability of nutrients can depend on factors like pH and E_h (see Box 3.2), which affect the chemical form and solubility of nutrients. Although phosphate and silicate are the major, commonly occurring forms of P and Si, nitrate can be converged to volatile ammonia or oxides of nitrogen which are readily lost from terrestrial ecosystems to the atmosphere. In both terrestrial and aquatic ecosystems the mineral nutrients are released from detritus by the decomposer system, which will be considered in more detail in Section 3.3.

Box 3.2   **E_h and pH**

Redox potential (E_h)

Oxidation and reduction can be considered solely in terms of electron transfer (see Box 1.1). The electrons liberated during the oxidation of one chemical species must be taken up by another species, resulting in its reduction. Therefore, oxidation and reduction must occur together in a reaction, termed a **redox reaction**. The species that is oxidised gives up electrons and so is termed a reducing agent. Conversely, the species that is reduced accepts electrons and so is an oxidising agent. This relationship can be seen in the oxidation of Fe(II) by oxygen, which is the sum of two half-reactions:

$$4Fe^{2+} \longrightarrow 4Fe^{3+} + 4e^- \qquad [3.1a]$$
$$O_2 + 4e^- \longrightarrow 2O^{2-} \qquad [3.1b]$$

$$4Fe^{2+} + O_2 \longrightarrow 4Fe^{3+} + 2O^{2-} \qquad [3.1c]$$

Box 3.2
(Continued)

In this reaction Fe(II) loses electrons and is oxidised, while oxygen gains electrons and is reduced. Similarly, when organic matter is oxidised by oxygen in aerobic environments the oxygen is reduced (e.g. by the gain of hydrogen to form water).

The tendency for a redox reaction to occur is reflected by the tendency for electrons to flow and produce an electrical current, which can be measured as a voltage, the redox potential (E_h). The direction of flow depends on whether oxidation or reduction is occurring and its size is a measure of how oxidising or reducing conditions are. At the oxic/anoxic boundary in sediments $E_h = 0$, and conditions become more oxidising as E_h becomes increasingly positive. Conversely, conditions become more reducing as E_h values increase in negative value. The value of E_h depends on temperature, the number of electrons involved and the concentrations of oxidants and reductants.

Acidity and alkalinity (pH)

In Box 2.5 an acid was defined as a substance that donates H^+ ions. The greater the concentration of hydrogen ions in a solution, the more acidic it is. The acidity of a solution or its converse, alkalinity, can be expressed in terms of its pH, which is defined as:

$$pH = -\log_{10}[H^+] \qquad [3.2]$$

(where $[H^+]$ = hydrogen ion concentration).

Water is considered to be neutral, but it is partially ionised into H^+ and OH^- ions. The concentration of H^+ ions in water is 10^{-7} mol/dm^3, so a neutral solution has pH = 7. Below this value a solution becomes increasingly acidic as pH falls, while above it conditions become increasingly alkaline as pH rises.

In aquatic systems stratification of the water column has significant effects on nutrient availability and the timing of bursts of primary productivity. It is convenient to consider these effects in the oceans in terms of the latitudinal variations in stratification.

*Low latitude
oceans*

At low latitudes there is a permanent thermocline but a seasonal thermocline does not develop and so surface and nutrient-rich deeper waters remain unmixed. Because the photic zone lies well within the permanent thermocline it becomes depleted in nutrients. Continued primary production is dependent upon the release of nutrients from detritus in surface waters. Productivity is, therefore, relatively low in tropical oceans but it is fairly constant, due to a year-round warm climate and ample sunlight

Table 3.2   **Classification and availability of essential elements in the ocean**

Biolimiting elements	P, N, Si	Usually almost totally depleted in surface waters
Biointermediate elements	Ba, Ca, C, Ra	Partially depleted in surface waters
Nonlimiting elements	B, Br, Cs, Cl, F, Mg, K, Rb, Na, Sr, S	Show no measurable depletion in surface waters

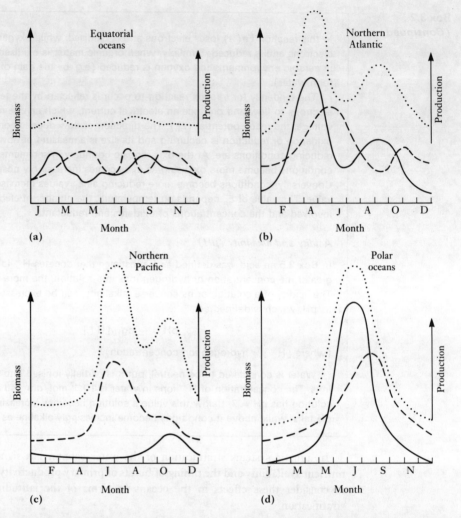

Figure 3.4   Approximate annual standing biomass profiles for phytoplankton (solid lines) and zooplankton (broken lines) in various regions of the oceans. Estimated phytoplankton production is indicated by dotted lines.

(Fig. 3.4a). In subtropical oceans in the mid-Pacific diatoms are responsible for approximately 80–90% of total algal production in the surface layers and their production is probably nutrient-limited. In deeper waters, but still above the thermocline (ca. 85–180 m), coccolithophores are the largest group of primary producers and their activity is probably light-limited (Venrick, 1982).

Middle latitude oceans   Increasing light intensity during the spring allows phytoplankton to assimilate the available nutrients resulting in a marked increase in productivity, giving rise to a spring phytoplankton bloom which can be seen as a film near the water surface. As the seasonal thermocline becomes fully established, at depths approximately corresponding to the depth of the photic zone (ca. 50–100 m), mixing between the water layers above and below it ceases. Biolimiting nutrients become depleted in the photic zone and phytoplankton production falls to a level dictated by the rate of nutrient recycling within the surface layer. While most organic material is recycled within the photic zone some can sink below the seasonal thermocline, in the form of dead organisms and faecal material from zooplankton, which exacerbates the shortage of nutrients. Nutrients are, however, abundant beneath the photic zone because they are released during bacterial decomposition of detritus, but their assimilation rates are relatively low in the absence of phototrophs.

In the autumn the surface water begins to cool, reducing the strength of the seasonal thermocline and, therefore, the density differences in the water layers on either side of it. Nutrients are once again brought into the photic zone as the seasonal thermocline breaks down and circulation occurs to greater depth in surface waters, giving rise to a second bloom. This bloom is usually smaller than that in the spring because of slightly lower light levels and nutrient supply. During the winter rough conditions result in complete disappearance of the seasonal thermocline and mixing down to the permanent thermocline is re-established, renewing the nutrient supply in surface waters ready for the next spring bloom.

In the northern Atlantic the two phytoplankton blooms are reflected in two peaks in biomass because there appears to be a lag between the onset of rapid primary production and increase in herbivorous zooplankton numbers (Fig. 3.4b). During the summer the zooplankton graze down the phytoplankton and their numbers decline before the second algal bloom occurs, whereupon zooplankton reproduction again follows the bloom. In contrast, in the northern Pacific phytoplankton biomass appears relatively constant, although there is the same overall pattern of primary production as in the Atlantic (Fig. 3.4c). The absence of a spring maximum in phytoplankton biomass arises because of a different strategy employed by the major Pacific herbivorous zooplankton (copepods), which use reserves of food stored over the winter to enable them to breed before the spring bloom. Consequently, large numbers of zooplankton are present and effectively graze down the phytoplankton as they appear. Zooplankton numbers decline somewhat in the late summer with the exhaustion of the initial phytoplankton bloom, resulting in the autumn bloom showing up as a slight increase in phytoplankton biomass.

High latitude oceans

In polar regions sufficient light intensity for photosynthesis is only available for five or six months of the year. There is no permanent thermocline at latitudes above 60° because there is insufficient insolation to create a significant vertical temperature gradient. As a result, nutrients are constantly being replenished in the photic zone, leading to a single large algal bloom during the summer (Fig. 3.4d).

Stratified lakes

In temperate lakes phytoplankton blooms occur in response to the seasonal availability of nutrients, which is controlled by stratification of the water column. The situation is, therefore, similar to seasonal thermocline development and decline in temperate oceans. During the spring surface water is at its most dense, having a temperature of only ca. 4°C, and so only small amounts of wind energy are needed to cause mixing of the water column. As the season progresses the upper layer of water is heated by insolation, stratification occurs when the thermal gradient is sufficiently developed (if the lake is of sufficient depth) and nutrients in the epilimnion then become depleted. Once firmly established, thermal stratification is not affected by wave motion during the summer as the extremely stable density gradient established within the thermocline prevents the transfer of energy to the hypolimnion.

In the autumn the intensity of sunlight is lower, heat is lost from the surface layers, the density difference between epilimnion and hypolimnion decreases and the cooling surface waters begin to sink. The end of stratification in the autumn can be quite abrupt, with overturning of the water column occurring in only a few hours, especially in strong winds. The nutrient-rich, oxygen-deficient waters of the hypolimnion mix with the epilimnion to provide the nutrients necessary for the following spring bloom.

Tropical lakes exhibit different stratification characteristics to their temperate counterparts because seasonal temperature variations are slight and so thermal gradients are low. However, thermal stratification can develop and is very stable because more energy is needed to mix warm tropical waters than colder temperate waters. Consequently, overturning of the water column may occur only as a rare event. Wind affects the mixing of water in all lakes, but in tropical lakes it can be the major factor. Wind strength tends to vary on a seasonal basis and so can lead to an annual breakdown in the thermocline in tropical lakes (i.e. warm monomictic lakes).

In very deep tropical and temperate lakes, mixing seldom occurs to the very bottom. Lake Tanganyika (East Africa) and Lake Baikal (Siberia) are examples of this type of lake, which are termed **meromictic**. Lakes in which circulation is established down to the bottom are classified as holomictic. Lake Baikal is the deepest (1620 m) and largest freshwater lake in the world, containing $> 10\%$ of the total standing fresh water. Its bottom waters are oxygenated, in contrast to those below ca. 200 m in Lake Tanganyika, which are permanently anoxic.

In shallow lakes at mid and low latitudes it is not possible for a steep density gradient to become established and so if stratification occurs it is weak and readily broken down, sometimes on a diurnal basis during cooling at night. In such polymictic lakes nutrient supply is virtually constant and so other factors, such as light and temperature, control any variations in phytoplankton production. Primary production can be higher in such lakes than in those where nutrients are immobilised within the hypolimnion for long periods.

3.2.4   Spatial variation in marine primary production

The previous sections have given a general idea of primary productivity in the oceans. However, production is not uniform over the surface of the oceans as a result of the effect of currents and wind action (see Box 3.1). This leads to some areas being particularly productive, while others are the oceanic equivalent of deserts. The general pattern can be seen in Fig. 3.5, and it has important implications for the likely location for deposition of organic-rich sediments.

The centres of the main subtropical gyres correspond to convergences (see Box 3.1), and the main thermocline is depressed, perhaps as deep as 400 m (Fig. 3.1). These surface waters are isolated from laterally adjacent waters by the current system. Nutrient supply is limited and primary productivity extremely low, as can be seen in Fig. 3.5. Such areas are termed **oligotrophic**. Conversely, in areas of divergence, phytoplankton production is continuously fed by nitrient-rich waters welling up from depth. These areas are extremely productive, as can be seen from Table 3.1, although their total area is limited. Among the most productive of these areas are the upwellings associated with the eastern boundary currents of the subtropical gyres

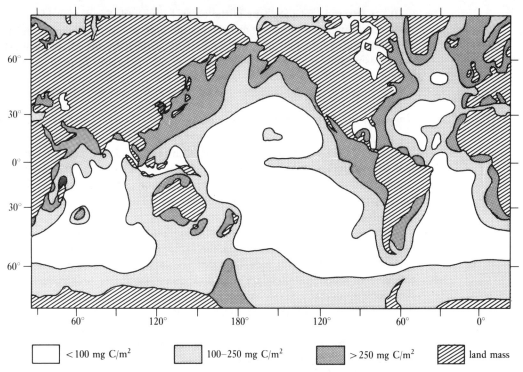

60°

30°

0°

30°

60°

60°   120°   180°   120°   60°   0°

| | <100 mg C/m² | | 100–250 mg C/m² | | >250 mg C/m² | | land mass |

Figure 3.5   Estimated annual primary production in the oceans. (After several sources, including FAO, 1972; Koblenz-Mishke et al., 1970.)

(see Box 3.1), along the western coasts of Africa and the Americas. The Peru upwelling is perhaps one of the best known and its high primary productivity results in rich anchovy fishing grounds. Occasionally these fisheries fail when upwelling is suppressed (during El Niño events). Highly productive areas with a rich nutrient supply are generally termed **eutrophic**. Divergences also occur off the west coasts of Australia and India, the east coast of Africa, around Antarctica, and in the open ocean along the equator and at ca. 50°N in the Pacific. The location of all these upwelling areas can be seen to correlate closely with the areas of highest productivity in Fig. 3.5. Other areas of high productivity in Fig. 3.5 correspond to nutrient-rich shallow seas, like the North Sea.

Lakes can also be categorised as eutrophic or oligotrophic on the same basis of nutrient availability and productivity. The water of oligotrophic lakes tends to be saturated with respect to oxygen, due to low biological oxygen demand. High rates of photosynthesis in eutrophic lakes may lead to the surface water being supersaturated with oxygen, while at depth anoxicity may develop because of the high oxygen demand during degradation of detritus.

Shallow seas and coastal areas generally benefit from river-borne nutrient supplies from adjacent land masses and so exhibit higher productivities than the open oceans. As water depth in continental shelf areas is generally not more than 200 m (Fig. 3.3), the waters are usually well mixed and do not suffer from nutrient limitation due to stratification. Consequently the average annual production on continental shelves on a unit area basis (ca. 270 g C/m²) is about three times that of the open ocean, although it is lower than that of upwelling areas (see Table 3.1).

It can be seen that the shape of the ocean basins and distribution of continental masses have an important effect on currents and high productivity zones. The degree of flooding of continental margins, which is controlled by the relative volumes of ocean basins and water, is also important in terms of the area of productive coastal waters.

3.2.5   Variation in phytoplankton populations

The main classes of phytoplankton are given in Table 3.3 and most are represented in both seawater and fresh water, although species vary between the two environments. Among the nanoplankton most of the Haptophyceae are marine, while the Chrysophyceae are mainly freshwater classes. The Euglenophyceae are also predominantly found in fresh water.

There are latitudinal variations in the major families of phytoplankton responsible for marine primary production. Of the three main primary producers, diatoms appear particularly important at high latitudes and along the equatorial belt, but in intermediate latitudes dinoflagellates and coccolithophores are more important. Planktonic cyanobacteria are relatively unimportant in the ocean but can be very important in eutrophic lakes, particularly in the tropics. Eutrophic lakes and tropical lakes in general tend to contain mainly larger species (e.g. Euglenophyceae and Chlorophyceae), while cold lakes and oligotrophic lakes tend to be dominated by small flagellates (e.g. Chrysophyceae).

In estuaries phytoplankton are less important than benthic and littoral (shoreline) communities, but benthic diatoms can be important. The larger plants (macrophytes) of saltmarshes and mangrove swamps may host a significant community of attached microalgae (**epiphytes**). The flow rate of rivers is generally too great to support phytoplankton communities.

During phytoplankton blooms in seasonally stratified water bodies the major families of algae present change as the season progresses, a process known as **succession**. In dimictic lakes the first to appear are the faster-growing diatoms, which later decline as silica becomes depleted. They are followed by other algal classes which require a different balance of nutrients and finally, if stratification persists, cyanobacteria may form dense blooms (Reynolds, 1984). A simpler

Table 3.3   **Algal classes present in the phytoplankton**

Class	Description
Cyanophyceae*	Cyanobacteria
Chlorophyceae	Green algae
Xanthophyceae	Yellow-green algae
Chrysophyceae	Golden algae (include silicoflagellates)
Haptophyceae	Yellow-brown algae (include coccolithophores)
Bacillariophyceae	Diatoms
Dinophyceae	Dinoflagellates
Cryptophyceae	
Euglenophyceae	Green flagellates

* Although bacterial, this class is sometimes included among the algae due to its previous classification as blue-green algae.

succession occurs in temperate oceans, with diatoms being followed by dinoflagellates. The succession probably results from varying requirements for light, temperature and specific nutrients between each family and also from grazing pressure. Within each family a succession of species can also be observed. Such successions are influenced by the presence of trace chemicals released by preceding species, some of which (e.g. vitamins) act as promoters, while others act as inhibitors of growth for potentially competing species.

3.3 Preservation and degradation of organic material

3.3.1 Fate of primary production in the water column

Much of the organic material produced by phytoplankton is consumed within the photic zone by herbivorous zooplankton. The most important marine zooplankton are the copepods, which are members of the macroplankton (Fig. 1.5). However, herbivorous protozoa also exist, which are members of the microplankton (see Fig. 1.5), comprising chiefly ciliates in fresh water and rhizopods (mainly radiolarians and foraminiferans) in the oceans. These protozoa can make a contribution to sedimentary material, such as the deep-sea oozes from *Globigerina* (a foraminiferan). Detritus, comprising the remains of various planktonic organisms and zooplankton faecal pellets, sinks down through the water column. It is colonised by bacteria which further break down the organic matter, releasing dissolved organic compounds and nutrients for uptake by phytoplankton. Most of the nutrients and organics are recycled within the water column, but some of the dissolved organics are adsorbed onto clay and other suspended mineral particles, which aids their transport to the sediment.

The amount of organic material that reaches the sediment is dependent on the depth of water through which it passes and the amount of primary production in the photic zone. The deeper the water, the longer the residence time and the longer the particles are exposed to degradation. Generally, the amount of organic material that reaches the bottom decreases by a factor of ca. 10 for every 10-fold increase in water depth. Zooplankton faecal pellets, being larger, fall more quickly through the water than phytoplankton remains (160 m/d for faecal pellets cf. 0.15 m/d for coccolithophores), which can be an important factor in preservation. It is thought that pellets > 200 μm in diameter are most important in the formation of organic-rich sediments (Suess, 1980).

In areas of high productivity zooplankton grazing is less efficient at removing the products of primary production from the water column and proportionally more reaches the sediment than in low productivity areas. Such areas are, therefore, potentially important in the formation of organic-rich sediments.

3.3.2 Sedimentary fate of organic material

Aerobic decomposition When detritus and colonised particles are deposited their fate depends on the rate of burial and amount of oxygen present at the sediment surface and within the interstitial spaces. In areas with sufficient oxygen concentrations, organic material is assimilated by suspension and deposit feeders (detritivores) or is decomposed by aerobic heterotrophic bacteria (decomposers). Decomposers inevitably form part of the diet of deposit feeders, being ingested with the detritus upon which they are working. The action of burrowing detritivores (e.g. polychaete worms) mixes the sediment (**bioturbation**) and aids oxygenation. Bacteria preferentially remove the more labile components from detritus and the residue becomes increasingly refractory. Much of the soluble products of the microbial breakdown of organic matter diffuses upward within pore waters to the sediment/water interface and is returned to the water column.

Decomposition of organic material (**mineralisation**) occurs rapidly under oxic conditions. The rate of oxygen supply to the sediment is a critical factor and is influenced by sedimentary particle size. The restricted size of pores in fine-grained silts and clays results in rapid reduction of water circulation within the sediment with increasing depth and, ultimately, oxygen can only enter pore waters by **diffusion** (a process caused by the random motion of molecules in a fluid resulting in the net movement of the molecules of a compound from regions of higher to lower concentration). The amount of organic matter in the sediment affects the balance between the rate of oxygen consumption during aerobic degradation of organic matter and the rate of diffusion of oxygen into sedimentary pore waters from the overlying water column. Pore waters in open marine, pelagic sediments in which oxygen consumption is low are often oxygenated to depths of 0.5 m or more (Murray and Grundmanis, 1980). In contrast, oxygenation is often limited to the top few millimetres of fine-grained shelf sediments, although oxygen penetration can extend to a few centimetres as a result of bioturbation and in the walls of well ventilated burrows (Revsbech et al., 1980).

As oxygen levels fall within the sediment and conditions become dysaerobic (ca. $0.3\%_{oo}$ oxygen, see Box 1.1) bioturbation ceases. If oxygen demand outstrips supply the sediment and sometimes the overlying water column become completely anoxic. In eutrophic lakes anoxicity usually develops within the hypolimnion. Under such conditions, the activity of **obligate** (i.e. strictly) aerobic bacteria (e.g. actinomycetes and certain species of *Bacillus*, *Pseudomonas*, *Corynebacterium* and *Flavobacterium*) is severely restricted.

Anaerobic
decomposition
Mineralisation continues under anoxic conditions, due to the activity of various anerobic bacteria, but the overall rate is slower. As well as obligate anaerobes, these bacteria also include **facultative** anaerobes (i.e. bacteria that are usually aerobes but which can function anaerobically). In the absence of molecular oxygen the anaerobes oxidise organic matter by using various inorganic oxidising agents (which act as terminal electron acceptors): manganese(IV), nitrate, iron(III), sulphate and bicarbonate. These processes release less energy to decomposers than the complete aerobic degradation of organic matter to carbon dioxide and water. Oxidising agents in anaerobic degradation tend to be utilised in order of decreasing energy return, as shown above, with each process taking the oxidation of detritus one stage further.

Some bacteria (e.g. species of *Clostridium*, obligate anaerobes, and *Bacillus*, facultative anaerobes) break down macromolecular components in detritus into simpler molecules by hydrolytic and fermentative processes. These products are the substrates for a range of other anaerobic heterotrophic bacteria which complete the mineralisation of organic matter. The most important of these bacteria are the nitrate reducers, sulphate reducers and methanogens (methane producers). While the term fermentation is sometimes applied to all anaerobic degradation processes, strictly it only describes reactions in which an internal source of electron acceptors is used, not an external source like nitrate or sulphate. An example is fermentation of glucose to carbon dioxide and ethanol, which can be considered to involve an internal redox reaction whereby part of the substrate is oxidised (to CO_2) and part reduced (to CH_3CH_2OH). A further example is shown in Eq. [1.6].

Hydrolysis of macromolecular components releases sugars, amino acids and long-chain fatty acids, which undergo fermentation to yield a relatively limited range of compounds upon which subsequent bacterial metabolism is based, as summarised in Fig. 3.6. The main products are short-chain volatile acids (e.g. formate, acetate, propionate, butyrate, lactate and pyruvate), together with alcohols (e.g. methanol and ethanol), methylated amines, carbon dioxide and water. Acetate (CH_3COO^-) is an important substrate and is produced, together with hydrogen and carbon dioxide, from other short-chain acids by acetogenic bacteria (e.g. species of *Clostridium* and *Acetobacterium*). Acetogenic conversion of H_2 and CO_2 to acetate also occurs.

Nitrate reduction (or denitrification) generally follows rapidly upon depletion of oxygen and yields carbon dioxide, water and nitrogen (via nitrite, NO_2^-). Nitrate reducers (e.g. species of *Pseudomonas, Bacillus, Micrococcus, Thiobacillus denitrificans*) are facultative anaerobes and use oxygen when it is in sufficient supply. The vertical extent of the nitrate reduction zone is usually very limited in marine sediments

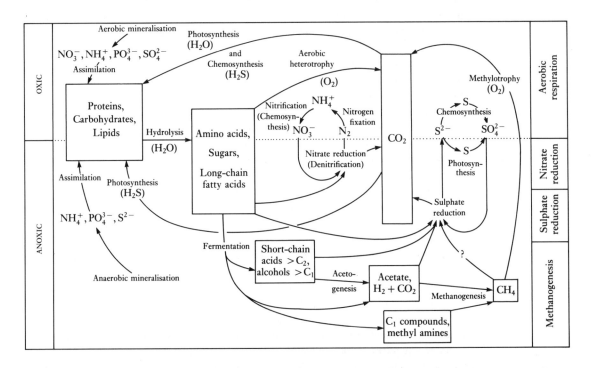

Figure 3.6   Generalised scheme of the role of bacteria in the carbon cycle and its coupling to the nitrogen and sulphur cycles. For clarity, the forms of N, S and P liberated at each stage of mineralisation are summarised on the left side of the diagram, where they contribute to the general mineral pools from which assimilation occurs. (After Fenchel and Jørgensen, 1977; Jørgensen, 1983a, 1983b; Parkes, 1987.)

because nitrate concentrations in pore waters are typically very low (only a few tens of $\mu mol/l$). Sulphate reduction becomes important when nitrate is depleted; its products are carbon dioxide, water and hydrogen sulphide. **Sulphate reducers** (e.g. *Desulfovibrio, Desulfobacter*) are obligate anaerobes. The depth of the sulphate reduction zone depends on the amount of organic matter present; it may be relatively shallow in organic-rich areas, where sulphate is rapidly depleted, but can occupy several metres in pelagic sediments with lower organic content. **Methanogens** (e.g. *Methanobacillus, Methanococcus*) are also obligate anaerobes and synthesise methane from the smallest fermentation products. Carbon dioxide and hydrogen are important substrates (CO_2 being an electron donor in the reaction with H_2), but some species can use simple compounds which are either C_1 compounds (e.g. methanol and formate) or readily yield C_1 units (methylated amines). In freshwater sediments ca. 70% of methane generation may result from utilisation of acetate (produced by acetogens), with the remainder deriving from CO_2 and H_2.

Bacterial communities and their interactions

Reliance of one group of bacteria on the products of another group is a general feature of bacterial communities, involving photosynthetic and chemosynthetic bacteria as well as members of the decomposer system. A much simplified representation of this behaviour is shown in Fig. 3.6, indicating the role of bacteria in the carbon, nitrogen and sulphur cycles. The aerobic and anaerobic respiration processes of bacteria release nitrogen and sulphur from detritus and convert them into forms assimilable by plants. Phosphorus is also released for uptake but remains in the form of phosphate throughout its cycle.

Aerobic and anaerobic communities are, therefore, interdependent and their relative positions in the water column and sediment reflect their reliance on particular substrates as well as oxygen levels. For example, photosynthetic bacteria such as the green and purple sulphur bacteria (*Chlorobium* and *Chromatium*, respectively) live just within the anaerobic zone, where there is sufficient light penetration for them to make use of the products of anaerobic degradation processes (e.g. H_2S and CO_2). These phototrophic sulphur bacteria oxidise sulphide to elemental sulphur and then to sulphate, but the first step in the oxidation is faster than the second and so sulphur accumulates. This sulphur can be oxidised to sulphate by *Thiobacillus denitrificans*, a chemosynthetic bacterium that is unusual in being a facultative anaerobe (a nitrate reducer). Immediately above the anoxic zone a group of chemosynthesisers (e.g. species of *Thiobacillus, Beggiatoa, Thiothrix*), all obligate aerobes, converts the sulphide produced by sulphate reduction back into sulphate. If the water immediately overlying the sediment is anoxic, conversion of sulphide to sulphate often does not occur, and when it does it is minor and due entirely to the activity of anaerobic photosynthesisers. Under such conditions there is a build-up of sulphide and an increase in acidity which results in the decalcification of calcareous tests and shells. If iron is present within the sediments it can combine with sulphide to form insoluble pyrite, removing sulphur from the system. If iron is not available sulphur can be deposited, a process that can occur during the formation of carbonate sediments.

Some nitrogen fixation, which converts N_2 into NH_4^+, is carried out by anaerobes such as the photosynthetic bacteria and some species of *Clostridium* and sulphate reducers. However, the most important nitrogen-fixing bacteria are the cyanobacteria (e.g. *Anabaena, Nostoc*) and some other aerobes (e.g. *Rhizobium* in plant root nodules and free-living *Azotobacter*). **Nitrification**, the conversion of ammonium to nitrate (via nitrite), is carried out by the combined action of the aerobic chemosynthesisers

Nitrosomonas and *Nitrobacter*. The nitrate produced diffuses downwards into the upper layers of anoxic sediment where it is used by the nitrate reducers.

Methane, produced by methanogens under anoxic conditions, may be oxidised by aerobic **methylotrophs** (e.g. *Methylomonas*) back to carbon dioxide if it rises into a suitable oxic environment. The carbon dioxide produced may then be reintroduced into the biological cycle by phototrophs and chemotrophs. Some anaerobic oxidation of methane appears to occur, which may be associated with sulphate reduction (Jørgensen, 1983a).

The availability of substrates and the competition for them between various anaerobes is important. Sulphate is plentiful in marine environments and so sulphate-reducing bacteria are abundant and account for ca. 50% of carbon oxidation in marine shelf sediments (cf. ca. 3% by nitrate reducers). The H_2S liberated during sulphate reduction may partially inhibit nitrate reduction in marine environments and sometimes a nitrate reduction zone may not be present (Jørgensen, 1983b). It also appears that methanogens do not compete effectively with nitrate reducers or, more importantly, with sulphate reducers for the main substrates (H_2, CO_2 and CH_3COO^-). Methanogens, therefore, are largely confined to the area below the sulphate reduction zone and may, in marine environments, be relatively insignificant in the overall reworking of organic matter, although methanogenesis can extend to considerable depth. In contrast, sulphate levels are usually significantly lower in lakes (and also in soils), resulting in methanogens and also nitrate reducers being much more important.

Anaerobic and aerobic degradation reactions probably proceed at about the same rate under identical conditions, but anaerobic degradation often appears slower because most of the labile organic components have already been removed leaving mainly recalcitrant material. Total microbial numbers decrease with increasing depth in the sediment, although significant bacterial populations can still be found in some sediments at 500 m below the sea floor (Parkes et al., 1990). Eventually, all biologically mediated degradation ceases, an important factor being the decreasing energy returns from oxidation of the increasingly recalcitrant organic matter. The remains of dead bacteria also contribute to the organic material present in the sediment and are degraded, or preserved, in the same manner as the rest of the detritus.

There are broad similarities between the decomposer communities of terrestrial and aquatic ecosystems, but while fungi are important in the degradation of woody material on land they are much less so in the decomposition of planktonic remains. Bacteria are important in all environments. The composition of bacterial communities depends on a number of factors such as acidity and salinity, in addition to the effects of oxygen levels described above. Bacteria can be active over a wide temperature range (e.g. ca. 0–45°C for most methanogens and *Desulfovibrio*), although most micro–organisms exhibit optimum growth rates within the range 20–35°C. Bacterial activity is generally suppressed when pH falls below ca. 5 and so fungi are more important than bacteria in acid soils. However, there are types of bacteria that are specially adapted to live under extreme conditions and are dominant in environments like hot springs and soda lakes (e.g. species of thermoacidophiles and halophiles, respectively). The role of bacteria in sedimentary environments is complex and not fully understood at present. Some types can grow in both aerobic and anaerobic environments and are able to carry out more than one of the functions shown in Fig. 3.6, and different species of the same genus may carry out markedly different processes.

3.3.3 Factors affecting sedimentary preservation of organic matter

There is some debate over the relative importance of productivity and anoxicity in the formation of organic-rich sediments (Demaison and Moore, 1980; Pedersen and Calvert, 1990). Sufficiently high productivity is clearly important if organic material is to reach the sediment in any quantity and stand a chance of preservation, rather than undergoing complete recycling within the water column. Seasonal algal blooms probably provide a better opportunity for sedimentary incorporation than a uniform, lower production rate throughout the year.

Anoxicity of sediment and bottom water is often a consequence of the degradation of large amounts of organic matter, but this relationship does not prove that anoxicity is a prerequisite for preservation. However, lipid-rich (i.e. hydrogen-rich), fine-grained sediments are generally finely laminated (i.e. non-bioturbated), consistent with anoxicity of sediments and bottom waters. Where the top layers of sediment and the overlying water are oxic any preserved organic matter tends to be hydrogen-poor, deriving from more refractory, lignified, higher plant material. Anoxicity drastically reduces the degree of decomposition of sedimentary organic matter by detritivores, even if the extent of bacterial degradation is not significantlly affected. It is possible, therefore, that the preservation of organic material in sediments is influenced by both anoxicity and the level of primary productivity.

Most soils are oxygenated and the organic matter they contain is efficiently recycled by organisms of the detrital food chain, resulting in little long-term preservation of organic material in soil. However, water-logging aids the development of anoxicity and the preservation of organic matter as peat.

At present, bottom waters and the upper layer of sediment throughout most of the oceans are oxygenated, as a result of the deep-water circulation pattern (see Box 3.1). In both marine and lacustrine environments oxygen consumption increases with depth under areas of high primary productivity, mainly due to microbial respiration during the degradation of organic matter. In oceanic environments an **oxygen-minimum layer** is produced beneath the photic zone, where oxygen demand is greater than supply and there is a maximum in carbon dioxide and nutrient concentrations because of the absence of phototrophic assimilation. In present-day oceans this layer is at ca. 300–1500 m. When upwelling occurs nutrients from this zone are taken up into the oxygenated photic zone, increasing phytoplankton production. The intensity of oxygen depletion within this zone is dependent on the residence time of the layer within the water column and the productivity of the overlying photic zone. When this zone intercepts the continental slope the bottom waters and sometimes the sediment of this area become anoxic (Fig. 3.7). An example is the accumulation of organic-rich sediments under the Peru upwelling.

Anoxicity is not always a consequence of *in situ* degradation of large amounts of organic matter. In parts of the open ocean, away from the coastal areas, an anoxic water layer is present that has no relation to those which impinge on continental margins. These oxygen minima occur at intermediate ocean depths and are associated with the influence of the Coriolis effect on the patterns of deep-water circulation (see Box 3.1) and not with high productivity.

Rapid burial can aid the development of anoxicity in sediments but can lead to the dilution of organic matter by large amounts of inorganic material. If the energy of the water system is too high deposition of fine-grained material is limited and any organic matter present is largely aerobically oxidised. This can be the fate of

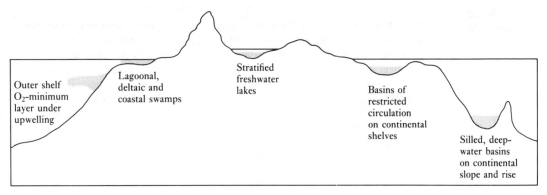

Figure 3.7   Important oxygen-depleted environments (stippled areas) associated with deposition of organic-rich sediments. (After Brooks et al., 1987.)

much sedimentary organic matter in productive areas, such as estuaries. Mangrove swamp and saltmarsh communities can also be prolific primary producers (see Table 3.1), and anoxicity can develop in the sediments trapped by the root systems of the macrophyte stands. However, the formation of organic-rich sediments in these environments is hindered by dilution with clastic material, a limited vertical extent of accumulation (if sea level remains constant) and subaerial exposure. Similar limitations can apply to the preservation of organic-rich sediments in freshwater swamps and marshes. Accumulation of significant thicknesses of organic-rich sediments in these areas requires water level to rise.

Coral reefs are also productive areas, but are confined to oligotrophic waters, where light penetration is maximum and coral polyps are not choked by detrital material. However, all nutrients are effectively recycled within the living reef system, providing limited potential for the formation of organic-rich sediments.

From the above it can be concluded that accumulation of organic-rich sediments is mostly associated with relatively high productivity in low-energy oxygen-depleted environments. Some general examples of such depositional environments are shown in Fig. 3.7. Basins are important sites for the potential large-scale accumulation of organic-rich sediments because anoxicity can develop in bottom waters during the degradation of large amounts of organic matter and thick sedimentary deposits can form (see Box 3.3). Examples are large eutrophic lakes and marginal marine basins in which the inflow of oxygenated bottom waters is obstructed by a sill or similar barrier (Fig. 3.7).

Box 3.3 | **Basins and sediments**

A **sedimentary basin** is a depression in the Earth's crust where a large thickness of sediment accumulates. These depressions can result from relative movement of adjacent crustal plates. Such tectonic movements give rise to a variety of basin types, which can be complex in formation and for which a variety of classification systems exist (e.g. Klemme, 1980; Selley, 1985). Generally, tectonic movement produces either compressional or extensional forces in an area of crust. The former can result in crustal down-warping and

(continued on next page)

Box 3.3
(continued)

basin formation. Extension results in crustal thinning and is generally associated with higher temperatures than in surrounding areas of normal crustal thickness as the hot mantle is nearer the surface. As tectonic activity subsides, cooling results in sagging of the crust and basin formation. Other mechanisms of basin formation exist, such as erosion and subsequent subsidence during cooling of a crustal bulge caused by a local hot spot in the mantle. Sediment loading can cause the development of a depression where a river delta progrades on to adjacent oceanic crust.

Inorganic sedimentary infill in basins generally results from erosion of material from surrounding areas of higher relief and its transportation by water (and sometimes wind) to the area of deposition. The composition of this **detrital** (or **clastic**) material varies, but aluminosilicate minerals are usually important. There are also biogenic sediments, resulting from the remains of organisms (e.g. carbonaceous and siliceous tests, peat) and **chemical sediments** formed by precipitation of minerals from solution (e.g. evaporites, some limestones and authigenic infills of pores by quartz and calcite crystalline growth).

The sedimentary infill in a basin can change with time, allowing the recognition of different bodies of sedimentary rock. Such a body is termed a **facies**, and it displays a set of characteristic attributes that distinguish it from vertically adjacent bodies. Various distinguishing attributes include sedimentary structures, mineral content and fossil content.

Tectonic processes control whether subsidence and progressive burial of organic-rich sediments occurs, leading to long-term preservation, or whether uplift halts deposition and possibly leads to erosion of the sedimentary accumulation.

3.4   Depositional environments

In the earlier part of this chapter we have considered factors directly related to the production and preservation of organic matter in sedimentary environments. This section is concerned with a more detailed examination of some of the environments where organic-rich sediments accumulate.

3.4.1   Lacustrine environments

Lakes contain only 0.02% of the water in the hydrosphere, yet their deposits of organic-rich sediments are of economical importance. There is a great diversity of lake types and sizes but an important feature of all lakes is that they are ephemeral features, on geological time-scales, and conditions in individual lakes may change over relatively short periods. This can lead to a variety of organic facies (see Box 3.3) in ancient lake basins reflecting different communities of organisms (i.e. autochthonous contributions), allochthonous sources and deposition conditions. An example of such a complex lake system is the ancient Green River Formation of Utah, Wyoming and Colorado, which contains huge reserves of oil shales.

Small lakes may be volcanic or glacial in origin, or they may occur as ox-bow lakes or coastal lagoons. Lakes formed in areas adjacent to glaciated regions (pluvial lakes) are now found in present-day dry belts (e.g. Salt Lake, Utah, USA) and were formerly much larger than at present. Very large-scale lakes are tectonic in origin (see Box 3.3) and are formed either in active tectonic areas such as extensional rift valleys (e.g. East African Rift) or as intracratonic sag basins (i.e. within a stable continental plate, like Lake Chad, Africa).

Lake size and morphology influence the thermally induced stratification of the water column and its stability, as there is a relationship between the depth of the seasonal thermocline and the maximum fetch of the water body: the longer the fetch (i.e. the distance over which wind acts on the water), the deeper the thermocline. The potential effects of stratification on productivity and anoxicity have been discussed in Sections 3.2 and 3.3.

Lake sedimentology is dependent on size, water chemistry and the amount of allochthonous, river-derived, clastic material. Organic material found in lakes ranges from algal remains through to degraded terrestrial material. Lakes can be classified according to whether they have an outlet and are hydrologically **open** or lack such an outlet and are hydrologically **closed**. However, some lakes have been both closed and open during periods of their history (e.g. Lake Kivu, Africa).

Open lakes Open lakes are characterised by having fairly stable shorelines, as inflow and precipitation are balanced by outflow and evaporation. Most of the allochthonous material in hydrologically open lakes is transported there by rivers. This input is, therefore, controlled by river drainage (or catchment) areas and the climatic influence on run-off. During the early summer the load can be heavy, but during the winter it may be only very minor, if much of the precipitation is held as snow and ice on higher ground. These conditions give rise to layered sediments which will be discussed more fully below. The depositional elements in a hydrologically open, freshwater lake are shown in Fig. 3.8.

In large lakes siliciclastic sediments (see Box 3.3) are normally concentrated around river mouths as beaches, spits and barriers fashioned by wave action. Deltas may also be present in relatively deep, freshwater lakes and resemble those found in marine environments. Nearshore deposition is characterised by accumulations of terrestrial plant material, deposited in interdistributary bays on delta tops, which

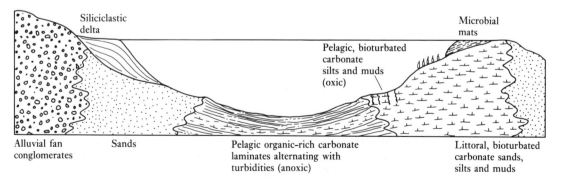

Figure 3.8 Generalised depositional components in a hydrologically open, freshwater lake. (After Eugster and Kelts, 1983.)

grade from peats to lignites (see Section 4.3.2). However, bacterial decomposer communities may sometimes actively rework the higher plant material deposited within the delta, their remains increasing the lipid content of the organic material. Delta fronts can become over-steep and collapse, depositing **turbidites** (accumulations of sediment, often thick, that have been transported down-slope in the form of a sediment–water slurry, also called a density flow) in deeper water. While this results in conditions suitable for preservation of organic matter, dilution by large volumes of clastic material tends to occur (see Box 3.3).

In offshore regions, clastic sediment is deposited by turbidity flows, pelagic rain (i.e. fall-out of material from the water column) and mass sediment flows. Organic material can be mostly autochthonous, of planktonic origin. It may be deposited in anoxic or poorly oxygenated bottom waters, due to water column stratification, in which bioturbation is inhibited. This often results in the formation of organic-rich shales and marls, which may be of great thickness and have a very high organic carbon content (Powell, 1986). Large, permanently stratified lakes with anoxic bottom waters are mainly associated with tropical climates.

The offshore deposition of many lakes is characterised by rhythmical sedimentation of alternating layers (or **varves**). This arises because fine-grained sediment may be suspended for long periods, depending on the degree of circulation and the availability of flocculating agents. The coarser-grained sediment layers result from fluvial summer deposits and the fine-grained layers from fall-out during the winter. The thickness and nature of the layers depend on the proximity of the inflowing river.

Sediment lamination is also caused by the deposition of calcium carbonate and is important in most freshwater lakes where clastic input from rivers is low. Calcium deposits may be chemical or biogenic (see Box 3.3). As with phytoplankton production, the abiogenic (chemical) precipitation of calcium carbonate is seasonal, occurring during the warmer months. During this period carbon dioxide is removed from water by planktonic photosynthesis, which causes a decrease in water acidity and precipitation of calcium carbonate. Subsequently, the water becomes more acidic and no longer supersaturated with respect to calcite (the main form of crystalline calcium carbonate), resulting in cessation of precipitation. In large, deep, seasonally stratified, anoxic lakes this seasonal variation gives rise to laminated, organic-rich sediments showing annual couplets. The lighter laminae represent the spring and summer production of carbonate with low magnesium content following phytoplankton blooms and the dark, carbonate-poor, organic-rich layers arise from the settling out of organic material and diatom remains during winter (Allen and Collinson, 1986). Most organic material in these lakes is preserved as carbonate laminates in the offshore, lacustrine, depositional region. Moving shorewards into shallower water, biogenic carbonate may form, consisting of shells or algal carbonates.

Closed lakes Hydrologically closed lakes can be divided into perennial saline lake basins and ephemeral salt-pan basins. Changes in water level in lakes with closed drainage cause substantial fluctuations in shorelines and much reworking of sediment in the nearshore zone. However, pelagic, organic-rich, carbonate laminates, interbedded with evaporitic minerals, can form in the centre of perennial saline lake basins when anoxic conditions exist due to stratification at times of high water levels (Eugster and Kelts, 1983). Stratification results from salinity differences and water below the halocline becomes depleted in oxygen.

In ephemeral continental **sabkhas**, evaporitic minerals (e.g. high magnesium calcite and gypsum) form on the edges of the water body and during periodic rains they may be washed into the lake and deposited. The organic-rich facies of these lakes may be formed during more arid periods as the growth of cyanobacteria is favoured. As mentioned in Section 3.2, mat-forming communities at the water line, dominated by cyanobacteria, can be highly productive in hypersaline environments. As well as colonial genera of cyanobacteria (*Lyngba*, *Microcoleus*) other organisms, such as diatoms and photosynthetic bacteria, are present in these communities.

3.4.2 Peat swamps and coal formation

Coal is a carbonaceous material formed by compaction and induration of plant remains which may have originally been deposited as peat in swamps. It is convenient to treat these swamps as a special type of lacustrine environment, as water balance is an important factor. Coal seams date back to the Middle Devonian, when extensive development of land plants occurred, but the first economic deposits originate from the Lower Carboniferous. The number of peat-forming plant varieties was small until the Lower Cretaceous, when a great increase in species diversity occurred.

A variety of peat types has been identified according to the main plant genera present. These include reed/sedge types (e.g. *Cladium*, sawgrass), woody types (e.g. *Taxodium*, cypress; *Rhizophora*, mangrove) and moss types (e.g. *Sphagnum*). Not only does the flora contributing to peats vary with environmental factors (e.g. climate) and over geological time, it also varies with time during the evolution of a swamp. Succession of plant communities can occur in swamps as a consequence of changing environmental factors relating to the building up of substrate (by dead plant material) above water level and subsequent drying out. Initially, therefore, water plants such as reeds dominate. As the ground builds up and dries out herbs and then scrub, characteristic of drier ground, can develop and eventually trees become established. If this succession occurs over the entire area of the swamp, the swamp disappears and peat accumulation ceases. However, succession may be cyclical, where the death of a stand of trees (e.g. as a result of fire) causes break-up of the substrate and reversion to more open water conditions. Different patches of the swamp can be at different stages in this cycle and, overall, peat accumulation continues. Horizons can then be recognised in the peat due to formation from different plant types.

For peat to accumulate there must be a balance between the production of plant material and the rate of decay. These factors are determined largely by climatic differences in that plant production and decay increase with rising temperature and humidity. At present, the largest peat deposits are forming under cool conditions at higher latitudes ($>45°$) in high moors and raised bogs. This is because plant growth is less dependent on temperature than is bacterial decay. Bryophytes (mosses) are particularly important in such areas, especially *Sphagnum*. Moss exudates and bacterial activity tend to increase the acidity of bog pools, producing an unfavourable environment for the decomposer bacteria. In tropical regions, such as the Mahakam Delta (Indonesia), large amounts of vegetation (including trees) are preserved because of extremely high rainfall, which leads to water-logging and anoxicity. The high rate of bacterial decay leads to oxygen depletion in the swamp and pools, thereby severely restricting further decay and resulting in the formation of peat deposits.

To create favourable peat-forming conditions precipitation and inflow must exceed evaporation and outflow. There must be a continuous rise in the groundwater keeping it at or above the sediment level to allow peat formation to continue. If the level of groundwater is too great the peat swamps drown and peat formation ceases. If the water level is too low the peat already formed is eroded and the organic material is decomposed as it becomes exposed to oxygenated shallow waters or even to the air. For enhanced preservation of peat it is necessary for the overlying water to be anaerobic, which is aided by retardation of water flow through the forming peat.

Much peat is deposited in **paralic** (i.e. marginal marine) areas and deltaic swamps. These swamps must be protected by bars and spits from major marine inundations and by levees from river floodwaters. As peat formation can be interrupted by the deposition of excessive fluviatile sediments, more favourable conditions are created if the river hinterland is of low relief, thereby reducing the river sediment load. Although most peats are autochthonous, some have an allochthonous element and are associated with enhanced mineral content. The latter peats may be formed from reworked peats or from vegetation that has been transported, often by rivers and wave action (e.g. Mahakam Delta). Coal formed under these conditions is detrital.

Okefenokee swamp

Some idea of ancient peat formation can be gained from the Okefenokee swamp of south Georgia, one of the most primitive swamps in North America. It covers ca. 1500 km^2 and has peat deposits up to 5 m deep, which are accumulating at ca. 0.5 mm/a. It is believed to be similar to swamps existing in the Tertiary. Beneath the peats are Pleistocene sands and muds, which originally formed the bed of a shallow lagoon at a time when the Atlantic reached much further inland. A few islands are present, representing offshore sand banks and bars built up by tidal action. The swamp is now, however, 30 m above sea level.

Initially, the formation of peat probably only occurred within shallow depressions but subsequently took place over wider areas, until only the tops of sand bars remained uncovered. In the present-day swamp there are two distinct environments: forests (covering most of the swamp) and open-water marshes (called prairies). The prairies are rarely flooded to >30 cm and only dry out during major droughts at intervals of ca. 25 years. The peat in prairies is thickest where it overlies depressions in the sands and muds of the ancient lagoon; it is red-brown and fibrous, mainly formed from roots, leaves and stems of water lilies, which are abundant in the prairies.

Much marsh gas is produced from the decay of peat in the prairies, but it can be trapped, causing large round areas of peat (batteries) to rise to the surface. Water plants on these batteries die when the surface dries out and they are replaced initially by grasses and sedges. These, in turn, are succeeded by shrubs and then large trees (mainly cypress), forming tree houses, which are anchored by tree roots to the underlying peat. Around the tree houses, in the transitional area between dry ground and open water, are fringes of vegetation such as maiden cane and sedges. However, the sedges rot away quickly and contribute little to the peat. Eventually tree houses may merge and form expanses of forest. Because these areas are relatively dry, oxygen is generally available for the degradation of litter, although peat still accumulates (ca. 1 m thickness). This peat is formed mostly from cypress remains (e.g. needles and wood) and is granular rather than fibrous and dark, due to oxidation and the presence of charcoal. Charcoal comes from forest and vegetation fires that can take place during droughts.

Major droughts can cause prairies to dry out and even the peat may then ignite, with the destruction of 30 cm or more of accumulation. The existence of prairies, therefore, is due mainly to fires which halt forest establishment. The relative abundance of prairies and forests can be seen in the vertical profile of peat types. A repeating sequence is observed, grading upward: cypress peat, charcoal zone, fibrous water lily peat.

3.4.3   Marine environments

Ocean environments can be divided into areas according to water depth: **continental shelf**, 0–200 m; **continental slope** (which includes the continental rise), 200–4000 m; **abyssal plain**, 4000–5000 m (see Fig. 3.3). Deeper areas are present, the troughs associated with subduction zones (destructive plate margins), but these are of relatively minor area and importance. Shelf, slope and continental rise are collectively termed the **continental margin** and, as we have seen, productivity is at its highest in this area of the oceans (Fig. 3.5). Basins in continental margins in which water exchange with the open sea is restricted by physical barriers are important settings for the formation of organic-rich sediments, together with upwelling areas (Fig. 3.7).

Marine shelf deposits   Mud-dominated, but not necessarily organic-rich, offshore shelf deposits are very abundant in the sedimentary record. Currently, thick shelf deposits are forming in areas that have low wave and current energies and where suspended sediment is at high concentrations. The type of sediments deposited is dependent upon climate, latitude, major wind belts and tidal energies.

Major periods for the widespread formation and preservation of organic-rich sediments in shelf deposits include the mid part of the Cambrian, the early mid Ordovician, the Devonian, early and late Jurassic and the mid and late Cretaceous. These periods may have coincided with minima in polar ice-cap development. The resulting larger volume of water in the ocean basins would have produced a large global (eustatic) rise in sea level resulting in marine **transgressions** and the extensive deposition of organic-rich, shallow, marine mudstones as the oceanic mid-water oxygen-minimum layer impinged on large, epicontinental, shelf areas. As the transgression of oceanic waters across the continents occurred large amounts of nutrients were liberated into epeiric (shallow inland) seas. Primary productivity would have been greatly enhanced generating large quantities of organic detritus. Water circulation would have been restricted, resulting in extensive oxygen depletion as bacteria oxidised the large amounts of organic material available. At times when polar ice-caps were well established, sea levels would be lower, the oceans would probably be well oxygenated and mixed, the shelf area would be much reduced as a result of marine regressions, and productivity and preservation of organic matter would have been much lower.

While variation in the size of the polar ice-caps can cause a change in sea level of up to ca. 200 m, tectonically related changes can be more significant. At times of high tectonic activity the mid-ocean ridges (constructive plate margins) occupy a larger part of the ocean basins. During the Cretaceous such activity may have resulted in a sea level rise of 600 m, leading to widespread marine transgressions, the effects of which are considered in more detail below.

In the present-day marine environment, shelves are also the setting for upwelling currents, which off the Namibian coast result in an oxygen-deficient zone measuring

340 km long by 50 km wide. Sediments deposited in upwelling zones have distinctive features including laminated non-bioturbated muds, phosphorites and uranium minerals (Allen and Allen, 1990).

The coastal shelf area has an intimate association with the land at the deltas of major rivers (e.g. Mississippi and Amazon). Peat and lignite formation is often a feature of the delta plain, although freshwater phytoplankton and bacteria that rework detritus may be locally important in lakes, swamps and abandoned channels. Brackish conditions may also occur, favouring those bacteria that rework organic debris (chiefly from higher plants) and giving rise to organic sediments with a high wax content. On the seaward slopes of the deltas a large proportion of organic material is of terrestrial origin and at least partially oxidised. Nutrients are abundant and productivity may be high, but preservation of organic material in high concentration is not favoured as it is often greatly diluted by organic-poor fluvial sediments derived from a hinterland that is not covered with lush vegetation. However, organic-rich sediments are presently forming in the tide-dominated portion of the Mahakam Delta plain. This is because there is a large amount of fine-grained organic-rich sediment in the delta which is derived from the fluvially dominated part of the delta plain that supports rich equatorial forests of mangrove and palm (Elliott, 1986).

Deposition of organic-rich sediments further down the shelf and on to the continental slope and rise often occurs as a result of turbidite flows, redistributing organic-rich sediments from delta fronts or from further up the shelf and slope (Summerhayes, 1983). While there is a certain amount of pelagic sedimentation, primary production is reduced away from the coastline as nutrient levels decline and detritus is largely recycled before it settles to the sea floor.

In hot arid areas where the continental shelf meets the land, marine sabkha-type environments exist. For example, on the Trucial coast of Abu Dhabi shallow marine carbonate sediments are reworked and bound by microbial mats. In the upper intertidal zone these mats are virtually undisturbed by predators and can grow and be preserved, eventually forming extensive organic-rich deposits.

Enclosed and silled basins   Apart from nearshore areas and deltas, continental shelves are not important as sites of organic-rich sediment accumulation except where basins are present. Whether or not the basin will contain organic-rich sediments depends on factors such as climate and water depth (200–1000 m is ideal) and hydrological factors. The nature of water exchange with the open sea is important when establishing whether a restricted basin has the potential to accumulate organic-rich sediments. A negative water balance occurs in arid hot climates, where surface seawater flows in to make good losses due to both evaporation and the sinking of hypersaline water which then flows out of the bottom of the basin. The bottom waters of the basin are, therefore, oxygenated and depleted in nutrients, providing unfavourable conditions for the formation of organic-rich sediments. Examples of such basins are the Red Sea, the Mediterranean Sea and the Persian Gulf.

When the water balance is positive, the outflow of relatively fresh water exceeds the inflow of denser, saline water. The latter tends to be relatively rich in nutrients but depleted in oxygen (due to the decay of organic matter sinking out of the overlying water column). A pronounced salinity gradient often occurs in the water column of the basin, allowing stratification to occur and the development of anoxicity in the lower layers. Nutrients are often trapped in such basins, resulting in high productivity and preservation of sedimentary organic matter in oxygen-poor conditions. Examples of such basins are the Black Sea, the Baltic Sea and Lake Maracaibo (Venezuela) (Demaison and Moore, 1980).

Production and preservation of organic matter in the Black Sea

The Black Sea is often used as a model for the preservation of organic-rich sediments of the type that can yield petroleum. It is the largest body of anoxic marine water in the world. Productivity is high in surface waters and degradation of the resulting detritus leads to anoxicity in the deep water of the isolated basin. Preservation of organic carbon is, therefore, high. Estimated production and preservation rates of organic matter are given in Table 3.4. Primary production is largely attributable to phytoplankton, but there are also small allochthonous inputs of detrital organic matter to the photic zone from rivers. Chemosynthetic bacteria make a significant contribution to autochthonous organic matter in anoxic waters lying below the photic zone. Most of the photosynthetic production and allochthonous inputs are recycled within the photic zone, while sulphate reduction accounts for most of the anaerobic degradation. However, ca. $4 \, \text{g} \, \text{C}/\text{m}^2$ is preserved in sediments annually.

Some interesting changes appear to have occurred in the water column of the Black Sea over the last decade (Repeta et al., 1989). The hydrogen sulphide interface has moved up into the photic zone, at ca. $80-100 \, \text{m}$ depth. Concentrations of bacteriochlorophyll-e (Fig. 2.22) and bacterial aromatic carotenoids (e.g. isorenieratene, see Fig. 2.21) reach a maximum at this depth, providing evidence for anaerobic bacterial photosynthesis by *Chlorobium* in this region of the water column. The biomass of photosynthetic bacteria (estimated at ca. $0.5 \, \text{g}/\text{m}^2$) may amount to half that of the phytoplankton, suggesting that the Black Sea has the largest and deepest community of photosynthetic bacteria presently in existence. It also suggests that such communities may have been more important than previously believed in similar ancient environments.

Cretaceous anoxic events

As mentioned above, during the Cretaceous there was widespread deposition of organic-rich black shales in marine environments. It was originally thought that this was due to anoxic bottom waters throughout the oceans (ocean anoxic events) resulting from diminished deep-water circulation, but this model has been superseded

Table 3.4   **Sources and fate of organic carbon in the Black Sea over the last 2000 years** (Data after Tissot and Welte, 1984)

	Annual flux $(g \, C/m^2)$
Inputs	
Photosynthesis	100
Chemosynthesis	~ 15
Rivers	5
Azov and Marmara Seas	2
Outputs	
Aerobic respiration	~ 100
Azov and Marmara Seas	3
Sulphate reduction	10
Dissolution	5
Sedimentary preservation	4

in recent years. During the mid to late Cretaceous, organic-rich sediments were indeed being deposited throughout the Tethyan Ocean and opening Atlantic, and on isolated seamounts and plateaux in the Pacific. The global climate was warm, influencing rainfall, vegetation and run-off. Surface and bottom waters were warm with only small temperature differences and ocean floor spreading rates were high. The increase in global temperatures would have caused a shrinkage of the ice-caps and a decline in the volume of cold oxygen-rich waters sinking at the poles. Thermoclines would have been less likely to form, although salinity differences may have become important in initiating stratification. These salinity differences would have been the main driving force for deep-water circulation and in low latitudes highly saline waters would have formed the main deep water mass in the oceans rather than the dense, cold and oxygenated waters of today. Because of higher temperatures in the Cretaceous these water masses would have had lower oxygen concentrations, although the oceanic circulation was probably not sluggish and enough oxygen would have been present in the bottom water currents to maintain oxidising conditions, especially in the Pacific Ocean.

It is possible that oceanic circulation was nearly as vigorous as it is today because higher levels of atmospheric carbon dioxide during the Cretaceous would have enhanced atmospheric circulation, increasing the velocity of the Trade Winds and intensifying current divergence and upwelling (Berner et al., 1983; Pedersen and Calvert, 1990) (see Box 3.1). Increased upwelling results in increased productivity and, ultimately, the deposition of organic-rich sediments. However, the warm waters of the Cretaceous did not contain high concentrations of oxygen and although water conditions were mostly oxic, depletion did occur in areas of high productivity in intermediate and deep water masses. Organic-rich sediments deposited during the mid Cretaceous were not synchronous and, therefore, were not global events. As oxygen levels in the ocean were easily changed locally, organic sediments were preserved under anoxic conditions in upwelling zones and in areas with large influxes of terrestrial material (e.g. where marine transgressions had caused an increase in the area of the continental shelf). An intensified and deeper oxygen-minimum zone (up to 3000 m deep), especially in areas of high productivity, would explain the deposition of organic-rich sediments where water depth was at an optimum: on large areas of continental shelf and slope where restricted basins may have occurred, on seamounts, on the Mid-Atlantic Ridge, etc. Redeposition of organic-rich facies by turbidite flows into deeper, oxic basins may also have occurred.

As was mentioned above, large-scale marine transgressions occurred during the Cretaceous, due mainly to increased activity of mid-ocean ridges and involving a sea level rise of perhaps 600 m above the long-term mean. High nutrient input from flooded continental areas would have resulted in high productivity in these areas, which probably also provided many basins in which anoxicity could develop and lead to the preservation of organic-rich shales. It has been noted that Cretaceous ocean anoxic events appear to correlate with transgressive periods and it is conceivable that significant amounts of terrestrial organic material could have been carried seawards, contributing to shale formation (Jenkyns, 1980).

Chapter 4

Formation of humic material, coal and kerogen

4.1 Diagenesis

4.1.1 Introduction

The preceding chapters have examined the main contributions to sedimentary organic matter, the types of organic compounds involved, and the environments associated with the large-scale production and preservation of this material. The transformations that the organic matter undergoes during its incorporation into sediments and subsequent lithification must now be considered. In organic geochemistry the term **diagenesis** is applied to the processes affecting the products of primary production that take place prior to deposition and during the early stages of burial under conditions of relatively low temperature and pressure. Biological agents are mainly responsible for diagenetic transformations, although some chemical transformations are possible and are often catalysed by mineral surfaces.

During diagenesis burial depth increases and the sediment undergoes compaction and consolidation. There is an associated decrease in water content and an increase in temperature. Biological activity eventually ceases as the later stages of burial, called **catagenesis**, are entered, where the alteration of organic matter is largely the result of elevated temperatures (ca. $>50°C$). The boundary between diagenesis and catagenesis is generally not sharp and corresponds approximately to the onset of oil formation, which is discussed more fully in Chapter 5.

4.1.2 Microbial degradation of organic matter during diagenesis

The transformations involved in the utilisation of primary production by heterotrophic organisms, particularly micro-organisms, during diagenesis strongly influence the form of the organic matter that may ultimately be preserved in sediments. Amino acids together with low-molecular-weight peptides and carbohydrates (sugars) can be absorbed directly by decomposer organisms and are usually rapidly recycled. They are, consequently, generally minor constituents in ancient sediments. Insoluble proteins and polysaccharides have to be broken down into their water-soluble constituent amino acids and monosaccharides, respectively, by fungal and bacterial extracellular enzymes before they are assimilated (Fig. 3.6). In this way, high-molecular-weight compounds forming storage (e.g. lipids and starch) or structural (e.g. exo-skeleton and cell wall) materials are transformed into assimilable components. As a result, the initial input of biological macromolecules (e.g. proteins, polysaccharides, lipids and lignin) often comprises no more than 20% of the total organics in the upper layers of sediments.

Degradation of organic material begins in the water column and continues after sedimentation. Different compound classes are degraded at different rates and only some compounds survive in a recognisable form. Proteins and nitrogen-containing compounds in general appear to degrade faster than the other major sedimentary organic compound classes, and the highest concentrations of amino acids are found at the water/sediment interface. The stability of individual amino acids towards diagenetic processes can differ. For example, in the shells of *Mytilus edulis* (a mussel) glycine progressively decreases in concentration relative to alanine with increasing sediment age.

Carbohydrates seem only slightly less readily degraded than proteins. Cellulose is degraded aerobically mainly by fungi in terrestrial environments and by bacteria in aquatic ecosystems. Anaerobic degradation of cellulose by bacteria (e.g. *Clostridium*) is also important in water-logged soils and deep-water sediments. The predominant sugars formed by hydrolysis of carbohydrates are hexoses and pentoses. Carbohydrates and proteins are not necessarily completely degraded and traces have been found in Precambrian sediments.

Lipids may also be recycled but some appear more resistant to degradation than proteins and carbohydrates and can survive diagenesis, undergoing only minor alterations. Lignins are the class of sedimentary organic compounds most resistant to biodegradation, although some fungi (especially the basidiomycetes, which include bracket fungi) and some bacteria (e.g. actinomycetes) can decompose them in terrestrial environments. It is interesting to note that it is the presence of microbial communities in their gut that permits terrestrial herbivores to digest higher plant material.

The general effect of microbial attack is, therefore, to reduce the concentrations of all compound classes with increasing sediment depth. However, association of even the most readily degradable substances with resistant structures can aid their preservation. For example, the proteins that form the matrix for mineralisation in invertebrate shells are relatively protected against degradation by their mineral covering. Highly cross-linked fibrous proteins such as keratin resist microbial attack but can be degraded by some fungi and most actinomycetes. Similarly, cellulose and hemicellulose from higher plants can be protected by lignified layers. Lipids can also be protected where they are incorporated in resistant coatings, such as in spores and pollens.

4.1.3   Geopolymer formation

The chemical residue from microbial degradation undergoes increasing condensation during diagenesis, finally yielding large amounts of brown organic material which becomes increasingly insoluble as hydrolysable components, such as carbohydrates and proteins, disappear. As we have seen in Chapter 3, certain conditions have to be met for the preservation and accumulation of organic matter in sedimentary environments, and the final form and properties of this matter depend on the source material and environmental conditions. At the end of diagenesis the polycondensed organic residue, or geopolymer, is called humin in soils, brown coal (or **lignite**) in coal swamps, and kerogen in marine and deep lacustrine sediments. These diagenetic products probably also contain varying amounts of largely unaltered refractory organic material.

4.2 Humic material

4.2.1 Occurrence and classification

Humic substances are found in soils, brown coals, fresh water, seawater and both marine and lacustrine sediments. The humics in soils are originally derived from plant matter. They can be leached, particularly from acidic soils in cold wet upland regions of the UK, and account for almost all the organic carbon in fresh waters, imparting the characteristic brown coloration to upland waters. In addition, most of the dissolved organic carbon (DOC) in seawater is humic or humic-like material and is sometimes described as 'yellow substances'. There are, however, significant differences between terrestrial and marine humics, which will be discussed later.

The term humic substances describes three groups of material: **fulvic acid, humic acid** and **humin**. This distinction is based on the traditional fractionation of soil humic material. Treatment of bulk humic material with dilute alkali dissolves the fulvic and humic acids, leaving a residue of insoluble humin. Acidification of the alkaline extract precipitates humic acid, leaving fulvic acid in solution. Fulvic acid comprises smaller and more oxygenated units than humic acid, and so has been proposed as possibly the oxidised degradation product of humic acid. Both humic and fulvic acids are thought to be intermediates in the diagenetic formation of humin. Humic substances can occur in different physical states. For example, humic acids are found in solution in fresh water, as solids or gels (colloids) in soil and as dry solids in coal.

4.2.2 Composition and structure

The complex polymeric nature and interaction between component chains of humic material make structural analysis difficult; however, compositional information can be obtained from elemental and functional group analysis. Elemental analysis of humic and fulvic acids from a range of soils is presented in Table 4.1. The atomic H/C ratio is quite low, being lower for humic acid (ca. 0.8) than fulvic acid (ca. 1.3), consistent with a higher aromatic content for humic acid. The atomic O/C ratio is also lower for humic acid (ca. 0.5) than fulvic acid (ca. 0.8), reflecting the higher content of polar groups in fulvic acid to which its greater solubility is attributed. There are similar differences between marine humic and fulvic acids. Nitrogen and sulphur levels are generally relatively low in humic substances but can be higher in marine than terrestrial humics.

Table 4.1   **Typical elemental compositional ranges for soil humic and fulvic acids** (After Schnitzer, 1978)

	Humic acid (wt%)	*Fulvic acid (wt%)*
C	53.6–58.7	40.7–50.6
H	3.2–6.2	3.8–7.0
N	0.8–5.5	0.9–3.3
O	32.8–38.3	39.7–49.8
S	0.1–1.5	0.1–3.6

Table 4.2   **Estimated abundance of functional groups (mequiv/g*) in soil humic and fulvic acids** (After Schnitzer, 1978)

	Humic acid	Fulvic acid
Total acidic groups	5.6–8.9	6.4–14.2
Carboxyl COOH	1.5–5.7	5.2–11.2
Phenolic OH	2.1–5.7	0.3–5.7
Alcoholic OH	0.2–4.9	2.6–9.5
Quinoid/keto C=O	0.1–5.6	0.3–3.1
Methoxy OCH$_3$	0.3–0.8	0.3–1.2

*mequiv/g is equivalent to mmol of each group per g humic substance, where 1 mol = 6×10^{23} molecules.

Oxygen is the major heteroatom in humic substances and occurs predominantly in the following functional groups: COOH, phenolic and alcoholic OH, ketonic and quinoid C=O, and OCH$_3$. The estimated abundances of these groups in soil humic and fulvic acids are given in Table 4.2.

The macromolecular structure of fulvic and humic acids probably consists of a flexible extended chain with only limited branching and cross-linking (Hayes et al., 1989). Attached to this backbone are smaller compounds, particularly sugar and amino acid residues, which can be released by hydrolysis. Concentrations of these sugars and amino acids are greater in humic than fulvic acids and are also greater in soil humic acids than freshwater humic acids. Sugars are usually more abundant than amino acids but their combined abundance does not normally exceed 20% of humic substance weight.

Units derived from lignin appear particularly important in the backbone structure of humic substances, some being largely unaltered lignin components while others show signs of microbial alteration. Some probable structural features of fulvic acid based on predominantly lignin-derived polycarboxyl-phenol components are shown in Fig. 4.1. Although single-ring aromatic systems dominate, there may also be small amounts of condensed (i.e. polycyclic) aromatic rings and furan derivatives. Aromatic units are often joined by ether (—O—) links and also sometimes by short-chain alkyl groups (particularly the C$_3$ group of lignin components). There is evidence for the involvement of polysaccharide-like material in the macromolecular backbone, suggesting that carbohydrates are incorporated during diagenesis. A degree of amino acid incorporation into the backbone may also occur.

Analysis of the degradation products obtained from pyrolysis and oxidative degradation studies suggest that terrestrial humic substances are highly aromatic (ca. 70% of total C). However, these techniques tend to overemphasise the importance of aromatic and phenolic units. Non-degradative techniques, such as ^{1}H and ^{13}C nuclear magnetic resonance, probably give more realistic aromatic carbon contents of 20–50% for soil humics, 20–35% for peat humics and <15% for marine humics (Hatcher et al., 1980). These data suggest that aromaticity can offer a means of discriminating between humic material from terrestrial and marine sources (Dereppe et al., 1980). In addition, they indicate that while aliphatic structures are more abundant in marine than terrestrial humics, they are important in all humic material (Hatcher et al., 1981). An important contribution to aliphatic structures are fatty

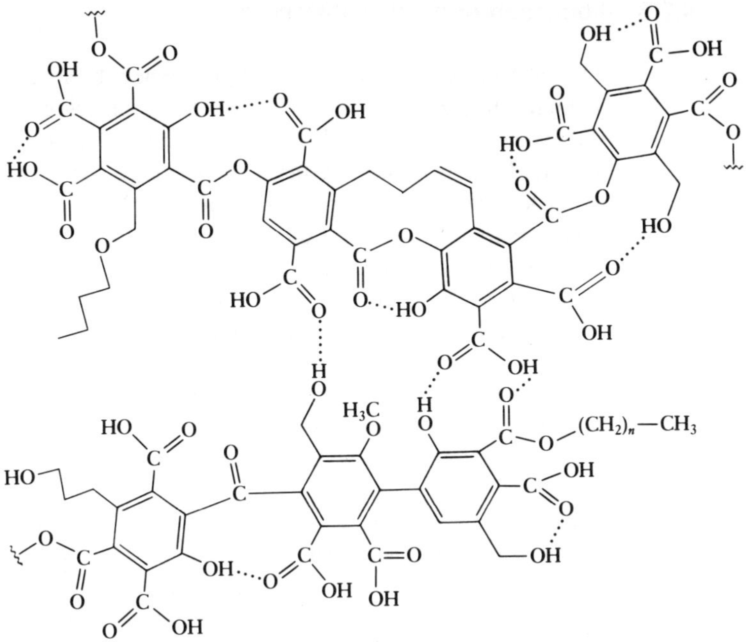

Figure 4.1   Some examples of structural units and bonding modes in fulvic acid (hydrogen bonding denoted by dotted lines).

acids, derived from microbial or higher plant sources, which are bonded by ester linkages to the macromolecular backbone. As a result of their greater ratio of aliphatic to aromatic structures, humic acids in marine sediments have higher H/C atomic ratios (1.0–1.5) than their soil counterparts (0.5–1.0) and also contain less phenolic constituents but often more sulphur.

Soil fulvic and humic acids are larger than their aquatic counterparts. Freshwater humic acids tend to have molecular weights of $\leqslant 10^3$, while soil humic acids generally have larger molecular weights, sometimes in excess of 10^6. Hydrogen bonding appears to play a role in molecular aggregation (see Fig. 4.1). An important factor influencing the shape of humic and fulvic acid molecules in solution is pH (see Box 3.2). At low pH fulvic acids are fibrous, but as the pH rises the fibres tend to mesh together to form a sponge-like structure by pH 7, while at higher pH the structure becomes plate-like. Humic acids behave similarly but precipitate when the pH falls below ca. 6.5.

The meshed structure formed by humic substances is capable of trapping smaller chemical species. For example, minor amounts of acyclic alkanes are found in most samples of humic and fulvic acids, and some of the fatty acids associated with humics may be similarly trapped components rather than bonded to the macromolecular backbone. Humic substances also usually contain a variety of metals, which are incorporated into the macromolecular structure. Metal ions can be surrounded by and bonded to suitable chelating groups, chiefly carboxylic acids, on humic molecules which stabilise the ions and allow them to be transported with the organic material. This important property of humic substances will be examined again later (Section 8.6.2) in relation to the environmental fate of heavy metals.

4.2.3   Formation of humic substances

The availability of various potential precursor substances in particular environments of humic formation appears to affect the composition of humic substances. The abundance of polycarboxyl-substituted phenols shows a positive correlation with terrestrial inputs, indicating that lignins and probably also tannins are important in humic formation in terrestrial environments. These components are resistant towards biodegradation and so their dominance in aerobic environments is not unexpected. Refractory lignified plant tissue, altered to varying degrees by microbial degradation, is the major substrate of soil humic material and contributes the bulk of the aromatic units. The by-products of the decomposer organisms and their remains are also likely to contribute to humic substance formation and may account for the major part of the aliphatic components (e.g. fatty acids and alkanes), supplementing the aliphatic contribution from higher plants (e.g. waxes). The higher aliphatic content of marine humics may be the result of an increased contribution from microbial lipids, particularly algal.

As diagenesis proceeds in naturally occurring humic material the amount of hydrolysable components, sugars and amino acids, decreases, suggesting either that they are lost from the humic structure or that they are being permanently incorporated into the macromolecular backbone. Condensation of the various units present in humic material progresses during diagenesis and yields the insoluble residue of humin in soils. Humic or humic-like substances are likely to make some contribution to kerogen, just as they do to brown coals. Sulphur levels generally appear to increase as diagenesis progresses, possibly involving incorporation of hydrogen sulphide into the macromolecules. For example, in marine humic substances sulphur levels increase with depth and in peat the humin fraction, which increases in abundance with depth, contains the greatest amounts of sulphur.

While carbohydrates are undoubtedly incorporated into the humic backbone evidence for amino acid incorporation is less conclusive, as the nitrogen content of humic material is low and mostly in the form of hydrolysable amino acids superficial to the macromolecular backbone. It is known that condensation reactions between sugars and amino acids can occur readily. A complex series of reactions is involved, collectively termed the Maillard reaction, producing brown polymeric substances called **melanoidins**. Functional lipids can be incorporated into the melanoidin structures which, upon artificial diagenesis, develop some of the properties of naturally occurring humic substances. However, there is little evidence to suggest that melanoidins make a significant contribution to natural humic material.

4.3   Coal

4.3.1   Classification and composition

Classification   Coals are usually classified as either humic or sapropelic. **Humic coals** are formed mainly from vascular plant remains (e.g. Westphalian coals of northern Europe). They tend to be bright, exhibit stratification and go through a peat stage involving humification (i.e. formation of humic material). The major organic components derive from the humification of woody tissue and have a lustrous, black/dark-brown appearance. In contrast, the less common **sapropelic coals** are not stratified macroscopically and are dull. They are formed from fairly fine-grained organic muds in quiet, oxygen-deficient shallow waters. Normally they do not go through a peat stage but follow the diagenetic path of hydrogen-rich kerogens, which will be considered in Section 4.4. Like such sediments, sapropelic coals contain varying amounts of allochthonous organic and mineral matter. The organic fraction contains autochthonous algal remains and differing amounts of the degradation products of peat swamp plants. Sapropelic coals are subdivided into cannel and boghead coals. **Boghead coals** (or Torbanites) contain larger amounts of algal remains together with some fungal material (e.g. the oil shale in the Midland Valley of Scotland). **Cannel coals** have a higher concentration of spores. Resin bodies are sometimes found in addition to the usual algal, vegetational and fungal material.

Petrological composition   The individual constituents of coal, its petrological components, which represent the preserved remains of plant material, are known as **macerals**. These are observed by transmitted or reflected light microscopy and can be recognised by their differing optical properties. There are three main groups of macerals: vitrinite, inertinite and exinite; they are summarised in Table 4.3 together with some general properties.

The major maceral group in humic coals is **vitrinite**, the lustrous, black/dark-brown constituents from the humification of woody tissue. Where the morphological structure of the woody tissue is preserved the vitrinite is termed telinite, which is transparent and orange-red with a vitreous lustre and any bodies present are filled with colloidal humic material or resins. Where the structure of the vitrinite is lost but it is still a translucent brown it is termed collinite. The term vitrinite is usually applied to the higher evolutionary stages of coalification beyond brown coals, while **huminite** is the commonly used name for this maceral group during diagenesis.

Among the **inertinite** macerals are opaque modifications of vitrinite, termed fusinite and semi-fusinite, which retain the morphological structure of woody tissue

Table 4.3   **Coal maceral groups**

Maceral groups	Macerals	General properties
Vitrinite	Telinite Collinite	Angular to subangular particles, sometimes show cell structure. Moderate transmittance. Fluorescence usually absent. Intermediate reflectance.
Inertinite	Micrinite Sclerotinite Semi-fusinite Fusinite	Angular, often cellular outline or granular texture. Opaque in transmitted light. No fluorescence. High reflectance.
Exinite	Alginite Sporinite Cutinite Suberinite Resinite Cerinite	Characteristic shape, e.g. algae, resins, spores. High transmittance. Intense fluorescence (at low maturity). Relatively low reflectance.

and are dull brown/black and friable. Other inertinite macerals are micrinite (granular or massive), which is dark brown, amorphous and derived from polymerised plant resins, and sclerotinite, formed from fungal remains (sclerotia).

Exinite is composed of lipid-rich, translucent, yellow/red macerals and is subdivided into resinite and cerinite (deriving from resins and waxes respectively), sporinite (from spore and pollen coatings), cutinite (from plant cuticles) and alginite (from algal bodies). Sometimes the term liptinite is used instead of exinite for this entire group of macerals in the context of coal, but when applied to macerals in kerogen it has a more specific use (see Section 4.4.2).

The division into three maceral groups is based mainly on reaction towards heating. Normally, the main product of heating vitrinites is a fused carbon residue and some gas is evolved. Inertinites are literally inert; they do not generate any hydrocarbons or appear to change in form, and it has been proposed that the fusain macerals (fusinite and semi-fusinite) may represent fossil charcoal from ancient vegetation fires. Exinites are generally transformed into gas and tar. With increasing maturity during coalification the recognition of macerals becomes more difficult as structural features become less distinct.

Various combinations of maceral groups occur in humic coals but four main lithotypes are recognised on the macroscopic scale: vitrain (comprising mainly vitrinite), durain (comprising exinite and inertinite), clarain (comprising vitrinite and exinite) and fusain (mainly fusinite). On the microscopic scale, macerals often occur in certain associations, recognised as microlithotypes. The three main microlithotypes are vitrite (mainly vitrinite), clarite (vitrinite and exinite) and durite (exinite and inertinite).

Chemical composition   The main elements in coal are carbon, hydrogen, oxygen, nitrogen and sulphur. Oxygen is present mainly in carboxyl, ketone, hydroxyl (phenolic and alcoholic) and methoxy groups, but the distribution of these functional groups varies with rank, as described below. Nitrogen is found in amines and in aromatic rings (e.g. pyridyl units). Sulphur is found in thiols, sulphides and aromatic rings (e.g. thiophenic units). Sulphur is a common constituent of coal in inorganic form, usually as pyrite. A variety of metals can be present in coal which were originally incorporated into the matrix of humic material during diagenesis.

4.3.2   Formation

There are two main phases in the formation of humic coals, **peatification** followed by **coalification**; coalification can be subdivided into a **biochemical stage** and a **geochemical stage**. The main agents during peatification and early coalification (biochemical stage) are biological, and these stages are equivalent to diagenesis. Late coalification (geochemical stage) is primarily the result of increasing temperature and pressure and may be equated with catagenesis. There is usually some overlap between the action of biological and physico-chemical agents and, therefore, the boundary between the two stages of coalification may be indistinct. The biochemical stage generally ends at depths of the order of several hundred metres, at the limit of penetration of percolating waters carrying micro-organisms.

The evolution of sedimentary organic matter with increasing burial and temperature is generally termed **maturation**. However, for coal another term, **rank**, is frequently used instead of maturity. In order of increasing rank, the main stages of humic coal formation can be termed: **peat, brown coal, bituminous coal** and **anthracite** (Table 4.4). Bituminous coals and anthracites are often collectively termed hard coals. Variations do, however, exist in the use of rank terminology, with lignite sometimes being applied to the brown coals as a whole and hard coals being substituted for bituminous coals. Various subclassifications of the main groups are also recognised but will not be considered here.

Peatification   Peat formation is an inefficient process; less than 10% of plant production in a typical peat-forming environment accumulates as peat, the rest being recycled by the associated microbial community or lost from the system (by mineralisation and leaching). The products of peatification are humic substances: brown hydrated gels with no internal structure. Oxidising conditions generally prevail at the surface of peat bogs but reducing conditions develop where there is a cover of stagnant water or with increasing compaction and depth. At the surface conditions are usually neutral or mildly acidic but with increasing burial depth acidity increases, bacterial communities change and their activities decrease and eventually cease. Conditions favourable for large-scale preservation, therefore, can develop just beneath the surface.

Peatification begins in the surface litter with the mechanical disintegration of plant material by invertebrates. This comminution greatly aids microbial degradation, mainly by increasing the surface area that can be attacked. Depolymerisation of polysaccharides by decomposers occurs during early peatification: hemicellulose constituents are rapidly removed, followed by conversion of cellulose into glucose units. Lignin is a more resistant biopolymer and its degradation tends to occur under aerobic conditions, mostly by fungi and some specialist bacteria, yielding large amounts of aromatic, phenolic and carboxylic acid (COOH) units. Microbial metabolites and the decaying remains of fungi and bacteria will invariably contribute to the organic matter (e.g. fungal carbohydrates).

As peatification proceeds the gradual biochemical transformation of lignin continues, involving depolymerisation and defunctionalisation (e.g. loss of methyl and methoxy groups). Random polymerisation and condensation of the products of the various biodegradation reactions then occur, with loss of functional groups, resulting in peat formation. During these biochemical changes gases are produced, such as CH_4, NH_3, N_2O, N_2, H_2S and CO_2. The CO_2 derives mainly from the breakdown of carbohydrates, particularly cellulose.

With increasing diagenesis the lignin and cellulose content of peat decreases but the content of humic substances increases. Cellulose is still found in peat but is absent from soft brown coal, which is formed at a more advanced stage of diagenesis. Lignin content can be up to 35% in peat but decreases to <10% in brown coal. Lipid material is confined largely to leaves, spores, pollen, fruits and resinous tissues (resins are particularly abundant in conifers). Although these components are very minor in higher plants they are concentrated during peatification as they are fairly resistant to degradation. Peats have a high water content (ca. 95%) and usually contain 5–15% bitumen (see Box 4.1), which derives from lipid components and comprises ca. 50% waxes, 20% paraffins and 25% resins.

Biochemical   Condensation of organic residues progresses during the early stage of coalification
stage of   with further loss of oxygen-containing functional groups. Bacterial action continues
coalification   to be an important factor. With the loss of functional groups the residue becomes

relatively concentrated in carbon and hydrogen. The final preserved organic-rich product of diagenesis, brown coal, contains ca. 50–70% C and ca. 5–7% H.

Aromatic units are important in the structure of coal. Initially they derive mainly from the aromatic units inherited from lignin but with increasing diagenesis the aromatic content of the macromolecular organic material increases generally. Modified lignin is a major constituent of huminite macerals. Cell wall carbohydrates do not appear to make a significant contribution to the aromatic network of coal. Aliphatic chains are restricted in abundance and in length, reflecting the paucity of aliphatic structures in the main coal precursors, lignin and cellulose. The aliphatic structures that are present derive primarily from lipid-rich constituents, such as plant cuticles, spores and pollen. At the end of diagenesis, in the resulting brown coal, all carbohydrate material has been removed by anaerobic microbial and geochemical processes.

As well as the macromolecular material, brown coals also contain small amounts of low-molecular-weight volatile substances (e.g. resin and wax constituents), encapsulation within resistant structures aiding their preservation during diagenesis. It appears that the metal ions inherited from humic substances are retained in the macromolecular structure during peatification and coalification.

Geochemical stage of coalification

With the termination of biologically mediated transformations the diagenetic stage of coal formation ends. As temperature and pressure become the dominant agents in the transformation of coal the geochemical stage of coalification begins. During the geochemical stage carbon content increases and there is a concomitant decrease in oxygen content. Hydrogen levels remain fairly constant for most of the geochemical stage but begin to decrease at an increasing rate at the highest levels of maturity. Nitrogen content is ca. 1–2% throughout, while levels of sulphur are variable but generally <1%. At the beginning of this stage aromatic nuclei are present, which appear to bear peripheral functional groups (e.g. methyl, hydroxyl, carboxyl, carbonyl, amino) and to be partly linked together by aliphatic rings and CH_2 (methylene) units. The changes in elemental composition during the geochemical stage reflect both the continued elimination of functional groups and the growth of aromatic nuclei that results from increasing aromatisation of cycloalkyl structures and condensation between individual aromatic nuclei. The aromatic carbon content of vitrinite increases from ca. 70% in hard brown coals to ca. 90% in anthracite. While exinite macerals initially have a much lower aromatic content than vitrinite, by the later stages of coalification their aromatic content parallels that of vitrinite. In inertinites the amount of carbon involved in aromatic structures is >90% at all ranks.

Methane, carbon dioxide and water are regarded as the main volatile products of coalification. Compaction continues with increasing burial depth, resulting in decreased porosity and increased density. Water content decreases with increasing rank; from up to 95% in peat to as little as 1% in anthracites. The production of methane and carbon dioxide increases but the abundance of other volatile components decreases during coalification. Some of these bulk changes with increasing rank are shown in Table 4.4.

Temperature is particularly important in determining the extent of geochemical reactions. It is possible to identify temperature intervals associated with the various stages of coal formation. The formation of brown coal appears to be associated with a temperature range of ca. 10–40°C, bituminous coal with ca. 40–100°C, and

Table 4.4 **Summary of changes in some bulk characteristics of coal with increasing rank** (After Teichmüller and Teichmüller, 1968)

	Rank stages	ASTM classification*	%C†	% volatiles†	% water	%R‡	Calorific value (kcal/kg)§
BROWN COAL			—60—		—75—	—0.25—	
	SOFT BROWN COAL	Brown coal					
			—53—		—35—	—0.3—	—4000—
	HARD BROWN COAL — Dull	Lignite					
			—71—	—49—	—25—		—5500—
	— Bright	Sub-bituminous coal					
			—77—	—42—	—10—	—0.5—	—7000—
HARD COAL	BITUMINOUS HARD COAL — Low rank	*Bituminous coal* — High volatile					
		Medium volatile	—87—	—29—		—1.1—	—8650—
		Low volatile					
	— High rank	Semi-anthracite					
			—91—	—8—		—2.5—	—8650—
	ANTHRACITE	Anthracite Meta-anthracite					
			—100—	—0—	—0—	—11.0—	

* American Society for Testing and Materials.
† As % of dry ash-free weight.
‡ Vitrinite reflectance (see Section 5.6.2).
§ Ash-free value.

anthracite with 100–150°C. The geothermal gradient (see Box 5.1) is therefore an important factor controlling the rank of a coal.

Structural changes during coal formation The extent of peatification and coalification reactions can be examined using a van Krevelen diagram, in which H/C vs. O/C atomic ratios are plotted (Fig. 4.2a). These plots can also differentiate between maceral groups, as shown in Fig. 4.2b. Loss of oxygen-containing functional groups during diagenesis can clearly be seen in the decreasing O/C ratios of the main humic coal macerals (huminite/vitrinite) and their vascular plant precursors and also for inertinite (Fig. 4.2b). This loss of oxygen is paralleled by hydrogen loss in the lipid-rich tissues and macerals (exinites), which contain low initial amounts of oxygen but relatively high amounts of hydrogen (Fig. 4.2b). For the main sequence of humic coal formation in Fig. 4.2a (corresponding to the trend for vitrinite in Fig. 4.2b) the end of diagenesis corresponds to an O/C atomic ratio of ca. 0.1. At this point the atomic ratios of different maceral groups begin to become similar.

A general idea of functional group composition of brown coal, corresponding to the end of peatification (65% C content), is shown in Fig. 4.3. Carboxylic acid groups are important in brown coals and so are hydroxy groups, which are divided between the more acidic phenolic and less acidic alcoholic species. Methoxy group content has decreased from its original level of ca. 1.5% in humic acids to <1% in brown coals.

During the geochemical stage of coalification the O/C ratio decreases further as the remaining functional groups are eliminated (Fig. 4.2a). Initially, up to the rank of semi-anthracite (Table 4.4), only a slight decrease in the H/C ratio occurs, with hydrogen content generally remaining in the range of 5–6%. The H/C ratio then

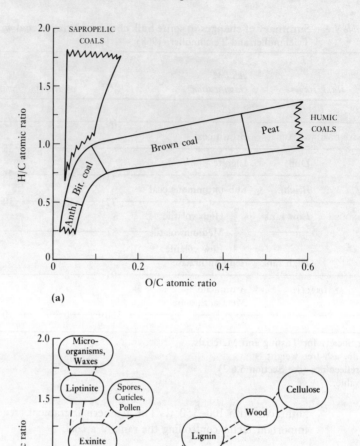

Figure 4.2   (a) Van Krevelen diagram showing the main evolutionary trends of sapropelic (cannel and boghead) and humic coals. Four stages are denoted for humic coals, in order of increasing rank: peat, brown coal, bituminous hard coal and anthracite. (b) Van Krevelen diagram showing the position of the main coal macerals and their diagenetic, evolutionary relationships to various components in living organisms (see Section 4.4.2 for an explanation of liptinite).

decreases at an increasingly rapid rate, reflecting the loss of hydrogen that occurs during aromatisation of cycloalkyl structures and progressive condensation of aromatic clusters. The decrease in oxygen content during coalification is illustrated in Fig. 4.3; a fall of ca. 23% is apparent between brown coal at the end of peatification and medium volatile bituminous coal (of ca. 85% C). From Fig. 4.3 it can be seen that all carboxylic acid groups have been eliminated during this phase of maturation

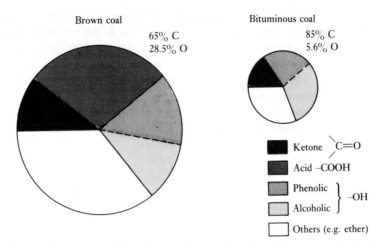

Figure 4.3 Distribution of oxygen-containing functional groups in coal at different maturity levels. (After Blom et al., 1957.)

along with remaining methoxy groups. Dehydration, decarboxylation and demethylation are, therefore, the principal reactions during early coalification and result in some aromatisation, while during late coalification aromatisation becomes the dominant reaction.

The complex polymeric nature of vitrinite, the major maceral of humic coals, imposes difficulties in the determination of its chemical structure. However, Fig. 4.4 presents a possible partial structure for vitrinite containing 83–84% C, based on general elemental composition and functional group analysis and bearing in mind the likely precursor molecules. At the macromolecular level the change in structure of vitrinite during the geochemical stage of coalification begins with the aggregation

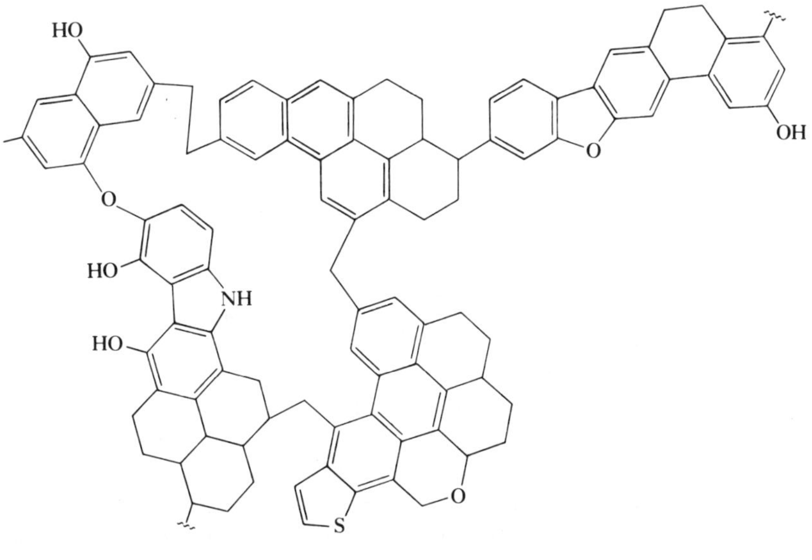

Figure 4.4 A partial structure for vitrinite (83–84% C). (After Heredy and Wender, 1980.)

of aromatic nuclei into clusters by the processes of aromatisation and condensation. These clusters form flat lenses (due to the inherently flat benzenoid ring systems involved), which are initially randomly orientated but gradually become more ordered, tending towards a parallel arrangement as pressure increases. Further condensation occurs between laterally adjacent lenses so that, by the anthracite stage, a generally parallel sheet arrangement has been attained, progressing towards the idealised structure of graphite (usually the most stable form of carbon). This is in contrast to the randomly orientated structure adopted in the bituminous coals of lower rank.

4.4   Kerogen

4.4.1   Formation

Geopolymer formation during diagenesis

Kerogen is the polymeric organic material from which hydrocarbons are produced with increasing burial and heating. It occurs in sedimentary rocks in the form of finely disseminated organic macerals and it is by far the most abundant form of organic carbon in the crust. In the classical theory it is believed that formation of kerogen proceeds via humic substances similar to the humic and fulvic acids found in soils (e.g. see Tissot and Welte, 1984). During early diagenesis in aquatic environments the organic material from primary production is broken down into smaller constituents by microbial action, which then undergo condensation reactions to give rise to humic substances. Microbial degradation and condensation follow in immediate succession, leading to a zone in the top few metres of sediment (and possibly also in the water immediately overlying it) where both processes are active at the same time. With increasing time and burial depth most of this humic material becomes progressively insoluble due to increasing polycondensation, which is associated with the loss of superficial hydrophilic functional groups (e.g. OH and COOH). Insolubilisation can begin quite early but continues to significant depths, as humic acids can still be found at several hundred metres depth in sediments that contain abundant terrestrial detrital material. The humin-like material that is formed continues to undergo condensation and defunctionalisation, resulting in kerogen. Lipid components are also incorporated into the kerogen structure. In this model of kerogen formation fulvic acids are viewed as the precursors of humic acids, in contrast to some theories of the evolution of soil humic substances (see Section 4.2.1).

At the end of diagenesis sedimentary organic matter is mainly composed of kerogen (Fig. 4.5), which is insoluble in organic solvents. There are also smaller amounts of organic material that is soluble in organic solvents and is termed bitumen (see Box 4.1). This bitumen comprises some small fragments of polymeric material (asphaltenes and resins) that have broken off the main kerogen structure, together with some free (i.e. not chemically bound to kerogen) relatively small molecules (molecular weights ca. < 600), which are mainly hydrocarbons and are of lipid origin (Fig. 4.5). Kerogen formation involves mild temperature and pressure conditions, and so the composition of the original organic input is an important influence on the chemical nature of the resulting kerogen. As for coal formation, microbial metabolism and the elimination of functional groups during condensation are accompanied by the evolution of volatiles such as methane, carbon dioxide and water.

Box 4.1

Bitumen

The term **bitumen** is usually applied to naturally occurring hydrocarbon deposits but excludes gases. Different types of bitumens can be recognised, all of which are soluble in carbon disulphide and the solid types are fusible. The term **petroleum** is used to describe naturally occurring liquid (i.e. oil) and gaseous hydrocarbon deposits. Oil can, therefore, be classified as a bitumen. Related to the bitumens are pyrobitumens, which are infusible solids that are insoluble in CS_2. On heating, pyrobitumens tend to decompose but can generate bitumen-like material. The different types of bitumens (in order of increasing melting point) and pyrobitumens are shown in Table 4.5.

Bitumen is composed of three main fractions (Fig. 4.5): asphaltenes, resins and hydrocarbons. **Asphaltenes** and **resins** are heavy N,S,O-containing molecules (molecular weight > 500), while the hydrocarbons are usually of lower molecular weights. The resins fraction of bitumen should not be confused with plant resins (e.g. from conifers). When bitumen is mixed with a light hydrocarbon (e.g. hexane) it separates into a soluble fraction, **maltenes**, and an insoluble fraction, asphaltenes. The maltenes can then be separated into hydrocarbons and resins by passage down a column of alumina, from which the hydrocarbons can be removed with a solvent like dichloromethane and the resins by subsequent application of methanol. Alternatively, aliphatic hydrocarbons can be eluted first with hexane, followed by aromatic hydrocarbons with toluene and then resins with methanol. Traditionally, the aliphatic hydrocarbons (or saturates) are described in terms of their paraffinic (acyclic alkane) and naphthenic (cycloalkane) constituents.

The classical model of kerogen formation described above has undergone reappraisal in the light of recent evidence. It appears that kerogen is formed at a relatively early stage of diagenesis, earlier than previously suspected, and that a significant fraction of it may derive from mixtures of selectively preserved, sometimes partly altered, resistant biomacromolecules, rather than from random recondensation of the constituents of depolymerised biomacromolecules (Tegelaar et al., 1989). Rapid formation of the recalcitrant polymeric material that constitutes kerogen is consistent with the extensive microbial degradation of marine and terrestrial detritus that generally occurs in the top metre or so of sediments and removes the more labile constituents (see Section 3.3).

Laboratory studies have shown that certain insoluble non-hydrolysable macromolecules which form protective structures, such as cutan (Nip et al., 1986; see

Table 4.5 **Classification of bitumens and pyrobitumens** (After Abraham, 1945)

Bitumens	Liquid petroleums (all crude oils and seeps)
	Mineral waxes
	Asphalts
	Asphaltites
Pyro-bitumens	Asphaltic pyrobitumens (generally < 5% oxygen content; e.g. oil shale)
	Non-asphaltic pyrobitumens (generally > 5% oxygen; e.g. kerogen and coal)

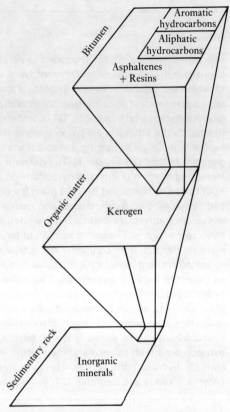

Figure 4.5   Composition of disseminated organic matter in sedimentary rocks. (After Tissot and Welte, 1984.)

Section 2.4.2), are resistant towards biodegradation. Although relatively minor components in organisms these resistant biomacromolecular structures become concentrated during the diagenetic degradation of more abundant and readily hydrolysable biopolymers, such as proteins and carbohydrates. For example, cutan is often the major constituent of fossil cuticles because it is more resistant towards degradation than hydrolysable cutin (Tegelaar et al., 1991). The degree of alteration that the resistant biomacromolecules undergo depends on the precise environmental conditions. There is also the possibility for less resistant material to be protected against biodegradation within coatings of resistant material (e.g. polysaccharides within higher plant cuticular membranes and lipids within microbial cell walls). Some of the non-hydrolysable biomacromolecules are initially soluble and a proportion may undergo cross-linking to form part of the insoluble kerogen matrix as diagenesis progresses.

Selective preservation seems a reasonable proposition as preserved parts of organisms, such as cuticles, spores and pollens, can be recognised in both coal and kerogen. If preserved remains can survive the conditions of coal formation it seems likely that they should survive the less oxidising conditions of kerogen formation. Some possible resistant biomacromolecular precursors of kerogen and their related macerals are given in Table 4.6. Messel oil shale kerogen contains significant amounts of selectively preserved, cell wall material of the alga *Tetrahedron* (Goth et al., 1988). This material, in both extant and fossil forms of the alga, is an insoluble and non-hydrolysable biopolymer (termed **algaenan**) that is rich in aliphatic structures.

Table 4.6 **Some resistant biomacromolecules and their related kerogen macerals** (After Tegelaar et al., 1989)

Resistant biomacromolecule	Maceral
Algaenan	Alginite
Cutan	Cutinite
Lignin	Vitrinite/fusinite
Polyterpenoids*	Resinite
Sporopollenin	Sporinite
Suberan	Suberinite
Tannin	Vitrinite

*Mainly from sesqui- and diterpenoids.

Similar algaenans have been found in other algae (e.g. *Botryococcus braunii*; Largeau et al., 1986). Such highly aliphatic, resistant biomacromolecules in algal cell walls and in plant cuticles are probably important precursors of the aliphatic network found in young kerogens with oil potential, and may also contribute significantly to the aliphatic content of humic material.

The contribution of bacterial remains to kerogen should not be overlooked. Resistant biopolymers have been identified in the cell walls of cyanobacteria and are believed to contribute to **amorphous kerogen** (Chalansonnet et al., 1988). It is probable that similar resistant macromolecular materials in other bacteria also contribute to amorphous kerogen and provide a degree of protection to membrane constituents (e.g. hopanoids and phytanyl lipids).

The classical and selective preservation models can be considered extremes (i.e. end members) of the scale of alteration that may be undergone by the biomacro-molecular precursors of kerogen. They are united in the scheme shown in Fig. 4.6, in which kerogen is seen to be the combination of resistant biomacromolecules, geomacromolecules, sulphur-rich macromolecules and incorporated low-molecular-weight biomolecules. The relatively minor amount of bitumen present at the end of diagenesis derives predominantly from preserved lipid components, shown on the lower right-hand side of Fig. 4.6. Quantitatively, the major pathways of the classical and selective preservation models are represented by, respectively, the routes to geomacromolecules (via LMW biomolecules and humic substances) and resistant biomacromolecules. Both models share the quantitatively less significant pathways to incorporated, low-molecular-weight biomolecules and preserved lipid components. The importance of the vulcanisation pathway to sulphur-rich macromolecules in kerogen varies, depending on diagenetic conditions, and will be discussed later. Partial alteration of biomacromolecules (e.g. the possible cross-linking of soluble components) is represented in Fig. 4.6 by the left-hand arrow to humic substances. Formation of kerogen at an early stage of diagenesis is readily reconciled with a model, such as that in Fig. 4.6, based primarily on selective preservation, in which the classical pathway to geomacromolecules plays, at most, a minor role.

Biomarkers   The preserved lipid components in Fig. 4.6, which are found in coals as well as kerogens, have also been called **geochemical fossils** and **biomarkers** (or biological marker compounds) because they can be unambiguously linked with biological precursor compounds and their basic skeletons are preserved in recognisable form throughout diagenesis and much of catagenesis. Many biomarkers originally possess oxygen-containing functional groups and undergo the same defunctionalisation

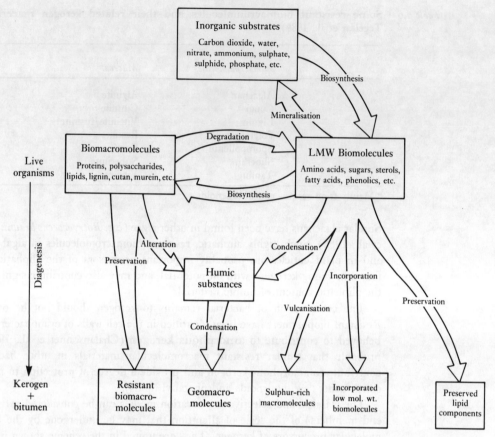

Figure 4.6   Integrated model of kerogen formation combining classical and selective preservation models. Kerogen is represented by the combined four boxes at the bottom, while the preserved lipid components box represents a constituent of bitumen.

processes as the bulk of the organic matter. Diagenetic products are, therefore, generally hydrocarbons, although small amounts of functionalised components (e.g. fatty acids) can survive diagenesis. In addition, unsaturated compounds (i.e. containing C=C bonds) tend either to become reduced (hydrogenated), resulting in the formation of aliphatic hydrocarbons (e.g. steranes and hopanes), or to undergo aromatisation. The latter process can occur, rather than hydrogenation, where a C=C bond is present in a cyclic system containing six carbon atoms, so that dehydrogenation (i.e. desaturation) within the ring system leads to an aromatic unit (e.g. A- and C-ring monoaromatic steroidal hydrocarbons, see Section 6.3.3). To summarise, then, during diagenesis biomarkers undergo defunctionalisation and either reduction or aromatisation. A proportion of the biomarkers may become a chemically bound part of the kerogen structure (incorporated LMW biomolecules in Fig. 4.6), but otherwise these compounds remain relatively unaltered, encapsulation within resistant macromolecular structures being an important factor in their preservation.

Sulphur incorporation   Incorporation of sulphur into macromolecules that are an integral part of kerogen (Fig. 4.6) can be important, depending on the sedimentary environment. Conditions favouring the activity of sulphate-reducing bacteria result in the production of sulphide, which is usually taken up rapidly by iron(II) ions, especially in some types of clastic and argillaceous sediments. Where there is a limited supply of iron(II) ions, free hydrogen sulphide and polysulphides (HS_4^-, HS_5^-, S_4^{2-} and S_5^{2-} are the most common naturally occurring sulphur-containing components in aqueous conditions) are produced, which can react with organic matter during diagenesis. The formation of sulphur-rich kerogens is, therefore, more likely in non-clastic sediments (e.g. carbonates, siliceous oozes and evaporites) where only very small amounts of iron are available. Inclusion of sulphur as sulphide (or perhaps even free sulphur in its native form) into low-molecular-weight unsaturated compounds appears to occur during early diagenesis and these compounds can then go on to become incorporated into the macromolecular structure of kerogen.

A summary of the possible mechanisms of sulphur incorporation is given in Fig. 4.7. Unsaturated precursor compounds that are potentially capable of incorporating sulphur include isoprenoidal alkenes from archaebacteria, bacterial hopanoids, long-chain (C_{37} and C_{38}) unsaturated ketones from coccolithophores (e.g. *Emiliania huxleyi*) and phytadienes from chlorophyll diagenesis. Addition of hydrogen sulphide across a C=C bond yields a thiol. Reaction of this compound with another unsaturated compound (i.e. intermolecular addition) can lead to the formation of sulphur-rich, high-molecular-weight substances (left-hand side of Fig. 4.7). Where there is more than one double bond present in the precursor molecule, addition and then loss of H_2S can lead to movement of double bonds along the chain (an isomerisation process). In addition, a thiol group formed from one C=C bond can react with an adjacent double bond in the molecule (intramolecular addition) to yield cyclic structures containing sulphur (right-hand side of Fig. 4.7). The size of the ring will depend on the relative positions of the double bonds. Initially, the cyclic centres are aliphatic, but undergo aromatisation with increasing maturity.

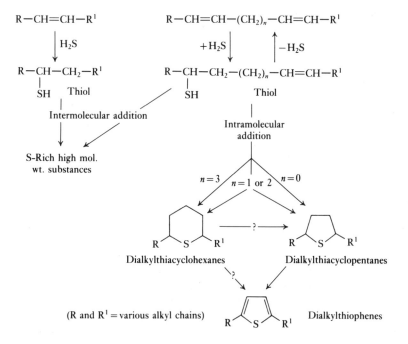

Figure 4.7   Possible mechanisms for sulphur incorporation into unsaturated substrates during diagenesis. (After Sinninghe Damsté et al., 1989.)

Incorporation of sulphur into individual geochemical fossils (e.g. hopanes) can also occur.

4.4.2 Kerogen composition

Because kerogen is disseminated in sedimentary rock it is often isolated prior to microscopic examinations of the kind used on coal. However, it is quite difficult to isolate kerogen quantitatively without alteration. The inorganic matrix is usually removed by successive treatment with hydrochloric and hydrofluoric acids. While microscopic examination reveals the presence of defined organic remains, such as algae, spores, pollen and vegetative tissue, these are usually only a minor part of kerogen and are fairly well dispersed. The same three maceral groups recognised in coals are found in kerogen: exinite, vitrinite and inertinite. However, for kerogen a subgroup of exinite, termed **liptinite**, is also distinguished and, therefore, a total of four maceral groups is assigned to kerogen. Liptinite macerals exhibit the usual properties of exinites, such as high UV fluorescence and high lipid content. The two maceral groups are differentiated by exinites having recognisable shapes, whereas liptinites consist of amorphous bodies. In addition, liptinites tend to have higher H/C ratios than exinites (Fig. 4.2b).

Carbon and hydrogen are the main elements of kerogen, but oxygen also appears to be important in the structure of kerogen and comprises 10–25% by weight in shallow immature sediments. For every 1000 C atoms there are ca. 500–1800 H, ca. 25–300 O, ca. 5–30 S and ca. 10–35 N atoms. Oxygen-containing functional groups include COOH, OH, CO, COOR and $CONH_2$. The aliphatic content of kerogen is higher than that of coal, reflecting the input of highly aliphatic planktonic and microbial lipids to the original sedimentary organic material.

As with humic substances and coal, structural analysis of kerogen is difficult and various oxidative and pyrolytic degradation techniques have been employed to identify structural units. Kerogen is a three-dimensional macromolecule formed from nuclei that are cross-linked by chain-like bridges. Both nuclei and bridges may bear functional groups. Lipids may be trapped within the structure, a molecular sieve property that is shared with coal. In general, the nuclei appear to be formed mainly from clusters (< 100 nm diameter) of 2–4 or more parallel aromatic sheets separated by gaps of ca. 30–40 nm. These sheets contain up to 10 condensed aromatic rings, some of which contain nitrogen and sulphur and possibly oxygen. The nuclei bear alkyl chains (mainly linear or with a few short branches), cycloalkanes and various functional groups. There are a variety of bridging structures between the nuclei, including aliphatic chains (linear and branched), oxygen- or sulphur-containing functional groups (e.g. —C(=O)—, —O—C(=O)—, —O—, —S—, —S—S—), and combinations of the two.

The composition of shallow, immature kerogen depends on the nature of the original organic matter incorporated into the sediments from which it is formed and also on the extent of microbial degradation. This composition determines the genetic potential of the kerogen, i.e. the amount of hydrocarbons that it can generate. For example, kerogen formed from organic matter with a large allochthonous higher plant contribution will have a lower aliphatic content than one formed largely from phytoplankton remains. The organic matter types from which kerogen is formed can, therefore, be described as humic and sapropelic; the former having poor oil potential while the latter may provide a good oil source.

4.4.3   Kerogen classification

Traditionally, four general types of kerogen are distinguished, types I to IV, which are broadly equivalent to the maceral groups liptinite, exinite, vitrinite and inertinite, respectively, although some differences do exist. As for coals, atomic H/C and O/C ratios for kerogens can be plotted on a van Krevelen diagram, generally allowing distinction of the main types at the end of diagenesis. Therefore, in Fig. 4.2b **type I kerogen** would plot in a similar position to liptinite and so would cannel and boghead (sapropelic) coals, which are hydrogen-rich. **Type II kerogen** plots in a similar position to exinite and **type III kerogen** near vitrinite, as does brown coal. **Type IV kerogen** plots with inertinite. Types II and III kerogen have higher oxygen content than type I, and the distribution of oxygen within the various functional groups differs for each type of kerogen (Fig. 4.8).

Type I kerogen   Type I, or liptinite-type, kerogen is relatively rare but has a high oil potential. It initially has a high H/C ratio ($\geqslant 1.5$) and a low O/C ratio (<0.1). It contains a significant contribution from lipid material, especially long-chain aliphatics. These lipids are predominantly either from the selective accumulation of algal material that may have undergone partial bacterial degradation, or from bacterial remains that result from extensive bacterial reworking of the primary input of organic matter to the sediment. An algal source gives rise to alginite macerals, whereas bacterial remains are often observed as amorphous bodies. The predominant macerals of type I kerogen are, therefore, liptinites. Compared with the other kerogen types it contains low amounts of aromatic units and heteroatoms (see Box 2.1). The limited amount of oxygen present is mainly in the form of ester bonds (Fig. 4.8).

Type I kerogen is dark, dull and either finely laminated or structureless. It is usually formed in relatively fine-grained, organic-rich muds deposited under anoxic conditions in quiet, oxygen-deficient, shallow-water environments (e.g. lagoons and

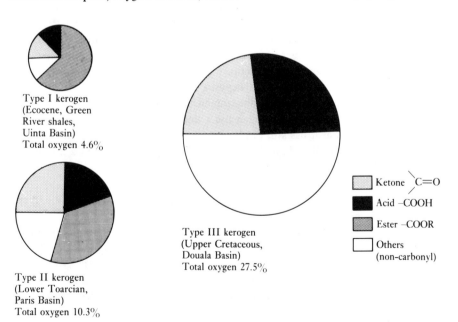

Type I kerogen
(Eocene, Green
River shales,
Uinta Basin)
Total oxygen 4.6%

Type II kerogen
(Lower Toarcian,
Paris Basin)
Total oxygen 10.3%

Type III kerogen
(Upper Cretaceous,
Douala Basin)
Total oxygen 27.5%

Ketone $\diagdown C=O$

Acid –COOH

Ester –COOR

Others
(non-carbonyl)

Figure 4.8   Distribution of oxygen-containing functional groups in examples of three different types of immature kerogen. (After Robin, 1975.)

lakes). For example, the tasmanites found in a number of oil shales were formed from algal remains in low-energy marine/brackish water environments, while the Carboniferous equivalent to the freshwater alga *Botryococcus braunii* formed the torbanites of Scottish oil shales. The Eocene Green River oil shale of Colorado, Utah and Wyoming is a further example of type I kerogen, although it is not homogeneous, due to varying environmental conditions, and its organic content varies from $<1\%$ up to 40%.

Pyrolysis of type I kerogen at $550-600°C$ yields more low-molecular-weight material than other types (up to 80% by weight from shallow, immature samples), indicating a higher oil yield. The main pyrolysis products are straight and branched acyclic alkanes (including isoprenoids, particularly those derived from the phytol unit of chlorophylls).

Type II kerogen

Type II, or exinite-type, kerogen is the most commonly occurring kerogen. It has relatively high H/C and low O/C ratios. Polyaromatic nuclei and ketone and carboxylic acid groups are more important than in type I but less so than in type III, while ester bonds are abundant (Fig. 4.8). Aliphatic structures are important and comprise abundant chains of moderate length (up to ca. C_{25}) and ring systems (naphthenes). Sulphur is often found in substantial amounts; in cyclic systems and probably also in sulphide bonds. Associated bitumens contain abundant cyclic structures (aliphatic and aromatic hydrocarbons, and thiophenes) and have a higher sulphur content than other types. A further classification of **type II-S** can be used for sulphur-rich type II kerogens (8-14% organic S by weight) where the atomic S/C ratio is >0.04.

Type II kerogen is usually formed in marine environments, from mixed autochthonous phytoplankton, zooplankton and microbial (mainly bacterial) organic matter deposited in reducing conditions. This material can be compared to exinite, but whereas in coals exines are usually well preserved, in marine environments the organic matter is usually substantially altered by reworking. Type II kerogen can also be formed from allochthonous material, where the macerals from lipid-rich, higher plant debris (e.g. spores, pollen and cuticles), together with plant membrane secretions (e.g. resins and waxes), become concentrated. Concentration of these macerals can sometimes be the result of sediment transport mechanisms, giving rise to lipid-rich coaly deposits. Exinite macerals are usually found in kerogens that also contain varying amounts of liptinites and vitrinites.

Type II kerogen has a lower yield of hydrocarbons upon pyrolysis than type I, but it has still produced oil shales of commercial value and sourced a large number of oil and gas fields. Pyrolysis products are often a mixture of those characteristic of types I and III kerogens.

Type III kerogen

Type III, or vitrinite-type, kerogen has a low H/C (<1.0) and a high O/C (up to $0.2-0.3$) ratio initially. Polyaromatic nuclei and ketone and carboxylic acid groups are important but there are no ester groups. A significant proportion of oxygen is in non-carbonyl groups (possibly heterocycles, quinones, ethers and methoxy groups; see Fig. 4.8). Only minor amounts of aliphatic groups are present. These are dominated by methyl and other short chains, often bound to oxygen-containing groups, but a few long chains are present, originating from higher plant waxes (liptinites) and cutin (exinites).

Type III kerogen is essentially formed from vascular plants and contains much

Table 4.7   **Kerogen classification according to pyrolysis products** (After Larter and Senftle, 1985)

Kerogen type	Major pyrolysis products	Major macerals
I	Paraffins	Alginite (*Botryococcus* torbanite)
I	Paraffins/naphthenes	Alginite (tasmanite)
I	Naphthenes	Resinite
II	Paraffins/naphthenes	Amorphinite (amorphous matter, marine origin)
II/III	Paraffins	Vitrinite/exinite
II/III	Phenols/paraffins	Sporinite
III	Phenols/paraffins	Vitrinite (marine/deltaic shales)
III	Aromatics/phenols	Vitrinite (coal swamps)
IV	Aromatics	Fusinite (inertinite)

identifiable vegetable debris. Vitrinite macerals, the structural woody (lignified) components of higher plants, are predominant. Microbial degradation during deposition is usually limited, possibly by rapid sedimentation and burial, but may give rise to varying amounts of liptinite. Vitrinite concentrations generally occur as coals or coaly shales, hence type III kerogen can be compared with coal in terms of its composition and behaviour with increasing burial.

Type III kerogen is less productive upon pyrolysis and is much less likely to generate oil than types I and II, but it may be a source of gas (especially methane) if buried deeply enough. Vitrinite is incapable of yielding oil, and any that is produced is generated from exinites and/or liptinites. Pyrolysis products are characterised by low-carbon-number phenols, particularly methoxy derivatives of lignin origin. Straight-chain aliphatic hydrocarbons (*n*-alkanes) are also present, dominated by the carbon number range characteristic of higher plant leaf waxes (ca. C_{23} to C_{35}).

Type IV kerogen   Type IV, or inertinite-type, kerogen comprises primarily black opaque debris largely composed of carbon in the form of inertinite. As this has no hydrocarbon-generating potential it is sometimes not considered as a true kerogen. It is probably formed from higher plant matter that has been severely oxidised on land and then transported to its deposition site.

Improved kerogen typing   The classical petroleum source rock is formed from marine organic matter deposited in a reducing environment and broadly corresponds to type II kerogen, with a high genetic potential. Most kerogens fall between types II and III in the van Krevelen diagram (i.e. between exinite and vitrinite in Fig. 4.2b). It is apparent that this classification system is an over-simplification of the variety of sedimentary organic material and depositional settings that can lead to kerogen formation. Detailed analysis of pyrolysis products by gas chromatography (see Section 7.4.3) has enabled a wider range of kerogens to be identified on the basis of dominant maceral and major pyrolysis products (Table 4.7).

4.4.4   Thermal evolution of kerogen

Structural changes   As in coal formation, with increasing compaction and burial depth bacterial activity declines and temperature increases, so that the low-temperature, biologically mediated processes characteristic of diagenesis are replaced by mainly thermally mediated changes of the kerogen structure. However, again as in coal formation, there is some overlap between the zones of diagenesis and catagenesis. As temperature and pressure increase with increasing burial the structure of immature kerogen is no longer in equilibrium with its surroundings and becomes unstable. Rearrangement of kerogen then occurs in a progressive manner to reduce the increasing molecular strain by the formation of a more ordered and compact structure, again very much along the lines described for coal. This is brought about by the elimination of bulky units which prevent the close packing of nuclei. Peripheral bulky groups on nuclei (such as non-planar cycloalkyl systems) and bridging units between nuclei (which include aliphatic chains) are ejected. Functional groups are chiefly associated with peripheral units and so they are progressively eliminated from the kerogen structure with increasing maturity. It is difficult to represent the complex structure of kerogen, but Fig. 4.9 attempts to show the various units present in type II kerogen at the ends of diagenesis and catagenesis. The elimination of bulky peripheral substituents and the increasingly compact and aromatic nature of the residual kerogen with increasing maturity are apparent.

A wide range of compounds is eliminated from the kerogen structure, including medium- to low-molecular-weight hydrocarbons, CO_2, H_2O and H_2S. Petroleum generation is, therefore, a necessary consequence of kerogen rearrangement, and will be considered in greater detail in the next chapter. Ultimately, it is possible for a graphite stage to be reached if sufficiently extreme conditions associated with high-grade metamorphism are attained. The changes associated with this stage are termed **metagenesis**. It is important to note that the initial kerogen structure will remain relatively unchanged even in ancient sediments unless the higher temperatures associated with catagenesis, usually related to deep burial, are attained.

Changes in chemical composition   Evolution of the different kerogen types with increasing maturity can be represented by a van Krevelen diagram, as in Fig. 4.10. In this diagram the evolutionary trends for types I, II and III kerogen broadly correspond to those of liptinite, exinite and vitrinite, respectively. Major components expelled from the macromolecular structure during diagenesis, catagenesis and metagenesis are shown. While type III kerogen is shown yielding oil during the first part of catagenesis, it should be remembered that not all type III kerogens can generate oil, and vitrinite alone is not able to do so.

As we have seen, progressive diagenesis is characterised by a decrease in oxygen content and a corresponding increase in carbon content, which results in a decrease in the H/C ratio and a more marked decrease in O/C. Oxygen loss is primarily due to the elimination of units containing C=O groups, particularly acid and ketone groups, but not apparently ester groups. Bonds involving heteroatoms are often weaker than C—C bonds and so they are the first to break. Some thermally induced bond breaking occurs towards the end of diagenesis, as temperature increases. For a typical oil-generating kerogen the main products evolved during diagenesis are carbon dioxide and water, together with asphaltenes and resins.

During catagenesis the compositional changes of types I and II kerogen are chiefly related to the decrease in hydrogen content due to the generation and release of hydrocarbons (acyclic and cyclic), involving the breaking of C—O and C—C bonds. Consequently, H/C ratios decrease significantly from (ca. 1.25 to 0.5). The aromatic character of kerogen increases as a result of both the loss of aliphatic components and the increasing aromatisation of naphthenic rings (i.e. aliphatic cyclohexyl groups are converted into benzene ring systems by dehydrogenation reactions). There is a relative increase in the amount of methyl (CH_3) groups as long-chain aliphatic components are preferentially removed. Residual C=O content is removed rapidly.

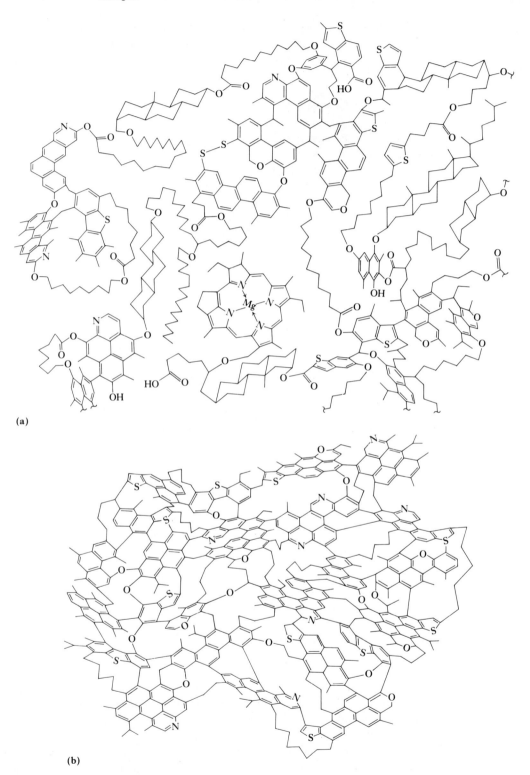

(a)

(b)

Figure 4.9   Structural characteristics of type II kerogen: (a) at the end of diagenesis (H/C = 1.25, O/C = 0.09), and (b) at the end of catagenesis (H/C = 0.73, O/C = 0.03). (After Behar and Vandenbroucke, 1987.)

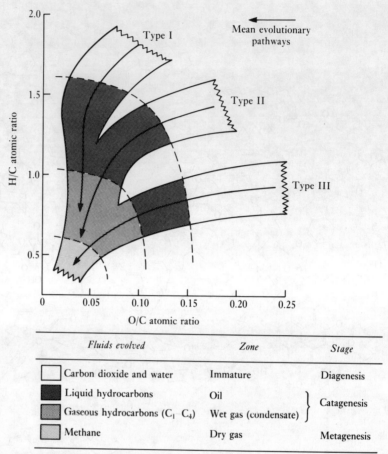

Figure 4.10 | Fluids evolved | Zone | Stage |
|---|---|---|
| ▢ Carbon dioxide and water | Immature | Diagenesis |
| ▦ Liquid hydrocarbons | Oil | ⎫ |
| ▨ Gaseous hydrocarbons (C_1 C_4) | Wet gas (condensate) | ⎬ Catagenesis |
| ▢ Methane | Dry gas | ⎭ Metagenesis |

Figure 4.10   Van Krevelen diagram showing the evolution paths of the three hydrocarbon-generating types of kerogen and the main fluids evolved at different stages of maturity (N.B. not all type III kerogens generate oil). (After Durand, 1980.)

The O/C ratio reaches a low level (ca. 0.05) and remains constant, indicating that the remaining oxygen is not affected, possibly being confined to stable structures like heterocycles. The fall in the H/C ratio during catagenesis is lower for type III kerogens, reflecting their reduced potential for oil generation, although quite large amounts of gaseous hydrocarbons (mainly methane) can be generated. However, oxygen-containing groups continue to be eliminated from type III kerogen, resulting in a further decrease in its O/C ratio.

Metagenesis is experienced by very deep samples or those near high geothermal gradients. By this stage hydrogen elimination has become slow (H/C $\leqslant 0.5$) in all types of kerogen and ca. 30% of the original carbon content has been lost. Methane is practically the only remaining hydrocarbon to be generated and during its evolution any remaining sulphur in the kerogen structure is lost, mainly as H_2S, sometimes in significant amounts. In extreme conditions the carbon content reaches 91% (by weight), with a corresponding H/C ratio of only 0.4. No aliphatic or C=O bonds remain; all carbon atoms are confined to aromatic systems. Major structural rearrangement of kerogen occurs as random layers of aromatic nuclei cluster and assume an orientation permitting maximum compactness.

Chapter 5

Generation and composition of petroleum

5.1 Petroleum generation

5.1.1 Introduction

The hydrocarbon-rich fluids (liquids and gases) evolved from kerogen during catagenesis and metagenesis are collectively termed petroleum (see Box 4.1). As might be expected, the amount and composition of the hydrocarbons generated depend on the composition of the organic matter incorporated into the kerogen structure. In the last chapter it was seen that type I kerogens are rich in acyclic, medium- to long-chain, aliphatic structures of the type generally abundant in oil. Type II kerogens contain relatively less acyclic aliphatic units than type I, but they are still capable of generating large quantities of oil. The predominant aliphatic units in type III kerogen are small chains (ca. $\leqslant C_4$), being derived from vitrinite macerals, and so hydrocarbon gases, particularly methane, are the main petroleum products. There may be some potential for oil production if components such as long-chain plant waxes have been incorporated into this kerogen. The influence of the distribution of acyclic aliphatic, cyclic aliphatic and aromatic units in the parent kerogens on the hydrocarbons generated at the peak of oil formation can be seen for examples of types I–III kerogen in Fig. 5.1. It should be noted, however, that the distribution and amount of hydrocarbons generated vary throughout catagenesis. Type IV kerogen does not contain any structural units capable of generating hydrocarbons.

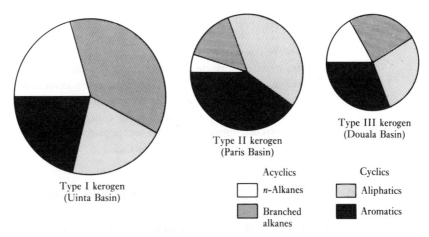

Figure 5.1 Distribution of hydrocarbons generated from different kerogen types at the peak of oil formation. Areas are proportional to amounts of hydrocarbon class per unit weight of organic carbon. (After Tissot and Welte, 1984.)

5.1.2   Hydrocarbons from coal

Because of the nature of the compounds that give rise to humic coals there is usually little potential for oil formation. Generally, there are few alkyl chains to be eliminated other than the predominantly methyl groups in vitrinite, from which large amounts of methane may be generated. Although the organic matter from which humic coals and type III kerogens are formed is similar, humic coals are much less likely to yield oil, even when liptinite/exinite content is relatively high. This is because the chemical and physical structure of coal endows it with a large capacity for absorbing liquid hydrocarbons. Any oil generated during coalification is usually retained within bedded humic coals which tend to exist as large continuous seams. In contrast, migration of oil out of the disseminated organic matter of type III kerogens occurs more readily (Hunt, 1991).

Some humic coals have been reported to be oil-prone, such as those within the Latrobe Group of the Gippsland Basin in south-east Australia. The oil potential of such coals probably arises because they contain significant amounts of lipid-rich material (liptinites and exinites) from a combination of two important sources: firstly, from lipid-rich exines, such as waxes, cuticles, resins and pollen, which become concentrated during diagenesis; secondly, from microbial biomass (liptinites) resulting from extensive reworking of higher plant material during diagenesis. These sources are suggested by the fact that many oils in this class are waxy, with n-alkane distributions dominated by members in the C_{23} to C_{35} range with an odd number of carbon atoms, reflecting a significant contribution from higher plant waxes. In addition, these oils often contain relatively large amounts of hopanes, signifying the importance of bacterial reworking, and their source beds often contain large amounts of amorphous kerogen which may derive from bacterial remains. The degree of bacterial activity and selective preservation of exines is likely to be controlled by factors such as the hydrological balance in the depositional environment, and suitable conditions may have occurred periodically during the formation of some humic coals.

Clearly, humic coals and type III kerogens have many common features based on their chemical composition. In addition, sapropelic coals (cannel and boghead), which have a high liptinite/exinite content deriving predominantly from autochthonous algal material, can be classified as typical type I kerogens having considerable oil potential. The ability of a particular kerogen or coal to yield hydrocarbons can, therefore, be considered in terms of the distributions of the four main maceral types, liptinite, exinite, vitrinite and inertinite. On this basis, humic coals and type I kerogens can be thought of as representing end members of a range of sedimentary organic material combinations, the composition of which can vary according to the depositional conditions and the type of contributing organisms. Whether or not hydrocarbons can migrate out of the source material is then a question of the chemical and physical properties of the source rock.

5.1.3   Variation in hydrocarbon composition with kerogen maturity

The amount and composition of hydrocarbons generated from a particular kerogen change progressively with increasing maturity. These changes are summarised for a typical type II kerogen (i.e. mixed exinite/vitrinite) in Fig. 5.2. During diagenesis the only hydrocarbons present are those inherited directly from living organisms

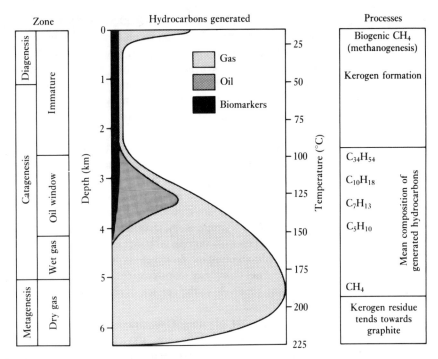

Figure 5.2   General scheme for hydrocarbon generation from a typical source rock. (After Tissot et al., 1974.) (N.B. The diagenesis/catagenesis boundary is shown at ca. 50°C, but if defined solely in terms of the onset of petroleum generation it could be considered to occur at ca. 100°C in this example.)

(mainly in the range C_{11} to C_{35}, such as those from cuticular leaf waxes) and also methane from methanogenesis. With increasing burial depth during diagenesis the gaseous end products of the anaerobic degradation of organic detritus, CH_4 and CO_2 (see Section 3.3.2), become less able to escape as water is expelled from the sediment and permeability is reduced with increasing overburden. Methanogenesis is the last truly biological process to operate at depth (at temperatures up to ca. 75°C).

Temperature steadily increases with burial depth until, during catagenesis, thermal energy is sufficient to cause certain hydrocarbon chains to break off and be expelled from the kerogen and the previously evolved asphaltenes and resins. As can be seen from Fig. 5.2, the size of hydrocarbons evolved decreases with increasing maturity, and on this basis hydrocarbon generation during catagenesis can be divided into two zones: the main zone of oil formation and the wet gas zone (see also Fig. 4.10).

During the main zone of oil generation (or 'oil window') there is significant production of liquid hydrocarbons of low to medium molecular weight, with C—O bond breakage in kerogen being important. The biomarker hydrocarbons, which were previously either trapped within or bound to the kerogen structure (see Section 4.1.1), are progressively diluted by the thermally generated hydrocarbons, which generally exhibit no specific structural or distribution patterns. The first liquid hydrocarbons to be evolved have relatively high molecular weights (mean composition ca. $C_{34}H_{54}$), but as temperature increases hydrocarbons of successively lower molecular weight are produced, as indicated by the mean compositions in Fig. 5.2. Oil generation is usually limited to a window of ca. 100–150°C (ca. 2.5–4.0 km depth in Fig. 5.2) (Mackenzie and Quigley, 1988). Lower temperatures have been proposed for the onset of oil generation (ca. 60°C), but such temperatures do not appear to be typical and may, in part, result from underestimation of maximum burial depths. Most extensional basins (see Box 3.3) experience some compression in the early stages of their evolution, sometimes resulting in uplifted regions which can suffer erosion. Under such circumstances it is possible to underestimate maximum burial depths and related temperatures (e.g. Paris Basin, Mackenzie and McKenzie, 1983).

Gas is produced at all depths down to the limit of hydrocarbon generation (ca. 230°C, ca. 6.5 km in Fig. 5.2). Near the surface it comprises methane from methanogenesis, while at greater depths methane and other hydrocarbon gases are produced in large quantities from the high-temperature fragmentation of kerogen. Gaseous hydrocarbons are considered to comprise C_1 to C_5 compounds, although C_5 members (i.e. various isomers of pentane) can be liquids at normal surface conditions. Oils contain varying amounts of dissolved gases and, similarly, gases can contain varying amounts of dissolved hydrocarbons ($\geqslant C_6$) which would normally be liquids. The term **gas condensate** is applied to gases from which liquid hydrocarbons condense at the surface during commercial recovery and the liquid is termed **condensate**.

In the later part of catagenesis cleavage of C—C bonds in kerogen becomes increasingly important. Only short alkyl chains remain on the kerogen structure, which yield light, gaseous hydrocarbons. These C_1 to C_5 gases are also produced by the thermally induced cracking of C—C bonds in previously released liquid hydrocarbons over a temperature range of ca. 150–180°C. This stage of catagenesis is termed the **wet gas** zone, during which gas condensates may be formed.

Towards the end of catagenesis the proportion of methane in the gaseous products rises rapidly with increasing temperature and kerogen evolution until, during metagenesis, it is the only hydrocarbon released. Methane generation ceases at ca. 230°C and the kerogen residue comprises virtually pure carbon (graphite). Thermodynamically, CH_4 and C are the most stable end products under the temperature (and to a lesser extent the pressure) regimes associated with metagenesis. The zone of methane generation during metagenesis is referred to as the **dry gas** zone. The whole zone of gas generation (wet and dry) is sometimes called the cracking zone, due to the thermal cracking of previously evolved hydrocarbons into smaller, gaseous products.

5.1.4   Reactions involved in hydrocarbon generation

An essential feature of hydrocarbon generation is the transfer of hydrogen from the kerogen nucleus, which becomes richer in condensed aromatic structures, to the expelled hydrocarbon fragments to form alkanes. Examples of reactions leading to this redistribution of hydrogen (or disproportionation) are shown in Fig. 5.3. Cyclisation and aromatisation of residual kerogen structures liberate the hydrogen required during dealkylation of kerogen (the cleavage of alkyl chains at C—O or C—C bonds). Ring-opening reactions can also occur and require addition of hydrogen atoms. Petroleum is not generated from inertinite as its predominantly aromatic nuclei lack sufficient aliphatic chains and hydrogen for disproportionation reactions to occur. In contrast, liptinite is hydrogen-rich, contains abundant alkyl chains and so can generate significant amounts of oil.

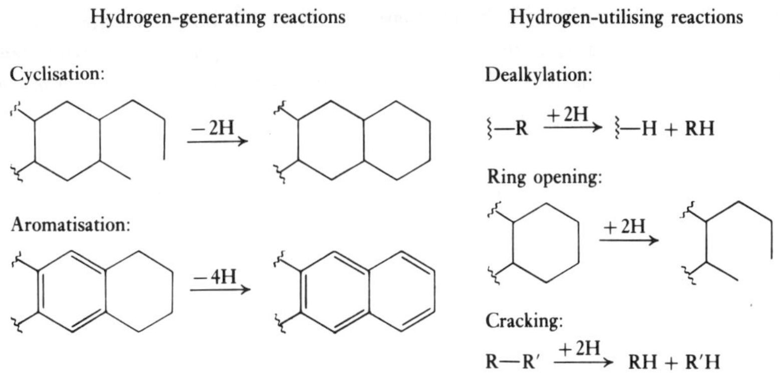

Figure 5.3 Examples of hydrogen transfer reactions during catagenesis.

As can be seen from Fig. 5.3, cracking of aliphatic chains requires hydrogen, which may again be supplied by increasing cyclisation and aromatisation of the kerogen structure. The production of small aliphatic chains depends on both the breaking of particular C—C bonds and the statistical likelihood of a particular fragment being formed. For example, C—C bonds towards the centre of a relatively long chain are generally slightly weaker than those towards the end. However, as the chain gets shorter, proportionally more energy is required to break the central C—C bond (although this energy difference does not change much for components $>C_8$). Also, the more ways there are of forming a fragment of a particular size, the more likely it is to be produced. Consequently, with increasing temperature, progressively smaller molecules are produced by cracking.

As the composition of kerogen varies according to the differing original inputs of organic matter, there is variation in the types of bonds that have to be broken during hydrocarbon evolution from types I, II and III kerogens. The predominant C—C bonds of type I kerogen are the strongest and so peak oil generation from this type of kerogen occurs at a slightly higher relative maturity level than for types II and III, which contain larger amounts of the weaker C—O, C—N and C—S bonds (Tissot and Espitalié, 1975; Waples, 1984). In particular, sulphur-rich kerogens (type II-S) contain a large proportion of alkyl chains attached to the kerogen nuclei by C—S and S—S bonds. These bonds are significantly weaker than the C—C and C—O bonds which are abundant in other kerogen types. As a result, oil can be generated from type II-S kerogens at considerably lower temperatures (ca. 80°C) than the normal type II (Orr, 1986). In addition to abundant sulphur-containing aromatics, the sulphur-rich oils produced contain large quantities of asphaltenes and resins, because breaking of the weaker S-containing bonds also results in relatively larger fragments than usual. The composition of the asphaltenes parallels that of the remaining kerogen with increasing maturity and so supports the proposition that asphaltenes are intermediates in the generation of hydrocarbons.

Isotopic fractionation
We have seen in Fig. 1.7 that organic matter, such as kerogen, preserved in sedimentary rocks is relatively depleted in ^{13}C. As might be expected, the hydrocarbons evolved from kerogen reflect this source-related isotopic signature. Methane generally exhibits an even lighter signature than the kerogen from which it is generated. This is because methane is generated by cleavage of C—C bonds in kerogen or during hydrocarbon cracking, and it requires less energy to break a

^{12}C—^{12}C bond than a ^{13}C—^{12}C bond. Hence, isotopically light methane is preferentially formed. Other hydrocarbons generated from kerogen are similarly relatively light, and so with increasing maturity the residual kerogen becomes isotopically heavier.

5.2 Importance of time and temperature in petroleum formation

5.2.1 Effects of time and temperature on hydrocarbon generation

Temperature   The importance of temperature in petroleum formation can be seen from the fact that most oil is generated from kerogen over a relatively narrow window of ca. 100–150°C. The general relationship between temperature and depth of burial is shown in Fig. 5.2, based on an average value for the geothermal gradient of 33°C/km (see Box 5.1). In basins where the temperature gradients are higher or lower, the depths for the peak generation of oil would be correspondingly shallower or deeper.

Box 5.1

Geothermal gradients, thermal conductivity and heat flow

As a general rule temperature increases with increasing burial depth because heat flows from the Earth's interior towards the surface. The temperature profile, or **geothermal gradient**, is related to the thermal conductivity of a body of rock and the heat flow through it by:

heat flow = (geothermal gradient) × (thermal conductivity of lithology)   [5.1]

Geothermal gradients are not necessarily uniform because of the varying thermal conductivities of different rock types. Thermal conductivity is dependent upon the mineralogical composition of the rock, its porosity and the presence of water or gas. Minerals have higher thermal conductivities than water and so non-porous rocks with a low water content are more conductive than porous rocks with a high water content. If the pores within the rock contain gas the conductivity is further reduced. Many evaporitic deposits have high thermal conductivities (halite and anhydrite ca. 5.5 W/m °C) while the values for shales with a high water content (and therefore high porosity) may be very low (ca. 1.5 W/m °C).

Differences in conductivity between adjacent lithologies can result in locally high temperatures and more mature kerogens. For example, where a salt dome is capped by organic-rich shale, the salt transports heat rapidly upwards through its structure until the cap rock is reached. The shale, being of lower thermal conductivity, is unable to dissipate this heat as quickly as the salt and its temperature rises, resulting in the kerogen in the shale reaching a higher level of maturity than if no salt dome were present.

The difference in temperature between the top and bottom of a shale of low conductivity is relatively great and so the temperature gradient is large. For rocks with high conductivities the temperature range is reduced and the thermal gradient is low. Generally, conductivity increases with increasing depth as rocks become more compacted, resulting in a corresponding decrease in the geothermal gradient.

Subsurface temperatures, and hence geothermal gradients, may be increased or decreased by the flow of fluids in basins and along fractures (as in a graben, which is a downthrown block between two faults). Regional tectonism usually causes a significant increase in heat flows and associated geothermal gradients.

Geothermal gradients can range from 10 to 80°C/km. The lowest values are found in convergent plate margins and Precambrian shields where orogenic (i.e. mountain building) events are of Palaeozoic age or older (heat flow values typically ca. $0.9-1.3 \, \mu cal/cm^2 \, s$). High geothermal gradients and heat flows (ca. $8.0 \, \mu cal/cm^2 \, s$) are typical along mid-ocean ridges and in rifted intracratonic areas or where the crust is thin.

In reconstructing the thermal history of kerogen in a sedimentary rock it must be taken into account that geothermal gradients can increase, decrease or remain about the same during the life of a basin. Gradients increase when the basin is affected by orogenic events that occur at the same time or just after sediment deposition. For example, the Tertiary Pannonian Basin (Central Europe) has been affected by the Carpathes alpine orogeny and the associated high heat flow and geothermal gradient (ca. 50°C/km) have resulted in source rocks of only Pliocene age reaching sufficiently high temperatures to generate oil.

Stable continental margins are areas where the geothermal gradient has decreased with time. Originally the sediments were deposited in areas of very high heat flows, close to the spreading ridge, as the margin developed upon rifting of the continent. However, as sea floor spreading continued the distance between these marginal sediments and the spreading ridge increased and so heat flows and geothermal gradients decreased. These conditions were prevalent during the Cretaceous period when an active phase of sea floor spreading was commencing and continental land masses, which provided copious sediments, were close to the spreading ridges (e.g. the Cretaceous organic-rich deposits off the Atlantic coast of South America). On stable platforms geothermal gradients have remained fairly constant as there has been little or no tectonic or magmatic activity since source rock deposition. The geothermal gradient is usually in the range 25–35°C/km for basins in such areas (e.g. Paris Basin, Palaeozoic basins on Precambrian basements in Australia and Mesozoic basins on Palaeozoic or older basements in west Canada).

Time   Time is also an important factor in petroleum generation. The higher the temperature experienced by a source kerogen the less time is required for oil generation. The amount of time needed increases exponentially with decreasing temperature. For example, at 60°C ca. 200 Ma would be required, but at 100°C only ca. 10 Ma would be needed. Petroleum-like hydrocarbons have been detected in hydrothermal vent mounds in the Guaymas Basin (California) and appear to originate from sediments < 5000 years old (from ^{14}C dating). Vent waters have recorded temperatures of ca. 200°C, but are believed to be expelled at temperatures of ca. 315°C and a pressure of ca. 200 bar (20 MPa) (Didyk and Simoneit, 1989). Temperature and time, therefore, appear to control oil generation.

5.2.2   Kinetic models of petroleum formation

Time–
temperature
index

Time and temperature alone can be used to predict the thermal maturity of *in situ* kerogen. One of the earliest, simple models, the time–temperature index (TTI; Waples, 1980), was based on the observation that the rate of a chemical reaction, such as bond cleavage in hydrocarbon formation from kerogen, approximately doubles for every 10°C rise in temperature (see Box 5.2). It is necessary to know the geothermal gradient (Box 5.1) and the burial history of a potential oil source rock in order to obtain the time and temperature information necessary to calculate the TTI.

Box 5.2

Reaction rates

The speed, or rate, of a reaction involving a single chemical species (a unimolecular reaction) depends on the concentration of the species involved. It is also dependent on temperature, and so the rate at a particular temperature is proportional to some power (x) of the concentration:

$$\text{rate} = k[K]^x \qquad [5.2]$$

where $[K]$ = concentration of the chemical species and k = a constant. The value of x can only be determined experimentally. Kerogen evolution can be approximated by reactions in which $x = 1$ (i.e. first-order reactions). The **rate constant**, k, is related to the activation energy (E_{act}; see Box 2.6) and temperature:

$$k = A \exp(-E_{act}/RT) \qquad [5.3]$$

where A = Arrhenius constant, E_{act} = activation energy, R = universal gas constant (8.314 kJ/mol K) and T = temperature (in kelvin, K).

From Eq. [5.3] reaction rates can be seen to increase with rising temperature. For most reactions near room temperature (300 K), E_{act} = ca. 50 kJ/mol, and so a 10°C rise in temperature results in an approximate doubling in reaction rate. However, the exponential dependence of reaction rate on temperature means that the rate increases by successively smaller amounts for every 10°C interval as temperature rises so that, for example, the increase in rate is only ca. 1.4 for a 10°C rise in the region of 200°C.

The first step is the construction of a depth vs. age plot for the geological section. An example is shown by line A in Fig. 5.4 for the base of a sediment stratum deposited in the late Jurassic at ca. 164 Ma b.p. (before present). Deposition then occurred at varying rates until ca. 108 Ma b.p., when there was a period (ca. 33 Ma) of uplift during the Cretaceous. Subsidence and deposition recommenced at the beginning of the Tertiary (ca. 65 Ma b.p.) and continued until the present. Lines B and C represent the bases of the Cretaceous and the Tertiary, respectively. Uplift and erosion caused the loss of some of the Cretaceous sediments (ca. 330 m) between ca. 108 and 65 Ma b.p. It can be seen that the three lines representing the bases of the Jurassic, Cretaceous and Tertiary remain parallel throughout.

The second stage is to superimpose isotherms representing temperature intervals of 10°C on the plot from geothermal gradient data. The surface temperature needs to be known, as do palaeogeothermal gradients if they are likely to have been different from the present gradient. In Fig. 5.4 a surface temperature of 20°C is used and a uniform geothermal gradient of 30°C/km assumed. However, different lithologies can have different thermal conductivities (Box 5.1) and this may be allowed for when plotting isotherms.

The next stage is to assign a numerical value for the incremental increase in maturation (represented by δTTI_i) that occurs in each 10°C temperature interval (represented by i). Each value of δTTI_i comprises a time function multiplied by a temperature function. The time function (δT_i) is simply the total time (in Ma) a particular stratum has spent in a given temperature interval. The temperature function of a given interval (r_i^n) takes into account the exponential dependence of maturation on temperature. The maturation rate increases by factor r for every 10°C rise in temperature. The most commonly chosen value of r is 2, representing a simple

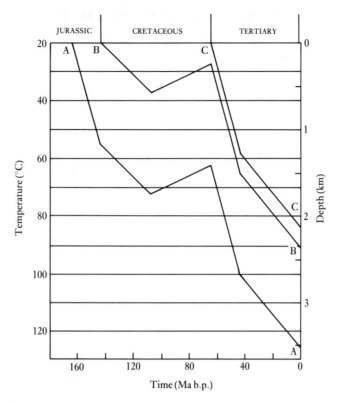

Figure 5.4   Burial history for three horizons with a superimposed temperature scale, forming the basis of a time–temperature index calculation.

doubling of rate every 10°C. However, values for r vary according to the types of chemical bonds in kerogen being broken, but they lie mostly within the range 1.6–2.5. The term n is an integer and for the temperature interval 100–110°C it is arbitrarily assigned the value 0. Above this interval n increases in steps of $+1$ for every 10°C rise in temperature, and below it n decreases in steps of -1. The maturity increase in an interval, i, is given by:

$$\delta \text{TTI}_i = (\delta T_i)(r_i^n) \qquad [5.4]$$

The total maturity increase for a stratum since deposition is then the sum of maturity increases for each interval:

$$\text{TTI} = \sum_i (\delta T_i)(r_i^n) \qquad [5.5]$$

Table 5.1 presents the interval and total TTI values for each of the three strata A, B and C in Fig. 5.4.

In general, oil generation (the 'oil window') corresponds to TTI values within the range 15–160. At greater values wet gas, containing $>5\%$ of ethane and higher-molecular-weight hydrocarbons, is produced and is preserved up to a maximum TTI value of ca. 1500. The limit for dry gas preservation is ca. 65 000. The present TTI values for horizons A, B and C in Fig. 5.4 are ca. 102, 9 and 5, respectively. Therefore, the Jurassic unit, bounded by A and B, lies almost entirely

Table 5.1   **TTI values for strata A, B and C in Fig. 5.4**

Temperature interval, i ($^\circ C$)	Time in interval, δT_i (Ma)			Temperature factor for interval, r^n	Maturity increase for intervals, δTTI_i		
	A	B	C		A	B	C
20–30	6	21 + 14	6	2^{-8}	0.023	0.137	0.023
30–40	5	47 + 5	5	2^{-7}	0.039	0.406	0.039
40–50	6	5	6	2^{-6}	0.094	0.078	0.094
50–60	14	5	8	2^{-5}	0.438	0.156	0.250
60–70	21 + 39	12	16	2^{-4}	3.750	0.750	1.000
70–80	13 + 6	17	17	2^{-3}	2.375	2.125	2.125
80–90	4	16	7	2^{-2}	1.000	4.000	1.750
90–100	6	2		2^{-1}	3.000	1.000	
100–110	17			1	17.000		
110–120	17			2	34.000		
120–130	10			2^2	40.000		
				Total TTI value	101.719	8.652	5.281

within the oil window, while the Cretaceous unit, bounded by B and C, has not yet reached sufficient maturity to evolve hydrocarbons.

More complex models   The TTI is based on a very simple model and can provide surprisingly accurate information. However, it tends to overestimate the effects of time and underestimate the effects of temperature in petroleum formation reactions and so is best suited to average heat flow conditions. With the use of computers it is possible to generate more rigorous kinetic models of oil generation which more accurately reflect natural systems (e.g. Wood, 1988). A problem in studying the kinetics of petroleum formation is whether laboratory simulations, which have to be carried out in a reasonable time (i.e. ca. 10^7 times faster than in nature), accurately reflect *in situ* processes.

Another model used to simulate petroleum formation treated kerogen as being formed from labile, refractory and inert components (Mackenzie and Quigley, 1988). **Labile kerogen** (i.e. liptinite and exinite) yields mainly oil, **refractory kerogen** (vitrinite) yields gas and **inert kerogen** (inertinite) produces no hydrocarbons. Labile and refractory kerogen, therefore, generate petroleum and may be termed **reactive kerogens**, while only H, O, S and N are eliminated from inert kerogen, yielding a residue of graphite. The resulting kinetic model provided results consistent with field investigations and indicates that:

(a) most oil is formed in the range 100–150°C;
(b) most gas is formed in the range 150–230°C;
(c) any oil left in the source rock undergoes cracking to gas in the range 150–180°C;
(d) oil is not expected to survive for geological time periods at > 160°C.

These guidelines are summarised in Fig. 5.5. Oil is often found at temperatures up to 160°C in the subsurface, but not in significant amounts above 170°C. This observation is in agreement with kinetic models which predict that the time taken for half an amount of oil to be converted to gas (i.e. its half-life) is < 1 Ma for temperatures > 170°C.

Kinetic models can be further refined by incorporating functions that describe

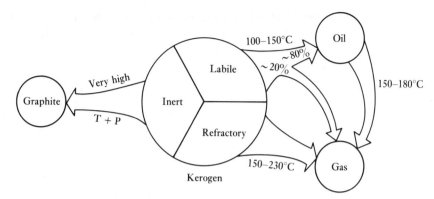

Figure 5.5   Summary of temperature ranges for petroleum generation from kerogen. (After Mackenzie and Quigley, 1988.)

in more detail the different types of reactions involved in petroleum generation. For example, these functions can describe the breaking of different types of kerogen bonds, the cracking of individual liquid hydrocarbons and the effects of clay catalysis on such reactions (e.g. Ungerer, 1990).

5.3   Migration of hydrocarbons

Petroleum is generally found in reservoirs at some distance from its source rock. The whole of the journey from source to reservoir is termed migration, but it is usually divided into primary and secondary stages. **Primary migration** is the expulsion of petroleum from the source rock into suitable adjacent carrier strata and **secondary migration** concerns its subsequent transport through the carrier rock to the reservoir. Important factors in oil migration are source rock and carrier rock **porosity** (a measure of the total volume of pores able to accommodate generated hydrocarbons) and **permeability** (the degree to which pores are interconnected, allowing flow).

5.3.1   Primary migration

Mechanisms of primary migration   As burial increases so does compaction of the generally fine-grained source deposits and much of the water originally associated with the sediments during deposition is expelled. By the time of the main phase of petroleum generation the source rock is relatively dense, with low porosity and permeability. The remaining pores may even be smaller than some of the petroleum molecules. Some pore water is still present, largely bound to mineral surfaces by hydrogen bonding, which further restricts the size of pore throats and severely hinders movement of hydrocarbon fluids out of the source rock.

The precise mechanisms by which primary migration occurs are not yet fully understood, although pressure and to some extent temperature are of importance. Different mechanisms may operate in different types of source rock (Stainforth and

Reinders, 1990). One possible mechanism is the movement of hydrocarbons through microfractures in the source rock that result from the development of over-pressure (e.g. Tissot and Welte, 1984). Over-pressure is mainly caused by the thermal expansion of pore waters that cannot escape quickly enough from the rock, although the generation of hydrocarbon fluids may also contribute. The resulting micro-fracturing relieves over-pressure by allowing the escape of hydrocarbons. These pressuring and fracturing processes probably recur in a cyclical manner. Upon entering larger pores, oil particles tend to coalesce as globules, forming stringers (elongated globules) upon expulsion through narrow pore throats. The general lack of solubility of petroleum components in water and the small amounts of unbound water remaining in the source rock suggest that a pressure-driven expulsion of this kind would occur as a discrete hydrocarbon phase. Depending on the amounts generated from kerogen, gases may be entirely dissolved in the liquid hydrocarbons or may form a separate gas phase.

Another possible mechanism for primary migration is the thermally activated diffusion of hydrocarbons, both liquid and gaseous, through the organic (i.e. kerogen) network of the source rock (Stainforth and Reinders, 1990). The rate at which the hydrocarbons diffuse away from their site of origin within the source rock (i.e. the region in which they occur at highest concentration) has been estimated to be of the same order of magnitude as their rate of thermal generation.

Diffusion and pressure-driven effects have been suggested to play a role in the hydrocarbon expulsion currently taking place from Kimmeridge Clay (Upper Jurassic) shales in the North Sea into adjacent sandstones, a process which generally appears to begin over a temperature range of $120-140°C$ (Mackenzie et al., 1988). At typical expulsion depths of >2.5 km enough hydrocarbons appear to be generated to form an interconnected hydrocarbon phase bleeding out from the shale into the sandstone pores. This flow is aided by higher pore fluid (i.e. capillary) pressures in the shale than in the more porous and permeable interbedded sandstones. The increased efficiency of expulsion close to the interface between shales and sandstones results in greater hydrocarbon concentrations in the margins of the sandstone than in the margins of the shale. The composition of these hydrocarbon fluids appears to be affected by molecular diffusion. While most of the hydrocarbons are expelled as a discrete phase from the shale by pressure-driven flow, there is a tendency for hydrocarbons to diffuse back into the shale margins from regions of higher concentration in adjacent sandstone margins (diffusion occurs down a concentration gradient and so does not have to coincide with the direction of bulk fluid flow). Because diffusion occurs more rapidly as molecular size decreases, it results in the shale margins becoming highly depleted in the weakly diffusive $>C_{10}$ hydrocarbons and enriched in the smaller, strongly diffusive $\leqslant C_{10}$ molecules.

In other sequences of shale source rocks and sandstones it has been noted that the expelled hydrocarbons are somewhat depleted in components $>C_{10}$ compared with those remaining in the source rock, and that the relative depletion increases with increasing molecular weight (Mackenzie et al., 1988). A possible explanation is that if a source rock is richer in small hydrocarbons ($\leqslant C_5$), under certain conditions of pressure and temperature during pressure-driven expulsion, separate gaseous and liquid hydrocarbon phases can be formed. The gas would be expelled first, having lower viscosity, and the larger hydrocarbons ($\geqslant C_6$) would escape from the source rock by dissolving in the gas. The observed trend in depletion of higher-molecular-weight hydrocarbons in the sandstone can then be attributed to the general decrease in solubility of hydrocarbons in gas with increasing molecular weight (e.g. $C_{20}H_{42}$ is less soluble than $C_{15}H_{32}$) (Mackenzie et al., 1988). Alternatively, a diffusion-driven expulsion from the source rock through the kerogen network could explain the enrichment of the more rapidly diffusing, lower-molecular-weight components in the expelled hydrocarbons without the need for separate gas and liquid phases to be formed (Stainforth and Reinders, 1990).

Expulsion
efficiency

Only if sufficient hydrocarbons are generated in the source rock will primary migration (i.e. expulsion of hydrocarbons from the source rock) occur, whether diffusion- or pressure-driven mechanisms operate. As might be expected, there is a lag between the onset of hydrocarbon generation and expulsion because a minimum degree of petroleum saturation within the source rock is required (possibly ca. 40%; Mackenzie and Quigley, 1988). In addition, the change in composition of hydrocarbons generated during catagenesis will lead to gradual compositional changes in the oil expelled from the source rock.

The proportion of generated hydrocarbons that is expelled from the source rock during the main stage of oil formation is strongly dependent upon the type and initial amount of kerogen present. Nearly all of the oil generated by rich source rocks may be expelled, but below a minimum initial concentration of reactive kerogen (ca. 0.5%) most or all of the oil generated may remain in the source rock (Mackenzie and Quigley, 1988). At higher maturity levels this residual oil is cracked to gas and can escape from the source rock more readily than the larger hydrocarbons. Expulsion of gas may be aided by microfracturing due to the increasing fluid pressure within the pore spaces as cracking proceeds. Similarly, liquid hydrocarbons trapped within the structure of humic coals tend to undergo cracking to gas and condensate at elevated temperatures. The fracturing of pores within the coal structure then allows these small hydrocarbons to migrate out of the source material.

A general model has been developed describing the types of hydrocarbons expelled from different types of source rocks. From this model it is possible to distinguish three end member classes for most source rocks, based on the type of kerogen present and its initial concentration (Mackenzie and Quigley, 1988). These two properties tend to govern the ratio of oil:gas in expelled petroleum from each class, as shown in Fig. 5.6. For each class the **petroleum generation index** (PGI; fraction of petroleum-prone organic matter transformed into petroleum) and **petroleum expulsion efficiency** (PEE; fraction of generated hydrocarbons expelled from source rock) are plotted as a function of temperature, assuming a general heating rate of $5°C/Ma$.

Class 1 source rocks contain predominantly labile (i.e. oil-prone) kerogen at concentrations $>10\,kg/t$. They begin to generate hydrocarbons as a liquid phase at ca. 100°C, which rapidly saturates the source rock and between 120 and 150°C 60–90% is expelled. The gas that is also generated is dissolved in the oil under typical conditions (gas:oil ratio 1:5 wt:wt, pressures $>30\,MPa$). Cracking of the minor amounts of fluid remaining in the source rock occurs at higher temperatures, resulting in the expulsion of a gas phase which is initially rich in dissolved condensate (typically, condensate:gas ratio 1:1 wt:wt and pressure 40–70 MPa at 150°C). Cracking is complete by ca. 180°C and dry gas evolution ceases around 230°C. Examples of this class of source rock are the North Sea Kimmeridge Clay and the Bakken Shale of the Williston Basin (North Dakota, USA).

Class 2 source rocks are essentially leaner versions of class 1, containing $<5\,kg/t$

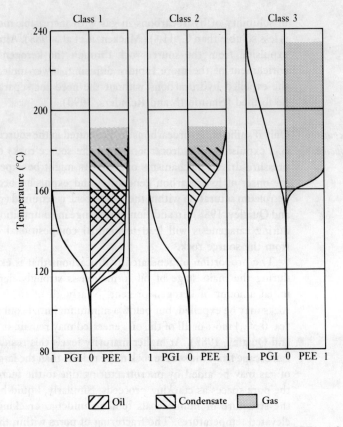

<figure>
Temperature (°C)

Class 1   Class 2   Class 3

240
200
160
120
80

1 PGI 0 PEE 1   1 PGI 0 PEE 1   1 PGI 0 PEE 1

Oil   Condensate   Gas
</figure>

Figure 5.6   Petroleum generation index (PGI) and petroleum expulsion efficiency (PEE) as functions of temperature for the three end member classes of source rocks. (After Mackenzie and Quigley, 1988.)

of mainly labile kerogen, and so exhibit similar variation of PGI with temperature. However, their leanness means that PEE is low and generated petroleum remains in the source rock until cracking commences at 150°C. Petroleum is then expelled as gas condensate, followed by dry gas. Continental shelf source rocks off Nova Scotia and in the Nile Delta are examples of this class.

Class 3 source rocks contain chiefly refractory kerogen, from which generation and expulsion of mainly dry gas occurs at >150°C. Examples of this class are Palaeozoic coals from Europe and North America. Some coals are mixtures of classes 1 and 3 (e.g. Mahakam Delta, Indonesia and Gippsland Basin, Australia) or of 2 and 3 (e.g. other Australasian coals). The behaviour of such mixtures can be estimated by adding the PGI and PEE curves together in the appropriate ratio.

5.3.2   Secondary migration

If suitably porous and permeable strata lie adjacent to the source rock, either above or below, the expelled oil may coalesce into larger stringers or globules within the carrier rock and may travel large distances in the direction of decreasing pressure gradient, until it escapes to the surface or is trapped by a suitable impermeable barrier (Tissot and Welte, 1984). The three important factors controlling secondary migration are buoyancy, capillary pressure and hydrodynamic flow. Oil and gas have specific gravities of 0.7–1.0 and <0.0001, respectively, compared with 1.0–1.2 for the aqueous pore fluids. Petroleum compounds, therefore, undergo buoyant rise in water-saturated porous rocks. Upward migration is usually retarded by layers of less permeable rock, so secondary migration generally occurs along permeable strata in the direction of decreasing pressure gradient. Extensive vertical migration generally requires suitable pathways produced by large-scale faulting.

Oil globules or gas bubbles must undergo distortion to pass through narrow pore throats, and this distortion is opposed by the interfacial tension between the hydrocarbon phase and the water lining the pore. This surface tension effect results in a capillary pressure that retards the flow of petroleum. Petroleum flow stops when capillary pressure exceeds buoyancy, due to a decrease in porosity and permeability. There is often hydrodynamic flow in sedimentary rocks, which can modify the buoyant flow of petroleum through carrier strata, depending on the relative direction of the two flows.

Secondary migration can involve distances of 10–100 km and occasionally more. The size of an oil accumulation is related to the area of source rock from which it was generated (the drainage area) and, therefore, to migration distance. These relationships can be used to obtain estimates of migration distances. As a general rule it can be assumed that a source rock contains 2% organic matter and 20% of the kerogen is transformed into oil, yielding ca. 4 kg of oil per tonne of source rock. If the expulsion efficiency is ca. 25%, then 1 t of source rock yields ca. 1 kg of oil. Typically, 20% of the volume of a reservoir rock is oil, so 1 t of reservoir rock contains ca. 100 kg of oil. If it is assumed that the thickness of a reservoir rock is comparable to that of the source bed, the ratio of the oil pool area to the drainage area is ca. 1:100. Using this approximation, a 3 km diameter oil reservoir requires a drainage area of radius ca. 15 km. A drainage area of several hundred square kilometres is implied for giant oil fields containing $>10^8$ t of oil, requiring secondary migration over several tens of kilometres. An example of such a giant field is the Athabasca tar sands, the area of which is several hundred square kilometres, implying secondary migration over 100 km or more. It must be stressed that the above calculations are only rough estimates and ignore factors such as the amount of oil lost during secondary migration, and ca. 2% of total available pore space appears to remain filled with petroleum after migration (Mackenzie and Quigley, 1988). Secondary migration distances are generally short in lacustrine sequences due to the limited scale of stratigraphic relationships between source rocks and reservoirs.

5.3.3 Traps and reservoirs

The secondary migration of petroleum in the subsurface is terminated by traps. Traps can be of a variety of types (e.g. Selley, 1985), which are usually divided into two classes: structural and stratigraphic. **Structural traps** are the commonest and are caused directly by tectonism. Anticlines are one of the most frequently encountered types of structural traps, other examples being faults and folds. In an anticline (an

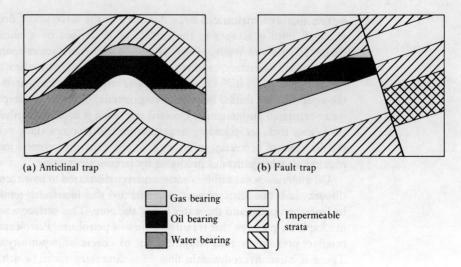

Figure 5.7 Examples of two types of structural traps: (**a**) anticlinal; (**b**) fault.

upward arch in the strata, Fig. 5.7a), petroleum is trapped below impervious strata because outward flow would require movement opposed to the direction of buoyant forces, although if sufficient petroleum is generated it can fill the anticline and spill over. In fault traps of the kind shown in Fig. 5.7b flow is opposed by large capillary forces in structures of very low porosity and permeability that cut across the carrier rock at the fault face. A variety of structural traps can be associated with salt domes (structures formed by plastic deformation of salt deposits under high pressure resulting in the formation of a dome that pierces the immediately overlying sediments). **Stratigraphic traps** are depositional features, such as barrier sand bars and islands (formed by wave and tidal action), channel sands (deposited in deltaic distributary systems), submarine fans (deep-water deposits from sediment gravity flows) and carbonate reefs. They are usually surrounded by less porous and permeable rocks, such as dense shales or limestones, which prevent the outflow of petroleum.

The rocks in which large volumes of petroleum are able to accumulate are termed **reservoir rocks**. They require suitable porosity (typically $10-25\%$) and permeability (typically $1-1000\,\mathrm{mD}$; $1\,\mathrm{mD}$ (milliDarcy) = ca. $10^{-9}\,\mathrm{m^2}$), with reasonable sized pores and an impermeable cap rock or seal to prevent escape of petroleum over geological time periods. They must also be in place before the onset of oil generation. Sandstones often provide suitable reservoir characteristics. More than 60% of all oil occurrences are in clastic rocks, while carbonate reservoirs account for ca. 30%.

Temperature and pressure conditions within the reservoir can be important in determining the hydrocarbon phases present. The solubility of gas in oil increases with increasing pressure, but decreases with increasing temperature. If the amount of gas generated exceeds the capacity of the oil to dissolve it, a separate gas phase will result which occupies the top part of the reservoir, due to its substantially lower density (as shown in Fig. 5.7). A light oil can dissolve more gas than a heavy oil. The permeability of the cap rock is important because gas can escape more readily than oil.

5.4   Petroleum composition

5.4.1   Gross composition of oils

Crude oils, like their associated bitumens in source rocks, contain hydrocarbons (aliphatic and aromatic), resins and asphaltenes (Box 4.1). Table 5.2 presents the typical range of elemental composition for crude oils. In terms of atomic ratios, for every 1000 C atoms there are around 1600–2200 H atoms and up to ca. 25 S atoms, 40 O atoms and 15 N atoms. Much of the nitrogen, oxygen and sulphur is associated with the resins and asphaltenes, and hence they are often collectively termed polar NSO compounds. However, a significant amount of sulphur can be present in compounds of medium molecular weight in the hydrocarbons fraction, usually in the form of thiophenic aromatics, deriving from the types of compounds found in sulphur-rich kerogens shown in Fig. 4.7. Other elements in oil include various metals, particularly nickel and vanadium, which are generally present in trace amounts and are mostly associated with the polar NSO compounds. The average oil contains ca. 57% aliphatic hydrocarbons, 29% aromatic hydrocarbons and 14% resins and asphaltenes; while sulphur, incorporated into thiophenic compounds, accounts for ca. 2% (by weight) of the aromatic hydrocarbon fraction.

The density of an oil is one of its most frequently specified properties. Most oils are lighter than water and so have specific gravities <1. An alternative measurement of oil density is often used, which is expressed as degrees of API gravity:

$$°API = \left(\frac{141.5}{\text{s.g. } 60/60F}\right) - 131.5 \qquad [5.6]$$

where s.g. $60/60F$ = specific gravity of oil at $60°F$ relative to water at the same temperature. By definition, the API scale is inversely proportional to density: light oils have API gravities $>40°$ and heavy oils have API gravities $<10°$.

The gross composition of any fossil fuel is important in terms of its uses and the resulting environmental impact. The most efficient fuels are capable of liberating the most energy from a given mass of fuel, i.e. they have the highest calorific values. Sulphur and nitrogen contents are also important considerations because the SO_2 and nitrogen oxides (NO_x) produced during combustion can lead to acid rain. In other applications of oil, e.g. as lubricants, pour point and specific gravity are important. Some of the basic uses of oil can be seen from the names given to the various fractions obtained upon distillation: gasoline $(C_4–C_{10})$, kerosine $(C_{11}–C_{13})$,

Table 5.2   **Elemental composition of crude** (After Levorsen, 1967)

Element	Abundance (wt%)
C	82.2–87.1
H	11.8–14.7
S	0.1–5.5
O	0.1–4.5
N	0.1–1.5
Others	$\leqslant 0.1$

diesel fuel (C_{14}–C_{18}), heavy gas oil (C_{19}–C_{25}) and lubricating oil (C_{26}–C_{40}). Although coal, oil and gas are all primary energy sources, oil is more valuable in terms of the uses to which many of its constituents can be put by the petrochemicals industry. There is increasing interest in coal as a source of similar compounds.

5.4.2   Hydrocarbons in petroleum

Major hydrocarbons   The hydrocarbon fractions of oils contain mainly aliphatic and aromatic hydrocarbons. However, as well as these true hydrocarbons there are also smaller amounts of aromatic hydrocarbon-like compounds containing a sulphur, oxygen or nitrogen atom. Major hydrocarbon types in crude oil are shown in Fig. 5.8.

The aliphatic hydrocarbons (also known as **saturates** or **paraffins**) are divided into acyclic alkanes (normal and branched) and cycloalkanes (or **naphthenes**). Normal (i.e. straight-chain) alkanes usually predominate, extending to C_{40} or more, but concentrations generally peak around C_6–C_8 and decrease above C_{10}. Usually found among the branched alkanes are *iso*-alkanes (2-methylalkanes) and possibly lesser amounts of *anteiso*-alkanes (3-methylalkanes); both groups exhibit similar carbon number ranges to the *n*-alkanes. Acyclic isoprenoidal alkanes may be important

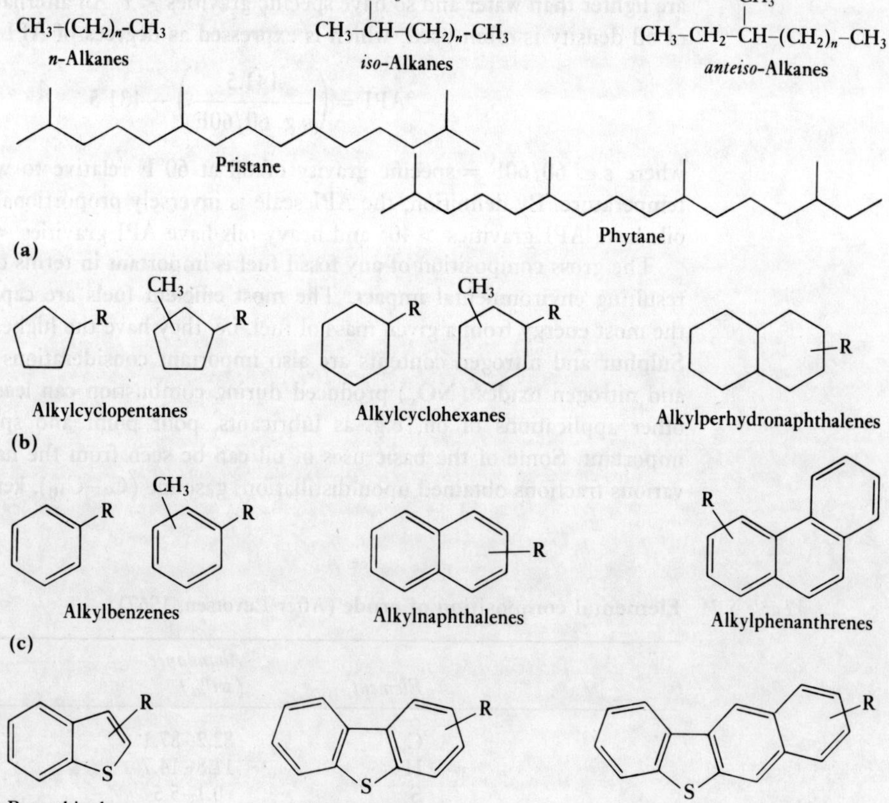

Figure 5.8   Major compounds in the hydrocarbon fraction of crude oils: (a) acyclic alkanes (paraffins); (b) cycloalkanes (naphthenes); (c) aromatic hydrocarbons; (d) sulphur-containing aromatics.

constituents, particularly pristane (2,6,10,14-tetramethylpentadecane) and phytane (2,6,10,14-tetramethylhexadecane) (Fig. 5.8a). Unfortunately, the term *iso*-alkane is sometimes used to describe all the branched alkanes and not just the 2-methylalkanes. Acyclic alkanes $<C_5$ are gases under normal surface conditions, while those up to C_{15} are liquids. The *n*-alkanes with >15 C atoms tend to be viscous liquids grading into solid waxes.

Major cycloalkanes (Fig. 5.8b) include cyclohexane and cyclopentane series with alkyl chains of similar carbon number range to the *n*-alkanes. Further series possessing additional ring-methyl groups are sometimes found. All these cycloalkanes are liquids, grading into solids with increasing alkyl chain length. Cycloalkanes with more than one ring system, such as the alkylperhydronaphthalenes in Fig. 5.8b, are generally present but their abundance tends to decrease as the number of rings increases.

Low-molecular-weight alkylbenzenes are generally the most abundant aromatic hydrocarbons and, again, alkyl chains often have similar carbon number ranges to the *n*-alkanes. As for the cycloalkanes, series of alkylbenzenes are sometimes observed with additional ring-methyl groups. Alkylnaphthalenes and alkylphenanthrenes are also usually present but larger polycyclic aromatic hydrocarbons, such as chrysene, are less abundant (Fig. 5.8c). Alkyl substituents on these polycyclic aromatic hydrocarbons take the form of methyl and ethyl groups, in contrast to the long alkyl chains present in alkylbenzenes. Aromatic compounds with fused cycloalkyl rings (naphthenoaromatics) may also be present. The short-chain alkylbenzenes are liquids, but the longer-chain members and also polycyclic aromatic hydrocarbons are generally solids under normal surface conditions.

Nitrogen, oxygen and sulphur compounds can be found in varying amounts in the aromatic hydrocarbon fractions of oils but they are usually less abundant than the major true hydrocarbons (benzene, naphthalene and phenanthrene derivatives). Sulphur-containing compounds like benzothiophene, dibenzothiophene, naphthobenzothiophene and their alkyl derivatives (Fig. 5.8d) are usually present in oils. Sulphur-rich kerogens can give rise to significant quantities of thiophenic compounds (see Fig. 4.7). Compounds containing oxygen and nitrogen are usually less significant components. Oxygen forms analogous compounds to sulphur, in which the furan unit (Table 2.1) replaces the thiophene unit, e.g. dibenzofuran. Nitrogen-containing compounds are usually the least abundant and, when present, include pyrrole and pyridine derivatives (Table 2.1) with additional benzene rings analogous to the sulphur aromatics.

The average hydrocarbon composition of crude oils is: acyclic (normal and branched) alkanes 33%; cycloalkanes 32%; aromatic hydrocarbons 35%; although, as noted earlier, the distribution of hydrocarbons expelled during catagenesis depends on the chemical composition of the source rock kerogen and its thermal maturity.

Crude oils can be classified according to the relative amounts of acyclic alkanes, cycloalkanes and combined aromatic hydrocarbons plus NSO compounds present. This classification is represented by the triangular plot in Fig. 5.9, and can be seen to distinguish between the main fields of marine and terrestrially sourced oils. The main classes of normal crudes resulting from this classification are:

(a) paraffinic oils, containing mainly acyclic alkanes and with $<1\%$ S;

(b) paraffinic–naphthenic oils, containing mainly acyclic alkanes and cycloalkanes, and with $<1\%$ S;

(c) aromatic–intermediate oils, containing $>50\%$ aromatic hydrocarbons and usually $>1\%$ S.

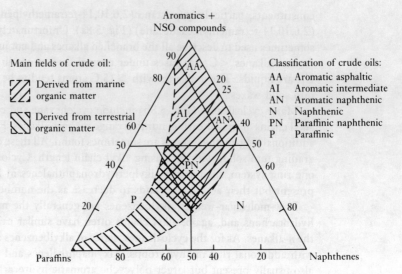

Figure 5.9   Scheme for classification of crude oils based on percentage content of paraffins, naphthenes and aromatics plus NSO compounds. (After Tissot and Welte, 1984.)

The relatively small amounts of hydrocarbons that are generated by most humic coals are characteristic of higher plant material. Aliphatic components in low-rank coals are dominated by wax-derived n-alkanes with an odd-over-even predominance in carbon number over the range C_{23} to C_{35}. This predominance disappears rapidly during the bituminous coal stage when lower-molecular-weight n-alkanes, probably arising from thermal cracking, become more abundant. Also during this stage of coalification cycloalkanes undergo aromatisation.

By the rank of bituminous coal, aromatics usually dominate the hydrocarbons of humic coals, the most abundant being benzene, naphthalene, phenanthrene and their alkyl derivatives. Aromatic sulphur-, oxygen- and nitrogen-containing compounds and their alkyl derivatives are also found. Among these compounds thiophenic components are often important, particularly alkyldibenzothiophenes (Fig. 5.8d). The alkyl substituents on all these aromatic compounds are generally restricted to methyl groups, as expected for compounds deriving largely from lignin components. Also present are small amounts of larger aromatic hydrocarbons formed from extensively fused benzenoid systems (e.g. White and Lee, 1980). Such polycyclic aromatic hydrocarbons are also produced during combustion of oil and coal, and will be considered further in Section 8.4.1.

Although the composition of natural gas accumulations can vary significantly, methane is generally by far the most abundant component (e.g. see Tissot and Welte, 1984). Lesser amounts of other hydrocarbon gases are usually present, chiefly ethane (C_2H_6), propane (C_3H_8), butane (C_4H_{10}), *iso*-butane, pentane (C_5H_{12}) and *iso*-pentane. Natural gas can also contain nitrogen, hydrogen sulphide, carbon dioxide and traces of helium.

Biomarkers   Also present among the hydrocarbons of oils are relatively small amounts (usually <1% by weight) of biomarkers, generally of lipid origin (see Section 4.1.1). By the end of diagenesis their functionalised precursor compounds have been transformed into hydrocarbons. While the bulk of hydrocarbons generated during catagenesis cannot readily be linked to specific precursors, biomarkers can. Pristane and phytane (Fig. 5.8a) are, therefore, included among the biomarkers and are often the most abundant of these compounds.

At the end of diagenesis a significant proportion of the biomarkers is present as discrete molecules trapped within the kerogen matrix. However, some biomarkers may also be incorporated into the kerogen structure and are released during catagenesis. Biomarkers include acyclic isoprenoids, steroids, terpenoids and porphyrins. Such compounds may provide a diagenetic source of hydrocarbons, but

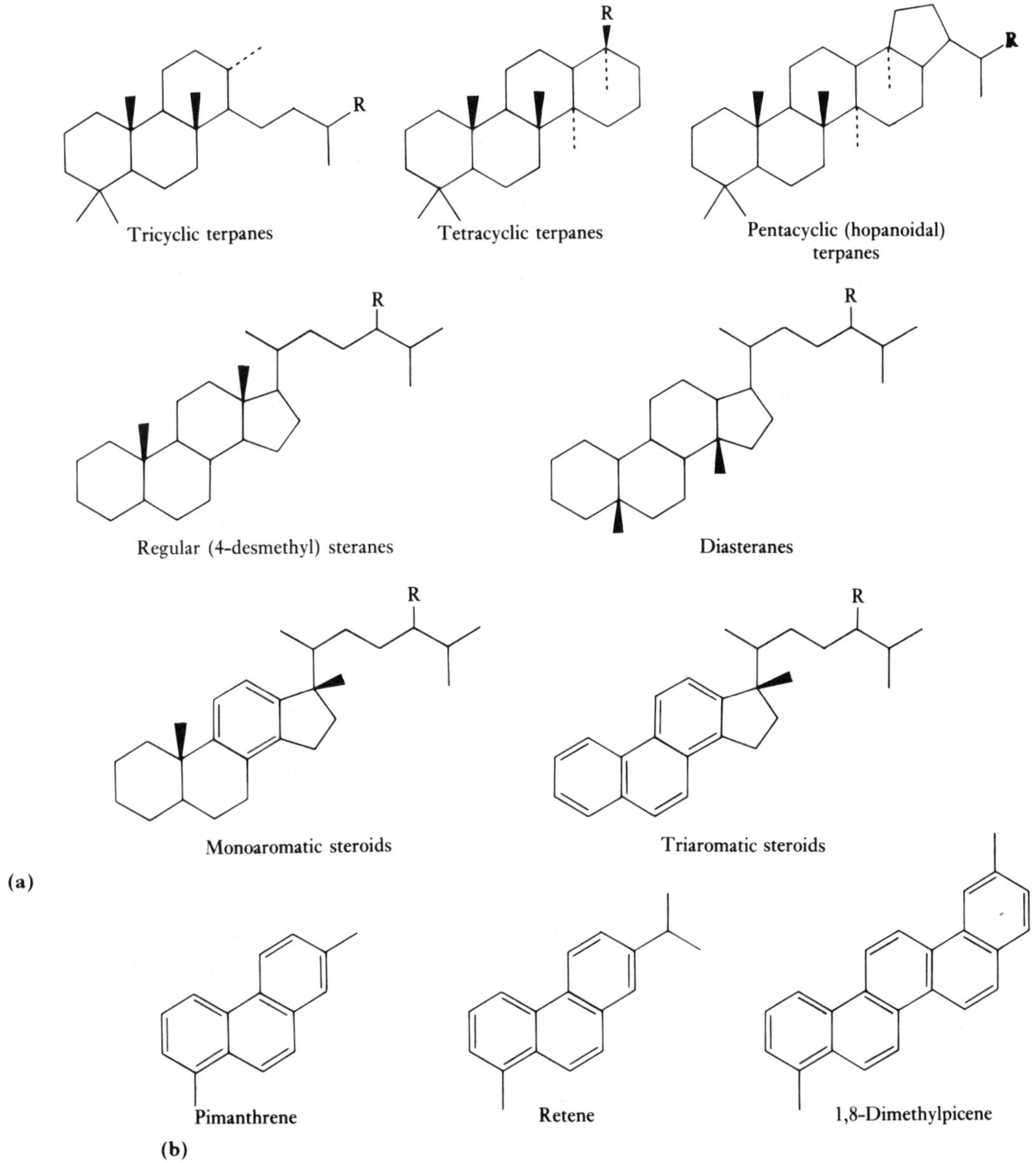

Figure 5.10 Some important biomarker hydrocarbons in (a) crude oils and (b) humic coals.

during catagenesis they are progressively diluted by new hydrocarbons. The transformations undergone by biomarkers and the information that can be obtained from them will be considered in detail in Chapters 6 and 7.

A great variety of terpenoidal and steroidal biomarkers is found in oils, but most of the commonly occurring compounds are given in Fig. 5.10a. The tricyclic terpanes occur generally in the range C_{20}–C_{26}, but occasionally the series extends to C_{40} or more. The members of this series from C_{23} upwards are represented by R = H to n-C_3H_7 in Fig. 5.10a, while the lower members are formed by successively shortening the alkyl chain bearing the R group. Major tetracyclic terpanes are usually C_{24}–C_{27} (R = H to n-C_3H_7, respectively, Fig. 5.10a). Hopanes are commonly found in the range C_{29}–C_{35} (R = H to n-C_6H_{13}, Fig. 5.10a), together with two C_{27} species called Ts and Tm (Fig. 2.18). Major steroidal species include C_{27}–C_{29} members of the regular (or 4-desmethyl) steranes, the rearranged steranes (or diasteranes) and the C-ring monoaromatic steroidal hydrocarbons (Fig. 5.10a). Two short-chain (C_{21} and C_{22}) members of these series are also common, their structures resulting from appropriate shortening of the alkyl chain bearing the R group in Fig. 5.10a. The corresponding triaromatic steroidal hydrocarbons are also usually found and, lacking the usual methyl group at the AB-rings junction, comprise C_{26}–C_{28} (R = H to C_2H_5, Fig. 5.10a) and C_{20} and C_{21} species. Somewhat less common are the 4-methylsteranes, which exhibit similar distributions to the regular steranes but with an additional carbon atom (due to the methyl group at C-4 in the A-ring).

In humic coals biomarkers are generally characteristic of higher plants. The dominant n-alkanes with odd carbon numbers in the range C_{23}–C_{35} can be considered as biomarkers because they can be unambiguously linked with higher plant waxes. Pristane and phytane, mainly deriving from chlorophyll, are also present. Depending on the types of contributing higher plants and diagenetic conditions, resin-derived diterpenoidal and triterpenoidal hydrocarbons can be important. The degree of aromatisation of these compounds depends on maturity. The tricyclic compounds pimanthrene and retene (Fig. 5.10b) are generally derived from pimaric and abietic acids, respectively (see Fig. 2.16c). Pentacyclic triterpenoids of the oleanane type (e.g. β-amyrin, Fig. 2.17) can give rise to picene derivatives (e.g. 1,8-dimethylpicene, Fig. 5.10b), which preserve the original methyl group substitution patterns, where these groups are not lost during aromatisation.

Many other acyclic isoprenoidal, steroidal and terpenoidal biomarkers are found in oils and coals. Some of the more important compounds will be discussed in subsequent chapters in relation to the geochemical information they can provide.

5.4.3 Comparison of crude oil composition with source rock bitumen

A number of factors can affect the composition of oil finally reaching the reservoir compared with the bitumen generated within the source rock. The main factors are migration, water washing, de-asphalting, biodegradation and thermal alteration.

Migration The minerals of the pore walls and their associated bound pore waters present a polar surface to the compounds generated from kerogen. The more polar oil constituents, the asphaltenes and resins and to a lesser extent the aromatic hydrocarbons, are attracted to this polar surface, become relatively concentrated in the interfacial layer and are less readily expelled from the source rock. Adsorption of the more polar constituents can continue throughout secondary migration, leading to a relative increase in concentrations of apolar hydrocarbons, a process often termed geochromatography (e.g. Mackenzie, 1984). Compared with the source rock bitumen, the oil that reaches the reservoir contains slightly less aromatics and significantly less resins and asphaltenes relative to the aliphatic hydrocarbons.

Water washing   Water washing can occur in the reservoir when there is infiltration of meteoric waters. It can also occur during migration, particularly where counter-current hydrodynamic flow is encountered. The effect is the removal of the more water-soluble components from oil, the same polar compounds preferentially adsorbed by pore water and mineral interactions during migration.

De-asphalting   Because asphaltenes are insoluble in light hydrocarbons (C_1–C_8), the infiltration of gas or light oil into a fairly heavy oil, either in the reservoir or during migration, can result in precipitation of asphaltenes (see Box 4.1).

Biodegradation   Oxygenated meteoric waters, carrying aerobic bacteria which are able to metabolise oil components, may reach the oil reservoir. The various components are not utilised at the same rate and there appears to be a general order of preference of removal. Long alkyl chains, particularly the unbranched variety, seem most susceptible to biodegradation. First to be degraded appear to be the *n*-alkanes, but long-chain alkylbenzenes are also affected at an early stage (Jones et al., 1983). The following hydrocarbon groups are then degraded in order of decreasing susceptibility: *iso-* and *anteiso*-alkanes; cycloalkanes; acyclic isoprenoids; C_{27}–C_{29} regular steranes; C_{30}–C_{35} hopanes; diasteranes; C_{27}–C_{29} hopanes; C_{21}–C_{22} regular steranes; tricyclic terpanes (Connan, 1984).

It is possible for complete removal of hydrocarbon classes to occur. While steranes and terpanes are less affected than other alkanes, they can also be totally removed under severe biodegradation. In general, aromatic components tend to be more resistant towards biodegradation than alkanes, and aromatic steroids appear only to be affected under the most severe conditions (Wardroper et al., 1984). Clearly, the properties of an oil can change significantly upon biodegradation, generally to the detriment of its commercial worth.

Thermal   Oil accumulations can continue to undergo thermal evolution, depending on the
alteration   depth of the reservoir and subsequent geothermal history. With increasing maturity oil composition changes and, as noted in Section 5.2.2, is degraded relatively rapidly above 160°C. For example, paraffinic–naphthenic oils (Fig. 5.9) are degraded to aromatic–naphthenic oils (with moderate S content, <1%), and aromatic–intermediate oils degrade to aromatic–asphaltic oils (with high S content, >1%).

5.5   Occurrence of fossil fuels

5.5.1   Temporal distribution of fossil fuels

Throughout the Earth's history there have been periods when conditions have been particularly suitable for the deposition of petroleum source rocks and coal. The

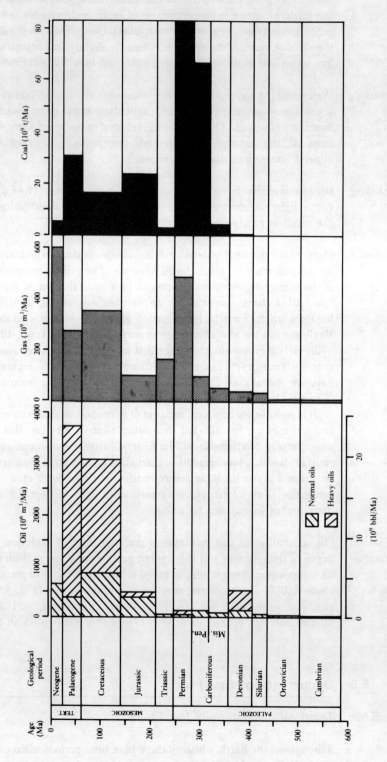

Figure 5.11 Relationship between amounts of fossil fuel reserves (proportional to area under graph) and source rock age. (Data after Bois et al, 1982.)

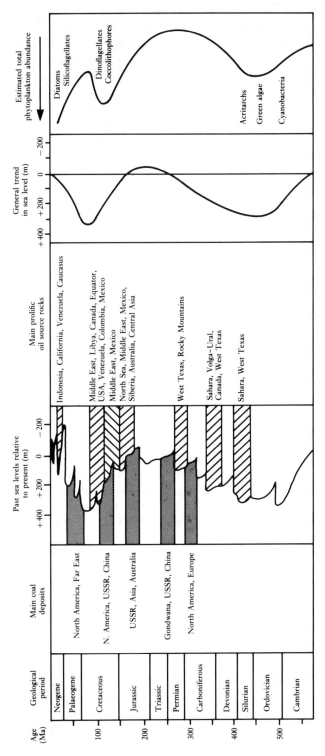

Figure 5.12   Relationship between marine transgressions and regressions and major depositional periods of coals and oil source rocks. (After Tissot, 1979.)

relationship between size of fossil fuel reserves and age is shown in Fig. 5.11, with the age for oil and gas reserves being that of the respective source rock. Two main episodes of coal formation can be distinguished: the first during the Carboniferous–Permian and a second smaller episode spanning the Jurassic to early Tertiary. Because of the influence of time on maturation, the majority of coals formed in the earlier episode are now bituminous coals or anthracites, whereas those from the Tertiary are mainly brown coals.

The amounts of oil shown in Fig. 5.11 include heavy oils (e.g. tar sands) as well as normal oils. About 25% of the total reserves are recoverable. It can be seen that there were two main episodes of oil source rock deposition: one during the Palaeozoic, peaking around the Devonian, and another during the Mesozoic, peaking around the Cretaceous. The Palaeozoic episode accounts for only ca. 13% of the total reserves of normal oils. These episodes correlate with periods of high sea level (see the general trend in sea level in Fig. 5.12) resulting primarily from the larger proportion of the ocean basins occupied by mid-ocean ridges during periods of more active spreading. As discussed in Section 3.4.3, such periods led to marine transgressions on to continental margins, creating suitable conditions for the production (e.g. high nutrient supply) and preservation (e.g. anoxic basins) of sedimentary organic matter. This is reflected in the correlation of phytoplankton abundance with organic-rich sediments in Fig. 5.12. The Palaeozoic peak in phytoplankton production was due to organic-walled organisms (acritarchs, green algae and cyanobacteria). The later productivity peak was dominated by calcareous nanoplankton (coccolithophores and dinoflagellates) initially and subsequently by siliceous plankton (silicoflagellates and diatoms) (see also Fig. 1.7). The smaller size of oil reserves for the Palaeozoic period compared with those of the Mesozoic is probably mainly due to the loss of Palaeozoic oils by thermal degradation or by escape from reservoirs due to later tectonic events. The surviving Palaeozoic oils are found in stable continental platforms that have escaped significant tectonic activity.

Superimposed on the general sea level trend is a cyclical rise and fall over shorter periods. The deposition of major coal deposits and oil source rocks is shown in Fig. 5.12 in relation to these cycles. It can be seen that the major prolific oil source rocks were deposited during global marine transgressions, while major coal deposits (accounting for ca. 95% of reserves) coincided with periods following regression. The latter is consistent with suitable conditions being predominantly found in lowland coastal swamps.

Gas reserves are less readily associated with their source rocks because gas is able to migrate more easily and over greater distances than oil. However, reserves are consistent with derivation from oil source rocks and coal (Fig. 5.12).

5.5.2   Oil reserves

Conventional oils

Conventional oils, or normal (i.e. liquid) crude oils, have API gravities of $\geqslant 20°$, while oils of lower API gravity are termed heavy oils. Proven conventional oil reserves amount to 175×10^9 m^3 (154 Gt or ca. 1100×10^9 bbl; 1 barrel (bbl) = ca. 0.16 m^3). Of this, the Arabian–Iranian province (Middle East) accounts for $>40\%$. Estimated total global reserves are ca. 600×10^9 m^3. Giant oil fields are particularly important in terms of global oil reserves. Only 33 supergiant fields (each containing

$> 5.5 \times 10^9$ bbl), out of a total of around 30 000 fields, account for 50% of oil reserves.

Most of the discovered oil has been generated from marine source rocks. In comparison, lacustrine source rocks are usually of much more restricted extent and hence, in absolute terms, produce less oil. For example, the largest lacustrine-sourced province, Songliao (China), accounts for $<1\%$ of global recoverable resources (ca. 8.5×10^9 bbl). Consequently, more effort has been put into the exploration of marine sourced oils. However, the discovery of smaller, lacustrine-sourced reserves may prove important in the future, as major fields become exhausted.

Heavy oils **Heavy oils** generally result from degradation within the reservoir. Sometimes a distinction is made between heavy oils (ca. $10-20°$ API) and extra-heavy oils or **tar sands** ($\leqslant 10°$ API). Estimated global reserves of heavy oils as a whole lie in the range ca. 450×10^9 to $1000 \times 10^9 \, m^3$ and are, therefore, comparable with conventional oil reserves. The largest reserves of heavy oils are in Venezuela (150×10^9 to $500 \times 10^9 \, m^3$) and Canada (220×10^9 to $420 \times 10^9 \, m^3$).

Heavy oils contain more asphaltenes (ca. 23% by weight) and resins (ca. 31% by weight) than conventional oils (ca. 14% by weight total NSO compounds), and they also usually contain more sulphur. The hydrocarbon content of the heavy oils is, therefore, lower than that of conventional oils and is more aromatic. Typically, heavy oils contain 30% aromatic and 16% aliphatic hydrocarbons (by weight), compared with 29% aromatic and 57% aliphatic hydrocarbons for conventional oils. By definition, therefore, heavy oils belong to either the aromatic–naphthenic class or the aromatic–asphaltic class (Fig. 5.9), depending mainly on the composition of the original oil and the extent of the degradation.

Much of the heavy oil reserves may not be recoverable. Bitumen has been recovered from the giant Athabasca oil sands (Canada), in which it forms a coating around sand grains and amounts to ca. 30% by weight of the raw material. Steam is used to recover the bitumen in giant tumblers, but economic operation requires relatively high oil prices.

Oil shales Any shallow rock that yields commercial amounts of oil upon pyrolysis can be classed as an **oil shale**. While oil shales are usually true shales, they can also be carbonates or marls. The organic matter in most oil shales appears to derive from marine or freshwater phytoplankton deposited in lakes, shallow seas, bogs or lagoons. Oil shales do not contain any significant amounts of free oil but can be considered as potential source rocks which yield oil upon artificial thermal maturation (pyrolysis at ca. $500°$C). Reserves are estimated at ca. $500 \times 10^9 \, m^3$ and so represent a potentially important source of oil, of similar size to conventional oil reserves. Reserves are concentrated mainly in North America (ca. $270 \times 10^9 \, m^3$), South America (ca. $130 \times 10^9 \, m^3$) and the former Soviet Union (ca. $50 \times 10^9 \, m^3$).

As for the heavy oils, economic recovery of oil from oil shales depends on the relative price of conventional oil. *In situ* retorting has proved successful for the late Cretaceous/early Tertiary lacustrine oil shales of Wyoming, Utah and Colorado (western USA). The organic matter content of these shales is ca. 15% by volume, chiefly comprising long-chain sapropelic hydrocarbons of the kind found in liquid crude oils. Retorting yields ca. 1 bbl of oil from 1.5 t of shale.

5.5.3   Coal

There is significantly more coal than oil (around six times as much if proved/recoverable reserves of coal are considered in terms of their equivalent in barrels of oil) and it is more evenly distributed on a global basis. Proven reserves total 1420 Gt (19 Gt anthracite, 1058 Gt bituminous coal and 343 Gt brown coal). Estimated total reserves are 10 754 Gt (28 Gt anthracite, 8093 Gt bituminous coal and 2633 Gt brown coal). Proven reserves, therefore, amount to 13% of the total possible reserves, while exploitable reserves are 5.5%.

5.5.4   Gas

Methane is the predominant hydrocarbon gas and can be associated with oil source rocks and coals, as it derives primarily from vitrinite. The 18 major oil provinces, which account for $>80\%$ of proven oil reserves, also account for ca. 70% of proven gas reserves, which total $105 \times 10^{12} \, m^3$. In terms of their equivalent in barrels of oil, this represents a similar amount to proven oil reserves. The largest gas reserves are in the Middle East and the former Soviet Union, each of which account for ca. 20% of the total. Estimates of total global reserves vary, but an amount in excess of $300 \times 10^{12} \, m^3$ seems likely.

5.6   Assessment of petroleum source rocks

It has been shown that the formation of petroleum depends on a number of factors. The type of hydrocarbons that may be generated during catagenesis depends on the type of organic matter preserved in the source rock, with higher plant material tending to be a source of gas, while more lipid-rich material can lead to oil generation. However, enough organic matter must be present for sufficient hydrocarbons to be generated so that expulsion from the source rock occurs (see Section 5.3.1). In addition, the temperature to which the source rock has been subjected is critical in determining what part, if any, of the hydrocarbon-evolving potential has been realised. These three factors – type of organic matter, amount of organic matter and thermal maturity – are important in the evaluation of potential petroleum source rocks. Some of the more commonly employed methods of assessing these factors are described below.

5.6.1   Amount and type of organic matter

The amount and type of organic material are interrelated in terms of petroleum generation. For example, a small amount of alginite (e.g. 0.5% of rock) is a much better potential source of hydrocarbons than a significantly larger amount of inertinite (e.g. 2.0% of rock).

Amount of organic matter

The amount of organic material in sediments and rocks is usually measured as total organic carbon (TOC) content, expressed as a percentage of the total substrate dry weight. It can be determined by grinding up a sample of source rock to a fine powder, removing carbonate by mineral acid treatment and then measuring the CO_2 evolved upon combustion. Average TOC values for shale and carbonate-type source rocks are ca. 2.0% and 0.6%, respectively. Typical minimum acceptable TOC values for various types of petroleum source rocks are ca. 0.5% for shales, 0.3% for carbonates and 1.0% for clastic-type rocks. The amount of hydrocarbons isolated from the bitumen extracted from finely ground rock samples can also provide a useful indication of whether any oil potential exists. Oil source rocks are generally considered to require a minimum hydrocarbon content of 200–300 ppm.

A pyrolysis technique known as **Rock-Eval** can provide much useful information on source rock samples. This reasonably portable piece of equipment is used to pyrolyse ground samples of source rock (ca. 100 mg) in an inert atmosphere under varying temperature conditions (Espitalié et al., 1977). The sample is firstly heated (at 10–50°C/min) to moderate temperatures (200–250°C), during which hydrocarbons already generated within the source rock (also known as the 'oil shows') are volatilised. This group of components is quantified using a flame ionisation detector (see Section 7.5.1) and constitutes what is termed the S_1 measurement. The temperature is then progressively increased (up to 550°C) until breakdown of the kerogen structure occurs (simulating catagenesis), and the resulting hydrocarbons are again determined by the flame ionisation detector to provide the S_2 measurement.

In the original design of the equipment the flow of components emanating from the thermal breakdown of kerogen was split into two streams, so that oxygen-containing components (e.g. CO_2 and H_2O) could be determined by a thermal conductivity detector (as they do not give a response on a flame ionisation detector), providing the S_3 measurement. Unless the temperature at which S_3 is measured is carefully controlled, problems can arise from the additional evolution of CO_2 from carbonate. Carbonate can be removed prior to analysis, but later versions of Rock-Eval and similar instruments have omitted the S_3 facility although they often include an S_0 measurement of free gases and an S_1' measurement of free liquid hydrocarbons.

The total amount of hydrocarbons that can be generated by a source rock, the **genetic potential**, can be represented by $S_1 + S_2$ as a proportion of the amount of source rock (expressed as kg hydrocarbons/t rock). Values <2 kg/t suggest no oil but possibly some gas potential, while values in the range 2–5 kg/t suggest a source rock with moderate oil potential and values >5 kg/t represent good oil potential.

Type of organic matter – optical methods

The type of organic material present in a source rock can be established by optical or physico-chemical methods. Optical methods are based on maceral recognition techniques developed originally for coal petrography (Section 4.3.1). Maceral examination can be carried out using reflected light microscopy of thin sections of whole rock or of isolated organic particles (the mineral matrix having been removed by acid digestion). Transmitted light microscopy can also be used for isolated maceral concentrates. Shape and degree of transmittance and reflectance, and also fluorescence under UV illumination, can be used to identify broad maceral groups (liptinite, exinite, vitrinite and inertinite) and types within these groupings, as in Table 4.3, providing information on sources of organic matter. More detailed analysis of organic particles can be undertaken by electron microscopy. An estimate of maceral distributions between the main groups provides an indication, together with TOC values, of the petroleum-generating ability of the source rock.

Type of organic matter – physico-chemical methods

Elemental analysis of isolated kerogen can be used to generate the H/C and O/C atomic ratios used in a van Krevelen diagram (Fig. 4.10), providing information on kerogen type and maturity.

Rock-Eval can yield very general information on the type of organic matter present by use of the S_2 and S_3 measurements in combination with TOC values. The **hydrogen index** (HI) is given by S_2/TOC (mg hydrocarbons/g organic carbon) and the **oxygen index** (OI) by S_3/TOC. A plot of HI vs. OI provides an analogue to the van Krevelen diagram, from which both organic matter type and maturity can be obtained. A minimum value of HI of ca. 20 mg/g is required for an oil source rock. The organic matter type application of Rock-Eval is becoming superseded by pyrolysis techniques in which individual products can be identified, providing more detailed information on kerogen constituents (see Table 4.7).

Carbon isotope distributions can be used to gain some information on type of organic material, although care and support from other data (e.g. maceral studies) are needed in interpreting the measurements. Kerogen from higher plant sources usually exhibits lower (by $3–5‰$) $\delta^{13}C$ values than that sourced by marine organisms.

Analysis of hydrocarbon distributions, particularly biomarkers, in source rock bitumen can give useful information on sources of organic material and also depositional environment, and will be discussed in Chapter 7.

5.6.2 Maturity of organic matter

Optical measurements of maturity

Macerals are affected by increasing maturity in a variety of ways. Perhaps the simplest is the change in colour, observed under transmitted light, of spores and pollens with increasing temperature (in a similar way to the charring of toasted bread). The carbonisation of these palynomorphs results in a colour change from yellow in immature samples, through shades of orange/yellow-brown during diagenesis and brown during catagenesis to black in metagenesis. Standardisation is required to remove the subjectivity of assessing colour change. Colour charts can be used and colours assigned a numerical value, such as in the **thermal alteration index** (TAI; Staplin, 1969).

Another property that changes with maturity in a progressive and defined way is the degree of reflectance of vitrinite macerals, which increases with increasing thermal maturity. Isolated macerals are usually immersed in oil to prevent stray reflections and illuminated with monochromatic light (546 nm, in the green region of the visible spectrum). Reflectance values are then expressed as a percentage by the term R_o. The method was initially developed for measuring the rank of coals, in which huminite and vitrinite particles are the main macerals. Calibration against other coal rank maturity parameters, such as volatiles content (which decreases with rank, see Table 4.4), has shown that **vitrinite reflectance** can provide an accurate assessment of the stage of coalification (McCartney and Teichmüller, 1972). Vitrinite reflectance has similarly found application in assessing thermal maturity in types II and III kerogen, but it cannot be used for type I kerogen because vitrinite is absent.

Vitrinite reflectance provides a measure of the thermal stress experienced by a source rock or coal. However, the maturity of a source rock is usually expressed in terms of the stage it has reached in the petroleum-generating sequence. As the temperature of oil formation can vary according to the type of organic matter present (Section 5.1.3), so can the associated vitrinite reflectance values. This is because the maturity measurement from vitrinite reflectance relates only to the thermal maturation of vitrinite macerals, not to the oil-generating macerals which can vary in the amount of thermal stress they require to generate oil. The main phase of oil generation usually occurs in the range $0.65-1.30\% \ R_o$, with peak generation occurring around $1.0\% \ R_o$. However, oil generation can commence around $0.5\% \ R_o$, which is usually considered to be the diagenesis/catagenesis and brown coal/bituminous coal boundary. A summary of vitrinite reflectance, thermal alteration index and time−temperature index values corresponding to the main stages of coalification and of petroleum generation can be found in Fig. 7.5, together with values for some biomarker maturity parameters. Wet gas generation from kerogen occurs in the range $1.3-2.0\% \ R_o$, giving way to dry gas $> 2.0\%$. The latter value is close to the boundary between bituminous coal and anthracite. Under advanced metagenesis, vitrinite reflectance values of around 11% can be associated with graphite formation.

A range of reflectance values is usually observed for vitrinite in a kerogen sample, and it is important to determine statistically the accuracy of the mean values used in maturity assessment and the significance of depth trends based on such values. Data can be recorded in the form of a plot of frequency (i.e. number of macerals exhibiting a given reflectance value) vs. reflectance (Fig. 5.13). In such plots two reflectance maxima can sometimes be observed for vitrinite, as in Fig. 5.13. That at lower reflectance represents the real maturity of the kerogen, while that at higher reflectance probably represents an input, during deposition of the sediment, of

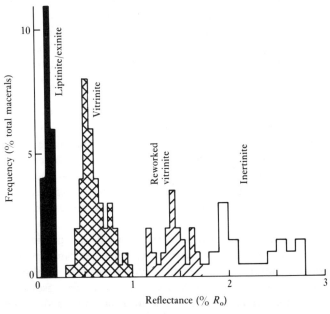

Figure 5.13   Example of vitrinite reflectance values of different macerals in a kerogen sample using a frequency plot (i.e. number of macerals of a given reflectance value vs. R_o).

reworked vitrinite from an eroded, older and more mature source. It can be difficult to identify all the finely disseminated vitrinite macerals in a kerogen sample, particularly if there is a paucity of other maceral groups for comparison of reflectance values. In addition, the reflectance values and morphology of vitrinite and inertinite macerals form a continuum, making distinction between these two groups difficult on occasion. It is, therefore, best to apply statistical evaluation to a large data set obtained from isolated and concentrated kerogen macerals.

Liptinite/exinite macerals fluoresce under blue/UV light and the fluorescence is characterised by its intensity and wavelength. Fluorescence is intense in immature samples but decreases during diagenesis and catagenesis, and by the end of the oil window it has usually disappeared. The intensity of fluorescence can, therefore, be used as a maturity indicator. In addition, the wavelength of fluorescence progressively increases (i.e. moves to the red end of the visible spectrum) with increasing degree of catagenesis. Fluorescence measurements are most accurate when vitrinite reflectance measurements are least accurate – i.e. the decrease in fluorescence is greatest where vitrinite reflectance is lowest and changes slowly – at the beginning of the oil window. These measurements are, therefore, complementary. In addition, while vitrinite reflectance cannot be applied to type I kerogen, fluorescence can.

Pyrolytic measurements of maturity

Rock-Eval data can be used to gain maturity information from kerogen samples. As hydrocarbons are generated in increasing quantity with increasing temperature, the S_2 measurement decreases while S_1 increases. Hence $S_1/(S_1 + S_2)$, the **transformation ratio (or production index)**, increases with increasing maturity. Migration of hydrocarbons out of the source rock will adversely affect this parameter in absolute terms, as will the presence of microreservoirs within the source rock, but a general increase with depth in a given well will still be observed.

During Rock-Eval analysis the temperature at which the maximum in the S_2 response is noted (T_{max}) increases with increasing maturity. This reflects the increasing thermal energy that is required to break the remaining bonds in kerogen associated with hydrocarbon generation during catagenesis. It is possible, therefore, to use T_{max} as a maturity parameter, but although it is independent of migration effects it is affected by the type of organic matter, as the thermal energy required to break the different types of bonds present varies (see Section 5.1.4).

Effect of maturity on identification of organic matter type

At maturity levels ca. $>1.5\%$ R_o it becomes increasingly difficult to differentiate between the types of organic matter from which the kerogen was originally formed. This results from structural changes in macerals and a trend towards similar chemical composition of the residual kerogens. The latter can be seen in the van Krevelen diagram in Fig. 4.10, in which the evolution lines for the three main kerogen types merge during the gas generation phase of catagenesis. Electron microscopy is required to distinguish types of organic matter at these maturity levels.

Chemical measurements of maturity based on bitumen

Bitumen abundance in source rocks can provide a rough estimate of maturity, although it is affected by the type of organic matter and migration. The corresponding measurement of volatiles content in coals is also used as a maturity indicator. Volatiles content is highest in peats and decreases with rank, from ca. 45% to 10% for bituminous coals. The distribution of hydrocarbons within bitumen provides more useful and dependable maturity information for oil source rocks, and will be discussed in Chapter 7.

A variety of maturity measures are used for coals, depending on rank. Moisture content (which decreases with increasing rank) and calorific value (which increases) are typically used for brown coals. Volatiles content and vitrinite reflectance are used for bituminous coals, and carbon or hydrogen content is used for low-volatile bituminous coals and anthracites. The carbon content of vitrinites increases to ca. 55% in peats, to ca. 75% in brown coals, to ca. 90% in bituminous coals and finally up to 100% in anthracites.

5.6.3   Stable isotopes and correlation of petroleums with source rocks

Because secondary migration can occur over significant distances it is possible, in a given exploration area, to have a number of reservoirs containing more than one type of oil and several potential source rocks. It is important to relate a family of oils with its source rock in order to determine where a source rock horizon has been buried sufficiently deeply to generate oil, to evaluate migration paths, and to aid exploration in adjacent areas where a similar source rock sequence may be found. Two of the most useful tools for correlating oils with each other and with source rocks are isotope distributions and biomarker distributions. The latter will be discussed in Section 7.4.2.

Oils and source rocks can be correlated by their carbon isotope distributions. Bulk comparison of kerogen with oils is not recommended, mainly due to the isotopic fractionation that occurs during petroleum generation (Section 5.1.4). Comparison of source rock bitumens with oils can also be misleading because the carbon isotope composition of hydrocarbons, resins and asphaltenes is different (the hydrocarbons are isotopically lighter), and the polar fractions are almost always significantly depleted in the oil by migration effects (Section 5.3). In addition, the isotopic signature of aliphatic and aromatic hydrocarbons differs (aromatics are slightly heavier by ca. $1-2\%_{oo}$), and migration also often causes a depletion of aromatic hydrocarbons relative to aliphatics in oils compared with the source rock bitumen. Consequently, comparisons are usually made of separate aliphatic and aromatic hydrocarbon fractions isolated from oils and source rock bitumens, a procedure that removes the effects of migration-related compositional variations.

Stable isotopes of elements other than carbon can be applied to correlation studies. Levels of ^2H (deuterium, or D) relative to ^1H and of ^{34}S relative to ^{32}S can be expressed as per mil values of δD and $\delta^{34}S$, calculated relative to standards in a similar way to $\delta^{13}C$ values (see Eq. [1.7]). A combination of such isotopic measurements can be quite informative for correlation purposes.

Plots of δD vs. $\delta^{13}C$ can be used to distinguish methane from different sources. Figure 5.14a shows the characteristic isotope distributions for major sources of methane in the subsurface, with atmospheric methane and the main field for oils included for comparison. Methane formed from kerogen at elevated temperatures can be called thermogenic (T, Fig. 5.14a) and has a different isotopic composition from bacterial (methanogenic) methane. The latter is mainly generated by fermentation of acetate (F, Fig. 5.14a) or reduction of CO_2 (R, Fig. 5.14a). Conditions within a sediment can cycle (e.g. on a seasonal basis) between those favouring fermentation (warm) and those favouring reduction (cold and acetate-depleted), producing methane of a mixed signature (Schoell, 1988). Geothermal methane (G, Fig. 5.14a), generated by the pyrolytic action of a magmatic heat source on organic

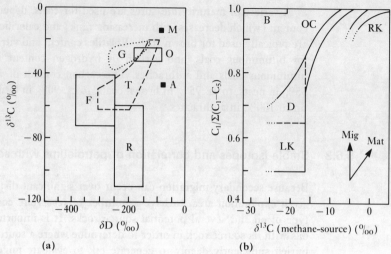

Figure 5.14   (a) Deuterium and ^{13}C concentrations in naturally occurring methanes. (A = atmospheric methane, R = methane from CO_2 reduction, F = methane from acetate fermentation, T = thermogenic methane, G = geothermal methane, M = abiogenic methane. Field for oils, O, shown for comparison.) (After Schoell, 1984a, 1988.) (b) Proportion of methane in gaseous products vs. isotopic fractionation (i.e. enrichment in ^{13}C in methane relative to its source) during methane generation from kerogen. (LK = gas from labile kerogen, RK = gas from refractory kerogen, D = gas displaced from solution in oil, OC = gas from oil cracking. Field of biogenic methane, B, shown for comparison. Mig and Mat arrows indicate trends resulting from migration and maturation, respectively.) (After Clayton, 1991.)

matter, can be distinguished isotopically from other sources, including thermogenic methane. Methane can also be produced by abiogenic processes in the mantle (M, Fig. 5.14a) and is isotopically heavy compared with other methanes, as expected. Examples of this type of methane have been obtained from sediment-free, mid-ocean ridge, hydrothermal systems.

Another plot used in distinguishing sources of methane is $\delta^{13}C$ of methane vs. amount of methane relative to other hydrocarbon gases (which can be expressed in a number of ways, such as $C_1/(C_2 + C_3)$) (e.g. Schoell, 1984a, 1984b). Modelling of the isotopic fractionation processes involved in gas generation from labile and refractory kerogens suggests that quite detailed information can be obtained on sources of gas, including mixed sources, providing the isotopic signature of the potential parent kerogens is known (Clayton, 1991). Figure 5.14b summarises the relationship between the composition of generated gas and the isotopic signature of methane relative to its source kerogen, and also indicates the way in which increasing maturity and migration affect these parameters. The merger in trends exhibited by gas from labile kerogen (LK, Fig. 5.14b) and gas from oil cracking (OC, Fig. 5.14b) is a consequence of some of the former being present in solution in oil (D, Fig. 5.14b) during the initial phase of oil cracking, resulting in a mixture of gases from these two sources.

Chapter 6

Molecular evaluation of Recent sediments

6.1 Biomarker distributions inherited from organisms

6.1.1 Introduction

Early on in the sedimentary process – in water column particulates and in the top of the sediment column – some organic compounds exist as largely unaltered constituents of the source organisms. Depending on the specificity of these compounds, it should be possible to gain an idea of the types of organisms contributing to the sedimentary organic matter and even to estimate their relative contributions. While considering the composition of organic material in organisms in Chapter 2 some examples of the specificity of compounds to groups of organisms were given. This specificity is not restricted to lipids, although certain lipid classes have received most attention because of their general ease of analysis. It is possible to assess sedimentary contributions in contemporary environments from the microscopically identifiable remains of organisms, particularly inorganic skeletal material (e.g. siliceous and carbonaceous tests). Unfortunately, these components may not survive in older sediments, but molecular evidence may, particularly in the form of biomarkers: lipid-derived compounds that can be traced to particular biological precursor molecules. The application of biomarker chemotaxonomy to ancient sediments is, therefore, important and requires an understanding of how biomarkers can be used as source indicators for extant organisms in **Recent** (i.e. Quaternary; $\leqslant 2$ Ma old) **sediments** and what changes these compounds undergo in the longer term. This chapter is concerned with the biomarker distributions inherited from various organisms and the transformations that occur during diagenesis. Chapter 7 will deal with the changes that occur during catagenesis.

The role of biomarkers as source indicators in contemporary sediments is also important in studying *in situ* microbial communities. Such a method provides advantages over conventional microscopic methods, which require the isolation and culturing of bacteria in the laboratory and which can result in a severely distorted impression of the distribution and activity of various members of the community. This is because the culturing conditions required to provide sufficient numbers of bacteria for identification are unlikely to be identical to *in situ* conditions and will prove more favourable to some species than others.

Analysis of sedimentary biomarkers involves extracting the sediment with a suitable combination of organic solvents. This procedure removes the **free lipids**, but a proportion, the **bound lipids** fraction, remains bonded to insoluble polymeric material in the sediment. Bound lipids are subsequently extracted following hydrolysis of the sediment residue, which breaks the bonds between the remaining lipids and insoluble matrix. The free lipids usually contain a substantial proportion of compounds in which two classes of components are chemically bonded together, such as fatty acids and fatty alcohols in wax esters. In contrast, the hydrolysis step during extraction ensures that the bound lipid fraction does not contain such combined components. The amount of combined components in the free lipids fraction depends on the degree of hydrolysis that has occurred during diagenesis, a process that releases individual components (e.g. fatty acids and sterols from steryl esters). Because of this, hydrolytic cleavage of the remaining combined components in the free lipids is normally undertaken in the laboratory, enabling distributions within a compound class (e.g. total fatty acids) to be examined. With increasing burial, the amount of free lipids usually decreases during the formation of insoluble kerogen, while changes in bound lipids reflect changes in kerogen composition. It can be informative, therefore, to analyse free and bound components separately.

In young sediments it is usually possible, from biomarker distributions, to distinguish between contributions from major groups of organisms (e.g. bacteria, phytoplankton, vascular plants), which broadly correspond to the division (phylum) level (e.g. algae, spermatophytes). Distinction is also usually possible at the next taxonomic level (e.g. algal classes of Chlorophyceae, Phaeophyceae and Dinophyceae; spermatophyte subdivisions of angiosperms and gymnosperms; see Box 1.2), but becomes increasingly difficult as the species level is approached. This is because there are relatively few examples of compounds exclusive to a family of organisms, demonstrating the link between the extents of biochemical and physiological diversification as evolution progresses. However, a distinctive combination of biomarkers may be sufficient to identify a contribution from a family (or even species), providing it can be reasonably ensured that such a combination may not arise from the combined contributions of other organisms. Absolute quantification of inputs of sedimentary organic matter using biomarkers alone is difficult, as it requires information on their abundance relative to total organic matter in a given organism (which can vary, depending on environmental conditions) and the relative stabilities of biomarkers in the sedimentary environment.

6.1.2 General differences between major groups of organisms

The more obvious compositional differences occur between the major groups of organisms, such as the presence of lignin, composed of polyhydroxyphenol units (Section 2.5.1), in only the higher plants. Another characteristic feature of higher plants is the use of waxes as protective coatings on leaves. In other organisms such coatings are usually biopolymers of carbohydrates or amino acids. The major wax components are saturated, straight-chain fatty acids with > 20 carbon atoms together with components of similar structure which are biosynthetically derived from these acids, including straight-chain alcohols (n-alkanols) and alkanes (n-alkanes). This biosynthetic relationship is reflected in an **even-over-odd predominance** (EOP) in carbon numbers for the fatty acids and alcohols, while the alkanes, which result from decarboxylation of fatty acids (i.e. loss of CO_2), exhibit a corresponding **odd-over-even predominance** (OEP) (see Eqs [2.7] and [2.8], Section 2.4.2). In contrast, in micro-organisms and multicellular algae the major saturated, straight-chain fatty acids and their biosynthetically related n-alkanes and n-alkanols generally have < 20 carbon atoms. A higher plant contribution to Recent sediments is, therefore, usually readily identified.

Contributions from photosynthesising organisms are characterised by various chlorophyll and related tetrapyrrole pigments. Chlorophyll-a is present in all plants and the cyanobacteria, while various bacteriochlorophylls are dominant in photosynthetic bacteria. Higher plants and green algae also contain chlorophyll-b, while brown algae contain chlorophyll-c. As demonstrated in Chapter 2, carotenoid accessory pigments can be useful markers for groups of photosynthetic organisms. Some non-photosynthetic micro-organisms contain characteristic carotenoids, such as the C_{50} bacterioruberin (Fig. 6.1) that confers the red colour to colonies of halophilic bacteria.

Various terpenoids can be specific to certain groups of organisms. Higher plants contain many types of terpenoids, some of which may be present in relatively large amounts in the more resinous angiosperms (e.g. resin acids and triterpenoids). Bacteria do not appear to contain sterols, unlike eukaryotes, and make use of other compounds in the role of cell membrane rigidifiers. Among these compounds are hopanoids, which are found in ca. 50% of extant bacteria but not in other organisms. Archaebacteria contain phospholipid ethers, and the biphytanyl ethers present in thermoacidophilic and methanogenic species are diagnostic.

In Section 6.2 the chemotaxonomic application of some compound groups will be examined in a little more detail, but before that it is useful to consider some of the variations that can occur in the lipid constituents of organisms.

6.1.3   Factors affecting the lipid composition of organisms

There are various factors that affect the biomarker distributions in organisms, some of which were mentioned in Section 2.6.1, such as the decline in storage lipids in zooplankton during times of poor nutrient availability. Light levels are important for plants because they affect the production of chloroplast pigments in green tissue, while root lipid content increases during periods of salt stress and drought due to a proliferation of internal membranes.

There is variation in the lipid content of different tissue types within organisms and so the degree to which different tissues are preserved can be important. For example, 18:1 and 18:2 account for ca. 75% of the fatty acids in the beans of soya plants, while its leaves contain only ca. 20% of these acids but ca. 55% of 18:3. There are also morphological changes in the life-cycles of plants, fungi and bacteria during which membranes, organelles and storage tissue appear and disappear. Lipids are important constituents of such tissues and so they also vary. For example, specialised pigments (e.g. carotenoids) are important during flowering but triglycerides are important as energy reserves when seeds are set. Most plant lipids (ca. 80%) are found in the chloroplasts (particularly as chlorophylls). In bacteria lipid synthesis increases rapidly just prior to cell division, while the lipid content of mycelium generally increases rapidly during the vegetative growth of fungi, and fungal spores also contain lipid-rich bodies which serve as energy reserves.

As temperature decreases higher plants and micro-organisms maintain the fluidity of their cell membranes by lowering the melting point of the constituent lipids. This can be achieved by increasing the proportion of unsaturated fatty acids and, in bacteria, by producing more *anteiso* (cf. *iso*) fatty acids of shorter average chain length. The nature of the substrate can also significantly affect the lipids synthesised by bacteria. High L-valine or L-leucine content leads to an increase in *iso*-branched acids, while *anteiso* acid production is favoured by high concentrations of L-isoleucine.

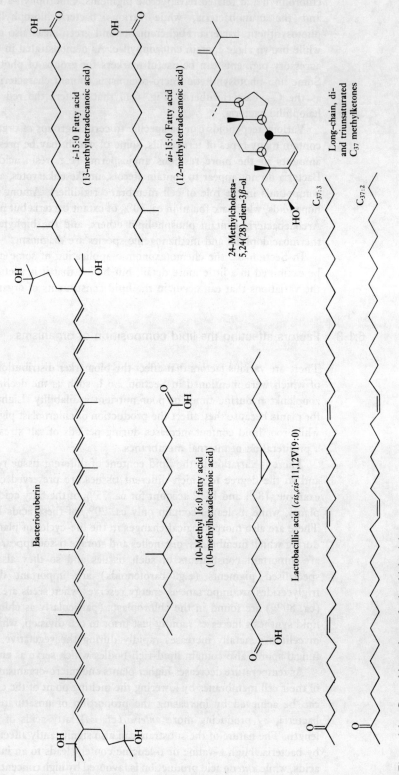

Figure 6.1   Some biomarkers of microbial origin.

Among the sulphate-reducing bacteria, *Desulfobacter* produces fatty acids with predominantly even carbon numbers when grown on acetate, while *Desulfobulbus* produces mainly odd-numbered acids from propionate ($CH_3CH_2COO^-$) but synthesises a mixture of odd and even acids when grown on lactate ($CH_3CH(OH)COO^-$).

6.2 Examples of source indicators in Recent sediments

As mentioned above, biomarkers are the most frequently used chemical indicators of sources of organic matter. Two of the most intensively studied classes of biomarkers will be examined: fatty acids and sterols. The units from which polysaccharides and lignins are formed (i.e. monosaccharides and phenolic compounds, respectively) can be analysed as readily as biomarkers and the source-related information that their distributions can convey will also be considered.

6.2.1 Fatty acids

Fatty acids are widely occurring compounds and fulfil a variety of roles, such as cellular membrane components (e.g. phospholipids), energy stores (e.g. triglycerides) and protective coatings (e.g. wax esters). They occur as either free fatty acids (i.e. unbound) or bound fatty acids (which are bonded through the ester linkage to other compounds). Most fatty acids in organisms are in the bound form. More than 500 fatty acids are known in plants and micro-organisms, but the most abundant are relatively few in number, palmitic acid (16:0) being the most common (see Table 2.3 for examples and Box 2.7 for notation scheme). In higher plants seven fatty acids account for ca. 95% of the total in combined leaf tissues and seed oil: 12:0, 14:0, 16:0, 18:0, *cis*-18:1ω9, *cis,cis*-18:2ω6 and *cis,cis,cis*-18:3ω3. The fatty acids of multicellular algae are generally similar to those of the higher plants. The three C_{18} unsaturated acids are also often abundant in unicellular green algae together with 16:4ω3 (Johns et al., 1979). All fungi contain 16:0, 18:0 and C_{18} unsaturated fatty acids.

As noted in Section 6.1.2, among the saturated straight-chain fatty acids, long-chain (generally C_{20} to C_{30}) components with an EOP in sediments are characteristic of higher plant detritus. Chain length is, therefore, a useful broad indicator of source type. More specific information can, however, be obtained from unsaturated, branched and hydroxy acids.

Mono-unsaturated fatty acids There is biological preference for the *cis* configuration at C=C bonds (although some clay-catalysed isomerism to the *trans* form can occur in sediments). The fatty acid 18:1 is generally either ω9 (oleic acid), which is common in animals, higher plants and algae, or ω7 (*cis*-vaccenic acid), which is particularly abundant in, although not exclusive to, bacteria. Like 18:1, 16:1 is also often an abundant acid in sediments and exhibits a similar differentiation between predominantly bacterially derived ω7 and algally derived $\omega 9$ isomers (although $16:1\omega 7$ is a major fatty acid of diatoms). This differentiation arises because of different pathways of fatty acid biosynthesis. The aerobic pathway has been described in Section 2.4.1 (and Fig. 2.11), during which unsaturated acids can be formed by the action of desaturase enzymes, resulting in $\omega 9$ compounds. In contrast, the anaerobic pathway operates in all anaerobic and many aerobic and facultatively aerobic eubacteria, again resulting in fatty acids with even numbers of carbon atoms. However, in this pathway enzymatic dehydration of an intermediate gives rise to a C=C bond with either cis-$\omega 8$ or $trans$-$\omega 7$ configuration. The $trans$ isomer can undergo the usual enzymatic reduction to saturated fatty acids but the cis isomer cannot, leading to $\omega 7$ unsaturated products. This biosynthetic route also appears to produce $\omega 5$ C=C bonds.

Not all monounsaturated fatty acids are biosynthesised as cis isomers with even numbers of C atoms. For example, $trans$-$16:1\omega 7$ and $trans$-$18:1\omega 7$ may be bacterial markers, while $trans$-$16:1\omega 13$ is produced by photosynthetic bacteria and some phytoplankton but not cyanobacteria. The fatty acid distributions of cyanobacteria can be quite variable, with some exhibiting major $16:0$ and $16:1\omega 7$, while others have abundant $18:1\omega 9$. The odd-numbered acids $15:1$ and $17:1$ with $\omega 6$ or $\omega 8$ C=C bonds are bacterial markers, produced by the anaerobic biosynthetic route. Among the sulphate-reducing bacteria, $17:1\omega 8$ appears to be characteristic of *Desulfobulbus*.

Poly-unsaturated fatty acids

Polyunsaturated fatty acids (PUFAs) are biosynthesised from saturated fatty acids by the action of desaturase enzymes (Section 2.4.1). The position of double bonds can again be useful in identifying source organisms; for example PUFAs such as $20:4\omega 6$, $20:5\omega 3$, $22:5\omega 3$ and $22:6\omega 3$ are generally characteristic of planktonic algae. Higher plants and green algae generally contain $18:3\omega 3$ among their C_{18} unsaturated acids. Of the three commonly occurring $16:2$ isomers in marine phytoplankton ($\omega 4$, $\omega 6$ and $\omega 7$) only the $\omega 4$ and $\omega 6$ isomers appear to be present in the macroscopic green, red and brown algae, while $\omega 4$ and $\omega 7$ are the common isomers in diatoms (Volkman and Johns, 1977). Abundant $16:3\omega 3$ suggests an input from green algae. Diatoms are different from other algae in exhibiting a characteristic distribution of $16:0$, $16:1\omega 7$, $16:3\omega 4$, $20:4\omega 6$ and $20:5\omega 3$, together with low amounts of C_{18} components. The presence of $16:3\omega 4$ appears to be particularly diagnostic of diatoms. Tetra- and pentaunsaturated acids are also found in other algae. Four $\omega 3$ acids, $18:4$, $18:5$, $20:5$ and $22:6$, appear to be characteristically abundant in haptophyceans and some dinoflagellates, but marine invertebrates can also be important contributors of $22:6$. In contrast to other organisms bacteria do not contain fatty acids with more than one C=C bond.

Iso *and* anteiso *methyl-branched fatty acids*

Branched acids can be source-specific and they are rarely unsaturated. They are formed by the incorporation of branched amino acids into the biosynthetic pathway, yielding *iso* and *anteiso* acids, as noted in Section 6.1.3. *Iso* and *anteiso* saturated fatty acids are found in fungi, molluscs and phytoplankton, but they are generally in higher levels in bacteria and are often observed in the C_{13}–C_{17} range. The C_{15} isomers (Fig. 6.1) are usually particularly abundant in bacteria and the ratio ($iso + anteiso$)/normal derived from C_{15} components can be used to assess relative bacterial contributions. Similarly, the $\omega 8$ isomers of *iso*-$15:1$ and *iso*-$17:1$ are bacterial markers (Perry et al., 1979), and *iso*-$17:1\omega 7$ is characteristically a major fatty acid in the sulphate-reducing bacteria *Desulfovibrio desulfuricans*.

Internally branched and cycloalkyl fatty acids

Internally branched acids occur naturally, such as the 10-methyl (numbering from the acid end) isomers of 16:0 and 18:0 found in fungi and bacteria. For example, 10-methyl 16:0 (Fig. 6.1) is characteristic of *Desulfobacter*. Cycloalkyl acids also exist, such as 17:0 and 19:0 cyclopropyl species. The position of the cyclopropyl group is important, as out of the more commonly occurring *cis*-11,12 and *cis*-9,10 isomers of 19:0 and the *cis*-9,10 isomer of 17:0, only the *cis*-11,12 isomer (lactobacillic acid, Fig. 6.1) appears to be a specific (probably aerobic) bacterial marker (Perry et al., 1979), as *cis*-9,10 isomers have also been found in some terrestrial plants.

Hydroxy fatty acids

The fatty acids of Gram-positive bacteria comprise mainly *iso* and *anteiso* fatty acids $<C_{20}$. In contrast, Gram-negative bacteria (which include most planktonic bacteria) are dominated by α-hydroxy acids (i.e. 2-hydroxy, the OH group is on the C atom next to the COOH group) and β-hydroxy acids (i.e. 3-hydroxy, the OH group is situated on the next but one C atom to the COOH group) of $<C_{20}$. However, β-hydroxy acids in the range $C_8–C_{26}$ have also been found in cyanobacteria (predominantly C_{14} and C_{16} in lipopolysaccharides) and some microalgae. *Iso* and *anteiso* isomers of both C_{15} and C_{17} β-hydroxy acids are probably bacterial markers.

Hydroxy acids are relatively uncommon in fungi but they are the most common substituted fatty acids in higher plants, particularly α-hydroxy acids. Many plants also contain di- and trihydroxy acids (Cardoso and Eglinton, 1983). Hydroxy acids of $>C_{20}$ with an ω-OH group (i.e. at the opposite end of the molecule to the COOH group) are vascular plant markers, derived from cutin and suberin. Shorter-chain ω-hydroxy acids may be derived from higher plants or bacteria. Higher plants also produce α,ω-diacids, although these compounds can arise from bacterial oxidation of other compounds (see Section 6.3.3). Mosses (Bryopsida, a class of the Bryophyta) contain monohydroxy acids, such as the 7-OH and 8-OH C_{16} diacids in *Sphagnum* (a genus of the Bryopsida).

6.2.2   Sterols

Sterols are found in both free and bound form (e.g. steryl esters and glycosides) in organisms. The carbon number distribution of regular (i.e. 4-desmethyl) sterols in young sediments appears to permit a degree of distinction between the contributions of some groups of organisms. The plankton in general contain dominant C_{27} and C_{28} sterols. Phytoplankton usually contain abundant C_{28} sterols (although diatoms can contain approximately equal amounts of C_{27}, C_{28} and C_{29} sterols) while zooplankton, of which the crustaceans are the dominant class (copepods account for ca. 90% by weight of the zooplankton), often contain abundant C_{27} sterols, particularly cholesterol (cholest-5-en-3β-ol, Fig. 2.20). In contrast, the major sterols in higher plants are C_{29} compounds, β-sitosterol (24α-ethylcholest-5-en-3β-ol) and stigmasterol (24α-ethylcholesta-5,22E-dien-3β-ol), although campesterol (24α-methylcholest-5-en-3β-ol), a C_{28} sterol, is also often abundant (see Fig. 2.20 for structures). Fungi contain $C_{27}–C_{29}$ sterols but ergosterol (Fig. 2.20), a C_{28} compound, often predominates. Terrestrial and marine invertebrates probably contribute a range of $C_{27}–C_{29}$ sterols.

A triangular plot of C_{27}, C_{28} and C_{29} sterols can be an aid to differentiating marine, estuarine, terrestrial and lacustrine environments (Fig. 6.2), based on the characteristic associations of contributing organisms. This approach to sterol

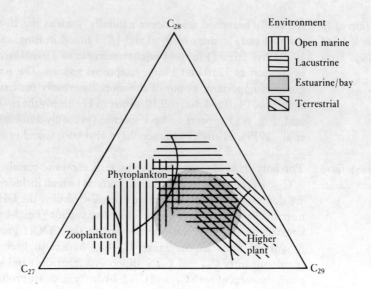

Figure 6.2   Sterol distributions (expressed as relative amounts of C_{27}, C_{28} and C_{29} components) in relation to source organisms and environments. (After Huang and Meinschein, 1979.)

distributions is very simplistic and may not always provide accurate indications of contributing organism groups. For example, it can be difficult to distinguish between marine phytoplankton and zooplankton on the basis of carbon number, and the former can also contain C_{28} and C_{29} sterols with similar structures to higher plant sterols. However, the configuration at C-24 can provide a degree of source specificity as it appears that higher plant-derived C_{29} sterols usually possess a $24\beta(H)$ configuration (i.e. 24α-ethyl, as in β-sitosterol and stigmasterol), whereas many algae (including the Chlorophyceae and dinoflagellates) biosynthesise sterols with the $24\alpha(H)$ configuration (Goad et al., 1974; Volkman, 1986). Unfortunately, C-24 epimeric pairs of sterols are not readily separated by the gas chromatographic techniques routinely employed in sterol analysis (see Section 7.5). In the absence of information on C-24 configuration it can be difficult to determine the precise origin of some sterols in Recent sediments. For example, the C_{29} compound 24-ethylcholest-5-en-3β-ol is found in higher plants and many unicellular algae and it is also often dominant among the sterols of cyanobacteria.

Among the phytoplankton many species of the Chlorophyceae contain Δ^7, $\Delta^{5,7}$ and $\Delta^{7,22}$ sterols (see Box 2.8 and Fig. 2.18 for numbering convention), but a few contain mainly 24β-ethylcholesta-5,22E-dien-3β-ol and 24β-methylcholest-5-en-3β-ol (i.e. the $24\alpha(H)$ epimers of stigmasterol and campesterol, respectively). Cholesterol occurs in variable amounts in most phytoplankton (including cyanobacteria) and is sometimes a major component. Diatoms contain a wide range of sterols but usually there are large amounts of brassicasterol (24β-methylcholesta-5,22E-dien-3β-ol, Fig. 2.20), cholesterol or 24-methylcholesta-5,24(28)-dien-3β-ol (Fig. 6.1). Cholesterol or brassicasterol is also often abundant in haptophycean algae (including coccolithophores). Dinoflagellates appear to be the most important source of 4-methylsterols in sediments (e.g. the C_{30} compound 4α,23,24-trimethyl-5α(H)-cholest-22-en-3β-ol, also known as dinosterol, Fig. 2.20).

6.2.3   Carbohydrates

The most abundant aldopentoses and aldohexoses resulting from hydrolysis of polysaccharides in sedimentary organic matter are usually lyxose, arabinose, rhamnose, ribose, xylose, fucose, mannose, galactose and glucose (see Fig. 2.4). The distributions of these monosaccharides can be used to differentiate between various higher plant sources and to distinguish marine from terrestrial sources (Cowie and Hedges, 1984). However, the variability in carbohydrate composition that can occur in algae, zooplankton and bacteria can make it difficult to differentiate these groups of the plankton.

Marine markers among the above pentoses are ribose and fucose. Ribose is present in many nucleotides and RNA, which are relatively abundant in the metabolically active organisms of the plankton. Fucose is used as a storage sugar by some plankton and bacteria, but it is rarely present in higher plants. In vascular plants structural polysaccharides predominate and the total level of carbohydrates is significantly higher than in marine organisms. Glucose is an important constituent of structural polysaccharides in vascular plants, leading to high glucose : ribose ratios (> 50). This ratio is usually lower in phytoplankton, but not always so because of the extremely variable glucose levels in phytoplankton. However, the amount of ribose or (ribose + fucose) relative to total monosaccharides excluding glucose is usually reliably greater for planktonic than higher plant sources.

Angiosperms (e.g. grasses and many deciduous trees) and gymnosperms (e.g. conifers) have characteristically different hemicellulose composition. Xylose-containing components are in greater relative abundance in angiosperms, while mannose-containing components are more abundant in gymnosperms. A plot of xylose vs. ribose, both expressed as levels relative to total monosaccharides excluding glucose, can allow distinction of marine, angiosperm and gymnosperm sources (Fig. 6.3a).

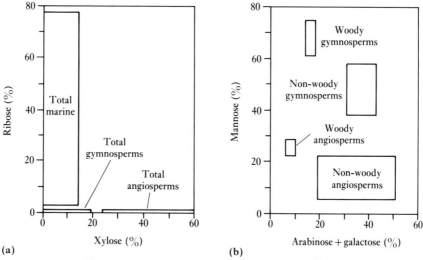

Figure 6.3   Plots of some monosaccharides (expressed as weight per cent of total amount of combined lyxose, arabinose, ribose, xylose, rhamnose, fucose, mannose and galactose) showing (**a**) differentiation of angiosperm and gymnosperm vascular plant sources from total marine plankton sources, and (**b**) distinction between different types of higher plant sources.

Non-woody vascular plant tissue (e.g. leaves and grasses) contains more pectin than woody tissue, resulting in higher relative abundances of arabinose and galactose. A plot of (arabinose + galactose) vs. mannose, both again expressed as levels relative to total monosaccharides excluding glucose, allows distinction of woody and non-woody angiosperm tissues and of the corresponding gymnosperm tissues (Fig. 6.3b).

Minor monosaccharides released upon hydrolysis of sediments can be useful source indicators. For example, a few specific methoxy (i.e. O-methyl) monosaccharides are found in phytoplankton, such as 3-O-methylxylose and 6-O-methylmannose in the coccolithophore species *Emiliania huxleyi*. Bacteria synthesise a wider range of these components together with deoxy monosaccharides.

6.2.4   Lignins

Lignin is an abundant component of vascular plants, and the phenolic units from which it is synthesised exhibit some compositional variation with plant type and are relatively stable towards chemical alteration. Lignin constituents are, therefore, potentially more useful indicators of different types of vascular plant sources than the less abundant lipids.

Alkaline oxidation of lignins from vascular plant tissue with copper(II) oxide (e.g. Hedges and Mann, 1979; Hedges and Ertel, 1982) yields four groups of structurally related products: *p*-hydroxyl, vanillyl, syringyl and cinnamyl groups (Table 6.1). There are sources of the *p*-hydroxyl group other than lignins, but the other three groups are diagnostic of lignin. Other than some rare exceptions for the cinnamyl group, all members of a group are either present or absent in the oxidation products of a particular vascular plant tissue. The levels of these groups in the woody and non-woody tissues of gymnosperms and angiosperms are presented in Table 6.2.

Plots of the amounts of any pair of the three oxidation product groups in Table 6.2 can be made. That of syringyl vs. vanillyl allows differentiation of the four types of vascular plant tissue and non-vascular plant material (Fig. 6.4a). When this type of plot is applied to sediment samples the proportion of total organic carbon that is contributed by vascular plants has to be determined, and this is not always a simple procedure. However, this determination is unnecessary if the ratio of the amount of one group of oxidation products to another is plotted along each axis. The most useful plot of this type in distinguishing the four main types of vascular plant tissues is syringyl:vanillyl vs. cinnamyl:vanillyl (Fig. 6.4b).

6.2.5   Carbon isotopes

In Chapter 1 it was noted that the most common pathway for photosynthesis, the C3 path (Box 1.3), results in a higher value for the $^{12}C:^{13}C$ ratio in the fixed carbon than in the CO_2 source. Some plants use a different pathway, involving carboxylation of phosphoenol pyruvic acid (PEP) instead of ribulose diphosphate, which subsequently forms a C_4 compound, oxaloacetic acid, instead of PGA (Box 1.3). Consequently such plants are termed **C4-plants**. The C4 path is a relatively recent evolutionary development of particular advantage in hot dry climates. Some plants, the **CAM-plants**, can use both paths. The C3 path operates in most higher plants,

Table 6.1   **Lignin oxidation products**

p-Hydroxyl group

HO—⟨benzene ring⟩—R

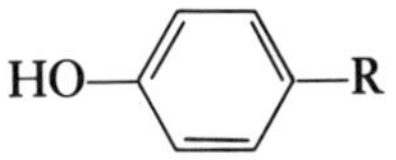 $R = -\overset{\displaystyle \mathrm{H}}{\underset{}{\mathrm{C}=\mathrm{O}}}$	*p*-Hydroxybenzaldehyde
$R = -\overset{}{\underset{\displaystyle \mathrm{CH_3}}{\mathrm{C}=\mathrm{O}}}$	*p*-Hydroxyacetophenone
$R = -\overset{}{\underset{\displaystyle \mathrm{OH}}{\mathrm{C}=\mathrm{O}}}$	*p*-Hydroxybenzoic acid

Vanillyl group

H_3CO
HO—⟨benzene ring⟩—R

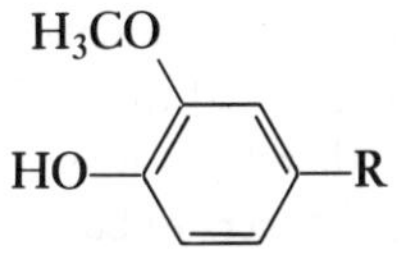 $R = -\mathrm{C}=\mathrm{O}$, H	Vanillin
$R = -\mathrm{C}=\mathrm{O}$, CH_3	Acetovanillone
$R = -\mathrm{C}=\mathrm{O}$, OH	Vanillic acid

Syringyl group

H_3CO
HO—⟨benzene ring⟩—R
H_3CO

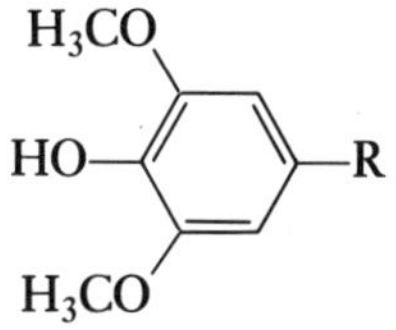 $R = -\mathrm{C}=\mathrm{O}$, H	Syringaldehyde
$R = -\mathrm{C}=\mathrm{O}$, CH_3	Acetosyringone
$R = -\mathrm{C}=\mathrm{O}$, OH	Syringic acid

Cinnamyl group

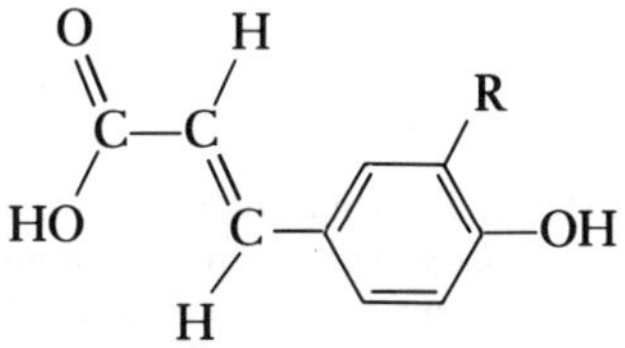 $R = -H$	*p*-Coumaric acid
$R = -OCH_3$	Ferulic acid

Table 6.2   **Lignin oxidation product distributions in different vascular plant tissues** (amounts are expressed as weight per cent of vascular plant organic carbon). (After Hedges and Mann, 1979)

	Syringyl	*Cinnamyl*	*Vanillyl*
Non-vascular plants	0	0	0
Non-woody angiosperms	1–3	0.4–3.1	0.6–3.0
Woody angiosperms	7–18	0	2.7–8.0
Non-woody gymnosperms	0	0.8–1.2	1.9–2.1
Woody gymnosperms	0	0	4–13

algae, cyanobacteria and some photosynthetic and chemosynthetic bacteria. Some bacteria use the C4 path, while others use different pathways generally involving acetyl coenzyme A. The balance between the rates of the different carboxylation reactions, the rates of dissimilatory reactions and the rates of diffusion of CO_2 into and out of the tissues controls the final isotope fractionation of the fixed carbon. Autotrophic organisms can be grouped according to the degree of isotopic fractionation, largely determined by the primary carboxylation step utilised (Table 6.3).

The range of $\delta^{13}C$ encountered in Recent marine sediments is -10 to $-34\%_{00}$, although the majority of samples lie in the range -20 to $-27\%_{00}$, and the mean value is $-24.5\%_{00}$. To a first approximation, therefore, it is possible to ascertain the relative contributions of different primary producers to bulk sedimentary organic matter from carbon isotope values. Such estimations for Recent sediments are more accurate if the contributing organisms can be identified and their isotopic composition measured for extant species growing in a similar environment.

While bulk carbon isotope measurements can give information on sources of sedimentary organic matter, potentially there is more information to be gained from

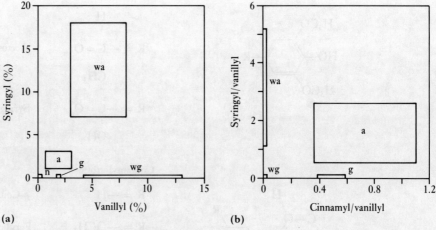

(a)   (b)

Figure 6.4   Plots of lignin oxidation product parameters permitting distinction of different higher plant tissues (a = non-woody angiosperm, wa = woody angiosperm, g = non-woody gymnosperm, wg = woody gymnosperm, n = non-vascular plant tissue). Amounts of syringyl and vanillyl components in (a) are expressed as weight per cent of vascular plant organic carbon.

Table 6.3   **Isotopic composition of major autotrophs and methanogens** (Data after Schidlowski, 1988)

Organism	$\delta^{13}C\%_{oo}$
Higher plants:	
C3	−23 to −34
C4	−6 to −23
CAM	−11 to −33
Algae	−8 to −35
Cyanobacteria	−3 to −27
Photosynthetic bateria:	
Green	−9 to −21
Purple	−26 to −36
Red	−19 to −28
Methanogenic bacteria	+6 to −41

the isotopic signatures of individual biomarkers. For instance, it should be possible to identify the source of a particular compound with more certainty and to evaluate its metabolic pathways. Isolating sufficient amounts of single compounds has been an obstacle to such measurements in the past, but with the advent of improved technology these measurements are possible from relatively small amounts of components, extending the application of this technique to biomarkers. For example, in a recent study (Rieley et al., 1991) it was observed that $\delta^{13}C$ values for leaf wax *n*-alkanes (C_{25}–C_{33}) varied from $-30.1\%_{oo}$ to $-38.7\%_{oo}$, depending on tree species. If such differences are genetically controlled the possibility exists that major higher plant species contributing to sedimentary organic matter can be identified.

6.3   Diagenesis at the molecular level

6.3.1   General diagenetic processes

Carbohydrates and lignins   As noted in Chapter 4, carbohydrates are relatively labile sedimentary components and readily undergo diagenetic changes; this can affect the interpretation of source indications based on monosaccharide distributions. For example, glucose as an energy storage carbohydrate is readily biodegraded, while bacterial cell wall components are less readily degraded. However, there appear to be differing degrees of susceptibility to biodegradation even among bacterial cell wall monosaccharides; for example fucose is less readily biodegraded than rhamnose.

    The lignified material of higher plants is more resistant towards degradation and can protect incorporated carbohydrates. The following order of increasing stability has been noted for hardwoods: hemicellulose < cellulose < pectin < syringyl lignin units < vanillyl and *p*-hydroxyl lignin units. The different relative stabilities of the polysaccharides are reflected in their major constituent monosaccharide units: xylose and mannose in hemicellulose; glucose in cellulose (often referred to as α-cellulose); and arabinose, galactose and galacturonic acid in pectin. Lignin and polysaccharide composition changes from the inside to the outside of woody cell walls and the middle lamellar region appears to be preferentially preserved, affecting the distribution of preserved components during diagenesis.

Biomarkers   During diagenesis biomarkers undergo the same main types of reactions as other biogenic organic compounds: defunctionalisation, aromatisation and isomerisation. Oxygen-containing functional groups predominate among the lipid components at the start of diagenesis and their loss involves reactions such as dehydration and decarboxylation. By the end of diagenesis these defunctionalisation processes lead to the formation of hydrocarbons, either saturated or aromatic. Non-aromatic, unsaturated hydrocarbons (alkenes) can be formed initially, such as sterenes from stenols, but they do not survive in the longer term. They may undergo hydrogenation to yield saturated hydrocarbons (alkanes) or, if the C$=$C bond is in a six-membered cyclic system, dehydrogenation to yield an aromatic system. For example, the diagenetic products of a sterol with a ring C$=$C bond (e.g. cholesterol) include steranes and C-ring monoaromatic steroidal hydrocarbons (see Fig. 5.10).

Biomarkers can undergo **isomerisation** (Section 2.1.3) during diagenesis and catagenesis. A major isomerisation process during diagenesis is the migration of double bonds in unsaturated biomarkers, such as sterenes. Configurational isomerisation at chiral centres can also occur during diagenesis but is mainly experienced at the higher temperatures associated with catagenesis and coalification. Both isomerisation processes involve the movement of hydrogen, either in the form of a hydrogen radical (H·, i.e. a hydrogen atom, which bears a single electron) or a hydride ion (H$^-$, i.e. a hydrogen atom with an additional electron) depending upon whether the hydrogen nucleus departs, respectively, with one or both of the electrons involved in its bonding to a carbon atom. Mineral surfaces (e.g. clay particles) are believed to play an important role in these processes, which are therefore geochemical rather than microbial.

Isomerisation at a particular chiral centre (**epimerisation**) converts the single, biologically conferred configuration into an equilibrium mixture of the two possible stereoisomers. The mechanism for the loss and subsequent re-addition of hydrogen at an acyclic chiral carbon atom is shown in Fig. 6.5. The intermediate involves a carbon atom bonded to only three groups and it has a trigonal planar geometry. The hydrogen species can re-enter the structure from either side and so there is an approximately equal probability that the original or the opposite configuration is formed. The same process operates at cyclic chiral centres, but the orientation of atoms on the groups surrounding the chiral carbon in the trigonal intermediate may partly hinder addition of hydrogen from one side (**steric hindrance**). This leads to a higher probability for the formation of one epimer, which will be present in greater abundance in the final equilibrium mixture. Configurational isomerisation is a reversible reaction (see Box 6.1) and so, ultimately, a dynamic equilibrium is achieved between the two epimers. The relative proportion of epimers in the equilibrium mixture depends on the effect of steric hindrance; it is approximately 50:50 for acyclic chiral centres because steric effects are about the same for both epimers. Isomerisation resulting in a 50:50 mixture of enantiomers (a racemic mixture) is termed **racemisation**.

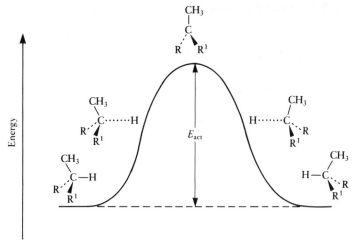

Figure 6.5   Energy profile for epimerisation at an acyclic chiral centre, showing the transition from a tetrahedrally bonded carbon atom to a trigonal planar geometry in the intermediate (E_{act} = activation energy).

Box 6.1

Reversible reactions

Reversible reactions do not go to completion and can occur in either direction. An important consideration is the activation energy for the reverse reaction. For the simple reaction in Fig. 2.10 the activation energy for the reverse reaction is the sum of ΔH and E_{act} (i.e. overall energy change plus activation energy for the forward reaction). If this is large the chance that a product molecule will possess the necessary energy for the reverse reaction to occur is small. However, as the numbers of product molecules increase and reactant molecules decrease, the probability of the reverse reaction occurring becomes greater, while the probability of the forward reaction occurring reduces. Eventually, the probability of each reaction occurring is equal, the forward and reverse reactions occur at the same rate and the proportions of reactants and products remain constant. This is a dynamic rather than a static equilibrium, with reactants and products being continuously interconverted.

Isomerisation at a chiral centre in a biomarker is an example of a reversible reaction in which reactant and product have similar energies. Initially, the concentration of reactant is high and that of product is low, so there is greater probability that a reactant will undergo hydrogen loss to form the trigonal intermediate. The ratio of product and reactant when dynamic equilibrium has become established reflects the balance between the probability of formation of the intermediate by each epimer and the probability of which epimer is most likely to be formed from the intermediate. Greater steric hindrance between the hydrogen atom and other atoms on surrounding groups for one epimer results in a higher energy for that epimer than the other and consequently a lower activation energy for intermediate formation. As a result, the equilibrium favours formation of a greater proportion of the less hindered epimer. For a single acyclic centre there is no difference in steric hindrance between epimers, the hydrogen can attack the intermediate from either side with equal probability, so that forward and reverse reactions have equal activation energies (Fig. 6.5) and a 50:50 mixture results.

6.3.2   Lipid diagenesis in the water column

It was noted in Chapter 3 that phytoplankton and their remains upon death sink only very slowly, allowing ample opportunity for predation by zooplankton and degradation by the microbial communities that are associated with particulate matter. Sediment traps enable falling debris of various sizes to be collected, depending on the mesh size used, at different depths in the water column. They provide information on the different communities present at various depths and on the transportation of organic matter from the photic zone to the underlying sediments, although it should be noted that **advection** (the horizontal movement of water bodies) is an important factor for suspended particulates. Sediment trap experiments have confirmed that only a small fraction of the primary production of the oceans reaches the sediment, with most being recycled in the photic zone. Only the larger particulates reach the sediment (generally $> 50 \ \mu$m diameter) and zooplankton faecal pellets can make a significant contribution. After removing certain components from their diet of phytoplankton, the zooplankton excrete by-products and unaltered phytoplankton remains. The distribution of lipids reaching marine sediments can, therefore, be significantly modified by zooplankton grazing. The role that zooplankton play in the transport of organic matter to sediments is particularly important in the open oceans, where water depth is around 4000 m. The faecal pellets of zooplankton are considerably bigger than phytoplankton remains and sink at a substantially faster rate, providing a more efficient system of transport for both labile and refractory organic components to the sediment.

Although living phytoplankton stay at a fairly constant depth in the water column, zooplankton migrate over quite considerable vertical extents, often feeding near the surface at night and migrating to depth during the day. The migration patterns are different for different species and can also vary with sex and growth stage for the same species. Concentrations of cholesterol (a dominant sterol in zooplankton) in particulates collected by sediment traps may sometimes reflect this behaviour, often reaching a maximum at a depth below the photic zone.

Crustaceans, such as copepods, cannot biosynthesise all the sterols they need from basic components and so they ingest and convert various algal sterols into components such as cholesterol. The modification of phytoplankton lipids in general by zooplankton during early diagenesis can be important. For example, the copepod *Calanus helgolandicus* when fed on a variety of phytoplankton in laboratory studies was found to be able to remove all algal hydrocarbons and a significant amount of fatty acids, particularly the polyunsaturated components (Prahl et al., 1984; Neal et al., 1986). There appears to be some selectivity towards the components that are digested and some phytoplankton lipids can reach the sediment unaltered after passing through zooplankton. For example, while *Calanus* seems to be able to digest most algal stenols, it appears to have a preference for certain positions of unsaturation, such as $\Delta^{5,7}$. In contrast, stanols and Δ^7 stenols appear not to be digested. Stanols are, therefore, likely to be more accurate indicators of phytoplankton inputs to sediments than most stenols (Harvey et al., 1987).

Coprophagous feeding (ingestion of faecal pellets) is not uncommon among zooplankton and appears to aid the uptake of algal hydrocarbons and polyunsaturated fatty acids by zooplankton. In addition, coprophagy seems to result in the preferential removal of C_{26} and C_{27} zooplankton sterols from faecal material relative to C_{28} and C_{29} algal sterols (Neal et al., 1986).

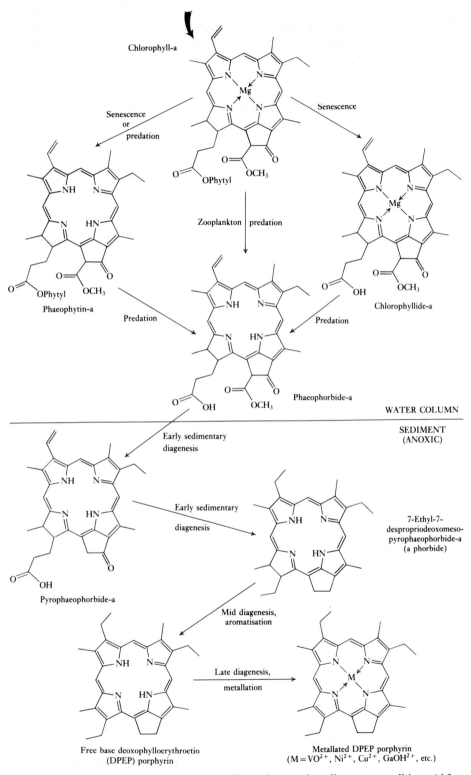

Figure 6.6   Possible diagenetic pathway for chlorophyll-a under anoxic sedimentary conditions. (After Barwise and Roberts, 1984; Baker and Louda, 1986.)

The predation of phytoplankton by zooplankton can also be important in determining the form in which cholorophylls reach marine sediments. Among the early diagenetic transformations of chlorophylls in the water column immediately below the photic zone are the loss of magnesium to give phaeophytins or the loss of the phytyl ester group to give chlorophyllides. Both processes occur naturally during **senescence**, the period of intercellular changes accompanying cessation of photosynthetic activity prior to death, as phytoplankton sink out of the photic zone. However, they also occur as the result of enzymatic hydrolysis during zooplankton metabolism, most efficiently in the case of *Calanus*, yielding phaeophorbides. These reactions are summarised in Fig. 6.6 for chlorophyll-a. Measurement of the levels of chlorophyll-a and its degradation products (chlorophyllide-a, phaeophytin-a and phaeophorbide-a) can, therefore, provide information on the physiological status of phytoplankton populations and the importance of grazing processes. For example, chlorophyll-a levels indicate phytoplankton biomass and productivity, while phaeophorbide-a concentrations in zooplankton faecal pellets captured by sediment traps provide information on grazing efficiency.

Carotenoids can also provide information on primary productivity and grazing rates, although the basic carotenoid structure appears to be less stable towards enzymatic degradation during herbivorous grazing than the porphyrin ring system of chlorophylls. For example, fucoxanthin is converted by enzymatic hydrolysis during metabolism by zooplankton (e.g. copepods and euphausiids) into fucoxanthinol (see Fig. 6.7) and so the relative amounts of these compounds provide similar information to the ratio of chlorophyll-a to phaeophorbide-a (Repeta and Gagosian, 1982, 1984). The concentrations of specific carotenoids in sediment traps can also give information on the fluxes of total organic matter in relatively anoxic environments such as the Peru upwelling, if the weight ratio of a particular carotenoid to total organic body mass is known for a phytoplankton species (e.g. 0.2% for fucoxanthin in diatoms).

Enzymatic hydrolysis is probably an important process in the diagenesis of a variety of other lipid classes, such as wax esters, triglycerides and steryl esters. In addition to the modification of phytoplankton biomarkers by zooplankton metabolism, all biomarkers in particulate matter are subject to alteration and supplementation by bacterial action. This action can vary in extent but is generally lower in the more rapidly sinking, larger particulates. The degree of oxygenation of the water column is also important to bacterial activity and the types of organisms present, as it is in the sediment. The effects of bacterial action in the water column can be considered an extension of that in the sediment. For example, the microbial hydrogenation of stenols to stanols can occur in anoxic sediments and beneath the oxic/anoxic boundary in the water column (where anoxicity develops in bottom waters).

6.3.3   Sedimentary diagenesis of lipids

Fatty acids   As noted in Chapter 4, when labile compounds are associated with refractory material they are protected to a degree from the effects of diagenesis. This may account for the apparently greater stability of long-chain fatty acids ($>C_{20}$) from higher plant material, which contains refractory lignified material, compared with the shorter-chain fatty acids ($<C_{20}$) from micro-organisms. Unsaturated acyclic components, such as polyunsaturated fatty acids, are degraded relatively quickly, both microbially and

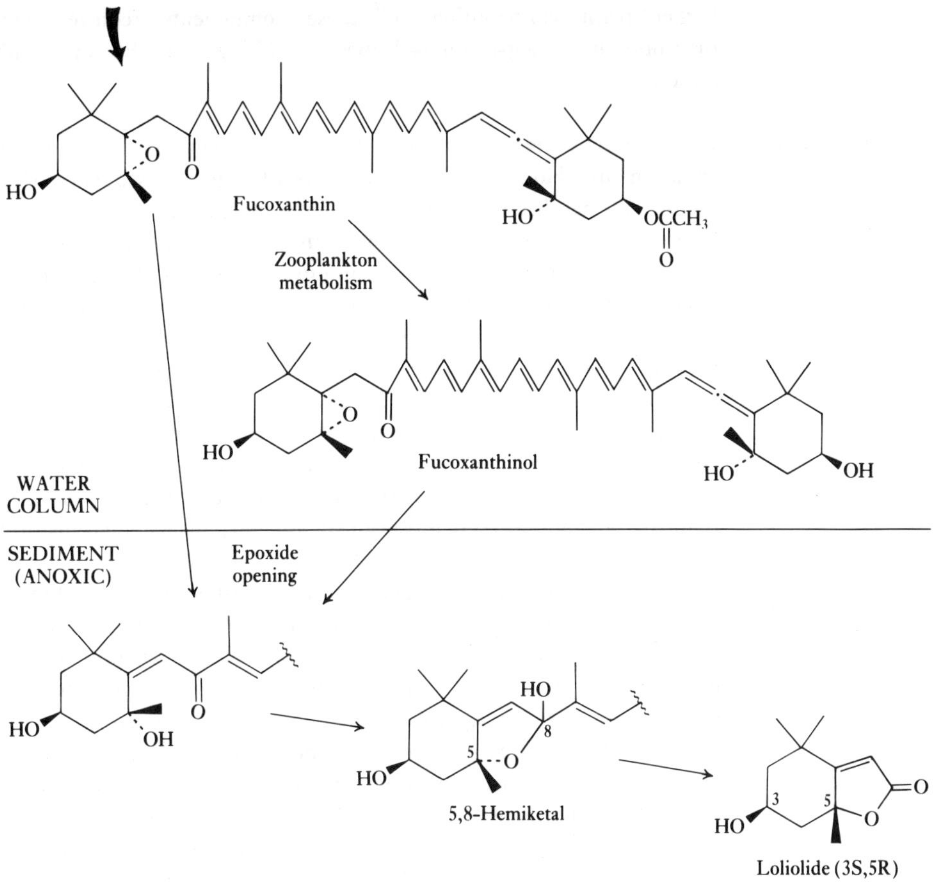

Figure 6.7   Possible early diagenetic pathways for fucoxanthin under anoxic sedimentary conditions. (Sedimentary intermediates in loliolide formation show only left side of molecule. Large arrow indicates biogenic input.) (After Repeta, 1989.)

chemically, during diagenesis. The more resistant saturated fatty acids, ω-hydroxy acids and α,ω-diacids, can survive in **ancient sediments** (i.e. sediments of greater than Tertiary age).

Fatty acids can be affected by microbial oxidation in sediments. For example, β-hydroxy fatty acids are produced from β-oxidation of saturated fatty acids. This microbial alteration of fatty acids requires care to be exercised when certain components are used as source indicators. For example, ω-oxidation of saturated fatty acids to ω-hydroxy acids and of ω-hydroxy acids to α,ω-diacids is performed by **yeasts** (unicellular fungi, mainly belonging to the ascomycetes) and bacteria. Consequently, while the long-chain ($> C_{20}$) ω-hydroxy acids are reliable indicators of higher plant sources, the short-chain components ($< C_{20}$) may be either microbial in origin or derived from higher plant suberin and cutin. Similarly, long-chain α,ω-diacids may be derived from vascular plant suberin and cutin or from microbial oxidation of long-chain ω-hydroxy acids (although in either event these diacids derive directly or indirectly from higher plant components). Confirmation of possible higher plant contributions of these components requires comparison with distributions of long-chain n-alkanes (exhibiting an OEP), an unambiguous source indicator.

Photosynthetic pigments   Photosynthetic pigments do not survive in recognisable form in oxic sedimentary environments. However, in anoxic settings the general sequence of reactions shown in Fig. 6.6 has been recognised for chlorophylls. As mentioned in Section 6.3.2, chlorophylls tend to undergo rapid demetallation and loss of the phytyl side chain in the water column, so that the main input to anoxic sediments are phaeophorbides, although some phaeophytins can also be present. These compounds then undergo defunctionalisation to yield pyrophaeophorbides (Fig. 6.6). Recently, pyrophaeophorbides with the carboxylic acid group esterified with a variety of sterols have been found in a number of marine and lacustrine sediments and in natural zooplankton populations. It would appear that these compounds are formed by enzymatic esterification of the hydrolysis products of algal chlorophylls. These processes probably occur during phytoplankton senescence or decay, or during grazing by zooplankton, and represent a previously unknown diagenetic transformation pathway for chlorophylls (Keely et al., 1992).

Decarboxylation of pyrophaeophorbides continues in early sedimentary diagenesis to yield completely defunctionalised phorbides (which can also be termed dihydroporphyrins; Fig. 6.6). The end of early diagenesis corresponds to *in situ* temperatures of not more than 25–30°C, which suggests that the sedimentary processes up to this stage are mainly microbially mediated.

As temperatures rise in mid-diagenesis the tetrapyrrole system of phorbides becomes fully aromatised, yielding free-base porphyrins, with deoxophylloerythroetio (DPEP) porphyrins predominating (Fig. 6.6). The long-term survival of porphyrins is enhanced by metal chelation during late diagenesis with species such as Ni^{2+}, VO^{2+}, Cu^{2+} and $GaOH^{2+}$. Nickel and vanadyl (VO) porphyrins are usually the most common and seem to be formed by different routes. Nickel porphyrins appear to be readily solvent-extractable from sediments, i.e. they are present as free species, while the vanadyl porphyrins are not extractable, i.e. they are bound to a polymeric matrix (e.g. kerogen). Nickel chelation, as measured by the ratio of nickel-complexed porphyrins to total porphyrins, increases with sediment depth during diagenesis, starting at ca. 45–55°C and finishing at ca. 60°C. Vanadyl complexation does not appear to follow a similar simple trend, although in immature sediments the distributions of vanadyl DPEP porphyrins more closely resemble the pattern expected from free-base porphyrin distributions than do those of the nickel DPEP porphyrins. The complexity of vanadyl chelation is probably due to both the fact that it occurs later than for other metals (laboratory modelling suggests that complexation occurs in the order Cu, Ni, VO) and the influence of the organic matrix to which the porphyrin is bound. Levels of vanadyl porphyrins increase with rising temperature, continuing into catagenesis. Late diagenesis, therefore, corresponds to the complete chelation of all free porphyrins and the onset of the evolution of vanadyl porphyrins (ca. 65°C).

Porphyrins of the DPEP type with carbon number ranges of C_{26}–C_{38} predominate during diagenesis in anoxic aquatic sediments and in sapropelic coals. However, in humic coals the main porphyrins are the etio type (see Fig. 6.8), $\leqslant C_{32}$ and complexed with Fe(III), Ga(III) or Mn(III). The reason for this difference is believed to be that plant porphyrins are largely degraded during the initially oxic depositional conditions

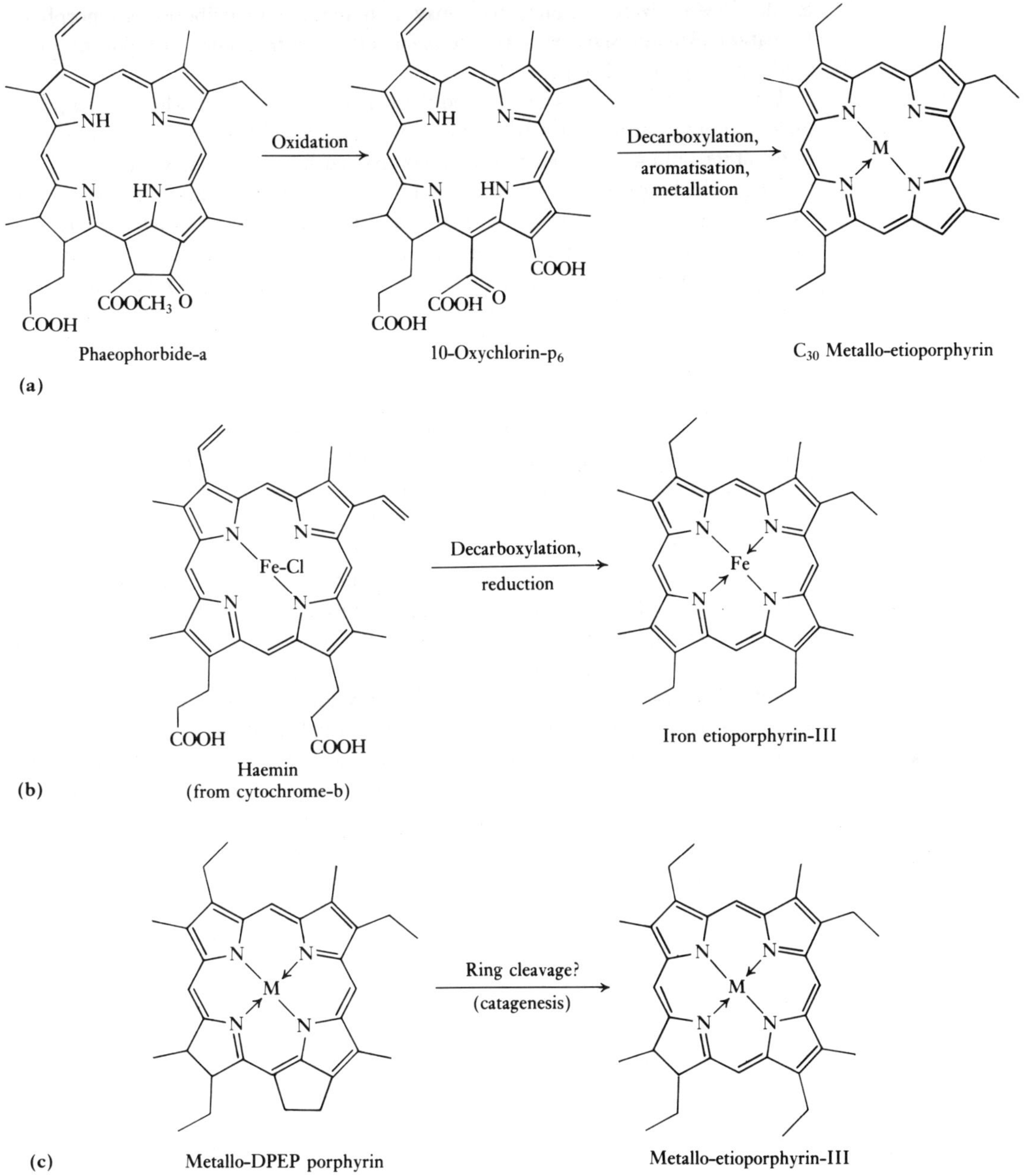

(a)

(b)

(c)

Figure 6.8   Possible origins of etio porphyrins in sediments and humic coals: (a) oxidative cleavage of
isocyclic ring of phaeophorbides during diagenesis; (b) haem source from microbial
cytochromes during diagenesis; (c) thermal cleavage of isocyclic ring of DPEP porphyrins
during catagenesis.

in peat bogs, and those that survive may undergo oxidative cleavage of the
five-membered ring that does not contain nitrogen (Fig. 6.8a). However, such cleavage
cannot lead to members $>C_{30}$, which are probably generated under anoxic conditions
from microbial cytochromes rather than chlorophylls (Fig. 6.8b) (e.g. Bonnett

et al., 1984). Bacteriochlorophylls may also make a contribution to porphyrin formation. Aquatic sediments and kerogens often contain some etio porphyrins in addition to the DPEP type.

Unlike chlorophylls, carotenoids do not generally survive the whole of diagenesis in recognisable form, even under anaerobic conditions. Defunctionalisation occurs, as for other lipid groups, but is often accompanied by chain fragmentation and ring opening, which destroy any recognisable link between a sedimentary component and its biological precursor. However, there are some exceptions in which the complete carotenoid structure is retained in the form of a saturated (e.g. lycopane and β-carotane) or aromatic (e.g. isorenieratane) hydrocarbon (Fig. 6.9). Methanogenesis appears to be the dominant oxidation process rather than sulphate reduction in marine and lacustrine sediments containing saturated or partially saturated carotenoids.

Occasionally the carotenoid structure is partially preserved, in the form of one cyclic system and part of the isoprenoidal chain. Examples shown in Fig. 6.9 include C_{11}–C_{31} 1,1,3-trimethyl-2-alkylcyclohexanes, which may arise from degradation of β-carotane (Jiang and Fowler, 1986), and C_{13}–C_{31} 1-alkyl-2,3,6-trimethylbenzenes, which are probably derived from degradation of isorenieratane (Summons and Powell, 1987). The position of methyl groups on the ring of aromatic carotenoids can provide information about source organisms. A 2,3,6-trimethyl substitution pattern (e.g. in chlorobactene and isorenieratene) is characteristic of the photosynthetic green sulphur bacteria (*Chlorobium*), while 2,3,4-trimethyl substitution is characteristic of the purple genera (*Chromatium*).

Individual carotenoids exhibit varying stabilities during early sedimentary diagenesis. Degradation rates increase in the order: β-carotene and diatoxanthin < peridinin and diadinoxanthin < fucoxanthin. Carotenoids containing 5,6-epoxide groups (i.e. a three-membered ring containing an oxygen atom), such as fucoxanthin and diadinoxanthin (see Fig. 2.21), rapidly disappear in anoxic sediments. The 5,6-epoxide group is highly reactive, due to the presence of an oxo group at C-8, and undergoes the chemical epoxide opening shown in Fig. 6.7. This is rapidly

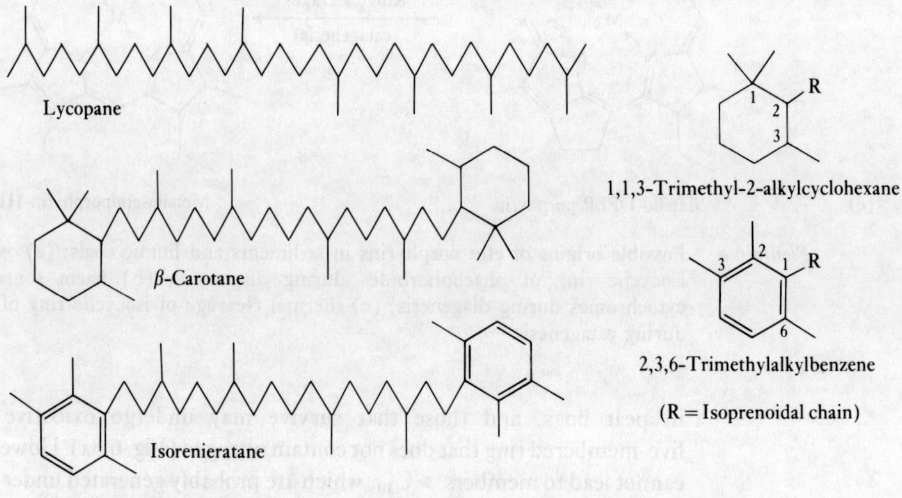

Figure 6.9   Some diagenetic products of carotenoids.

followed by fragmentation of a 5,8-hemiketal intermediate to yield loliolide. Loliolide has a 3S,5R configuration, as do all epoxide-containing carotenoids. In contrast, carotenoids containing a C=C bond in the 5,6 position (e.g. diatoxanthin, β-carotene, lutein and zeaxanthin) are degraded more slowly in anoxic Recent sediments, requiring microbially mediated oxidation (Fig. 6.10). The initial epoxide-forming step of this microbial oxidation can involve attack by an oxygen atom from either side of the ring system, ultimately yielding equal amounts of the epimers loliolide and isololiolide, where a 3-OH group is present in the original carotenoid (Fig. 6.10). The corresponding oxidation of β-carotene, which lacks an OH group at C-3, yields the 5R and 5S epimers of dihydroactinidiolide in equal amounts (Fig. 6.10). Other carotenoids (e.g. acyclics like lycopene, aromatics like okenone, and those containing a 4-oxo group like astaxanthin, see Fig. 2.21) do not undergo such 5,6-epoxidation reactions.

Steroids   The diagenetic fate of steroids has received much attention (e.g. Mackenzie et al., 1982). Our knowledge is largely founded on the identification of specific product–precursor relationships, based on the similarity of distributions in the

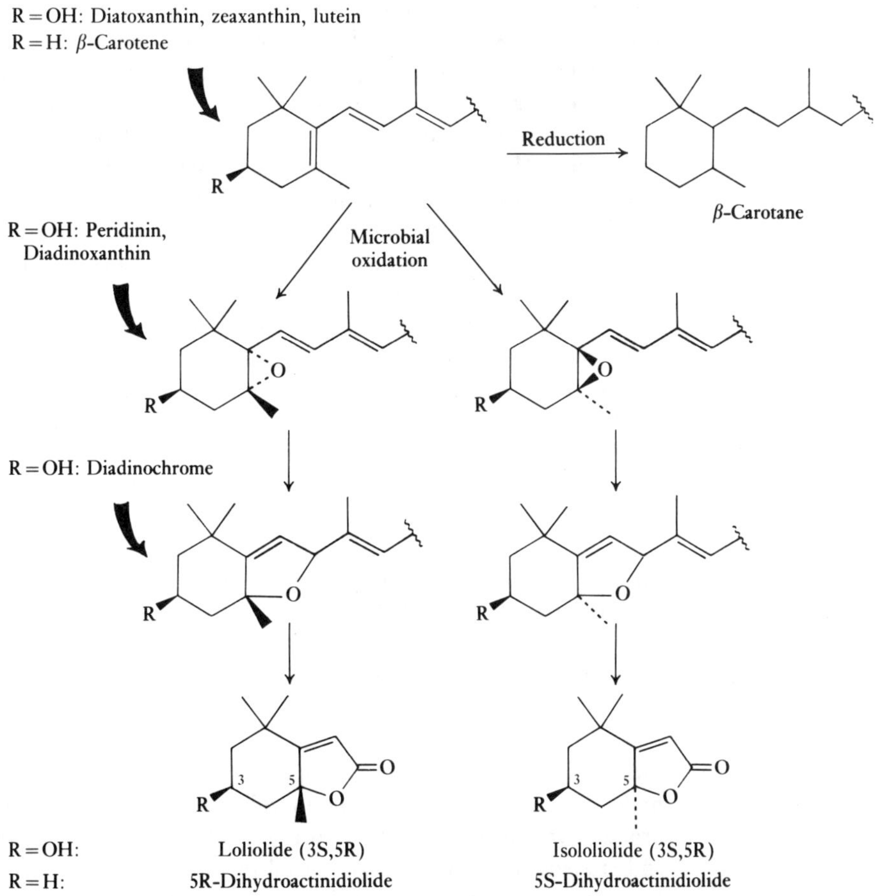

Figure 6.10   Diagenetic pathways of carotenes and β-cyclic xanthophylls in anoxic sediments (large arrows indicate biogenic inputs of specific carotenoids). (After Repeta, 1989.)

supposed product and precursor compound classes and on inverse abundance trends in product–precursor compound pairs with increasing burial depth. Where possible, transformations have been confirmed by simulated diagenesis in the laboratory using isolated or synthesised individual compounds. In this way, many diagenetic transformations have been recognised (e.g. de Leeuw and Baas, 1986; Peakman and Maxwell, 1988), but we shall consider the transformations of some of the more widely occurring steroids in sediments. Δ^5 Stenols are the predominant biogenic sterols and, along with stanols, constitute the major steroidal input to sediments. Free sterols appear to undergo diagenetic alteration much more readily than esterified sterols and so liberation of free sterols by hydrolysis may be considered the initial step in early diagenetic transformations. The subsequent transformations are summarised in Fig. 6.11 for the major products of diagenesis.

Hydrogenation (reduction) of free unsaturated sterols (stenols) to their saturated counterparts (stanols) occurs at an early stage of diagenesis. It is a microbially mediated process, occurring under anaerobic conditions, both in sediments and in particulate matter in the water column (Wakeham, 1989). The reduction may proceed via ketone (stanone) intermediates or by direct reduction of the C=C bond and yields predominantly $5\alpha(H)$-stanols (Fig. 6.11), with much smaller amounts of $5\beta(H)$-stanols. The ratio of stanols : stenols, therefore, increases during early diagenesis. However, a contribution of $5\alpha(H)$-stanols (e.g. cholestanol) can reach the sediment, resulting from partial heterotrophic transformation of algal sterols and from direct algal inputs. The sterols of phytoplankton contain about 5–20% stanols, although the exact amount can vary significantly with growth conditions (e.g. in the brown alga *Monochrysis lutheri* the stanol content is 0% in the exponential growth stage but rises to about 50% in the stationary phase).

The next diagenetic transformation, again under anaerobic conditions in the sediment or water column, is dehydration of stanols to **sterenes**. This process is also microbially mediated and converts the major $5\alpha(H)$-stanols into Δ^2 sterenes (Fig. 6.11), while the minor $5\beta(H)$-stanols are probably converted into Δ^3 sterenes. With increasing burial depth isomerisation of these sterenes to the more stable Δ^4 and Δ^5 isomers occurs. There are a number of divergent reaction pathways leading from sterenes to significantly different products (e.g. Brassell, 1985), the major products usually being **steranes**.

Reduction (hydrogenation) of Δ^4/Δ^5 sterenes to their fully saturated counterparts, **steranes**, results in mainly a $5\alpha(H)$ configuration. This reduction may also occur at the earlier Δ^2/Δ^3 sterene stage (Fig. 6.11). The configuration at most of the chiral carbons in the newly formed steranes is unaffected by diagenetic reactions and so is that inherited from the original biogenic steroid. At the end of diagenesis, therefore, steranes have predominantly a $5\alpha(H),8\beta(H),9\alpha(H),10\beta(CH_3),13\beta(CH_3),14\alpha(H),17\alpha(H),20R$ configuration. Isomerisation reactions involving chiral carbon atoms bearing a hydrogen atom (Section 6.3.1 and Box 6.1) occur later for steranes.

Rather than reduction the intermediate sterenes may undergo rearrangement of the steroidal skeleton, which is apparently catalysed by acidic sites on certain clay minerals (e.g. kaolinite and montmorillonite). The most important rearrangement products are **diasterenes** (Fig. 6.11), while minor products include spirosterenes and steroids that have lost part or all of the A-ring. These minor products result from alteration of the cyclic backbone of the precursor sterenes. In contrast, diasterene formation involves methyl group and C=C bond migrations, but the tetracyclic backbone remains unchanged. Diasterenes are more resistant towards reduction than

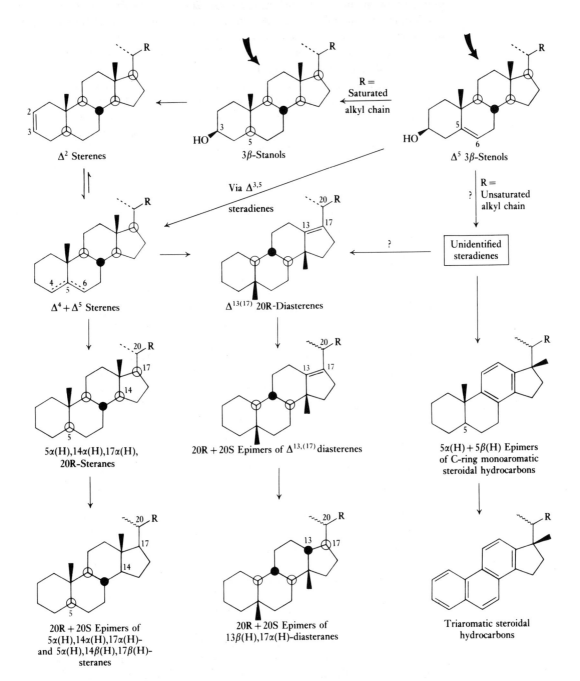

Figure 6.11   Summary of reaction pathways for major steroidal products of anoxic diagenesis (large arrows indicate biogenic inputs). N.B. The last row represents transformations during catagenesis. (After Brassell et al., 1984; Mackenzie, 1984; Peakman and Maxwell, 1988; de Leeuw et al., 1989.)

sterenes and so they survive for longer during diagenesis. They are affected by isomerisation at the C-20 position, the initial 20R isomer being converted into an equilibrium mixture of 20S and 20R epimers in approximately equal amounts before reduction occurs; this appears to be the opposite of the observed sequence for sterenes. Reduction yields mainly $13\beta(H),17\alpha(H)$-**diasteranes** (Fig. 6.11), with smaller amounts of $13\alpha(H),17\beta(H)$ isomers.

As an alternative to reduction it appears that biogenic stenols may undergo dehydration to yield diunsaturated hydrocarbons, **steradienes**. Double bond migration can then occur followed by aromatisation, yielding various monoaromatic steroidal hydrocarbons. The major products are C-ring monoaromatics and by the time they are formed isomerisation at the C-20 position has already occurred (Fig. 6.11). The intermediate steradienes for these compounds have not been identified as yet, but it is believed that the precursor stenols have an additional $C{=}C$ bond in the side chain on the five-membered D-ring. Minor aromatic products include A-ring monoaromatic steroids and B-ring monoaromatic anthrasteroids, both of which may derive from $\Delta^{3,5}$ steradienes by competitive reactions. These steradienes may also undergo rearrangement and/or reduction to yield saturated products (Fig. 6.11).

An important attribute of the diagenetic transformations of steroids is their sensitivity to temperature. For example, diasterene isomerisation at C-20 exhibits changes over small depth intervals of around 20 m. Therefore, the sequence of diagenetic reactions can be used to determine the extent of diagenesis in an immature sediment. The balance established between the products of the competing reactions occurring during diagenesis is affected by factors such as the geothermal gradient and the availability of mineral surfaces (particularly clays) for involvement in catalysed rearrangements. The products of subsequent catagenetic transformations will depend to a large extent on the hydrocarbon distributions present at the end of diagenesis.

In addition to 4-desmethylsterols, 4-methylsterols are abundant in nature and they follow similar diagenetic trends to their demethylated (or regular) counterparts, as summarised in Fig. 6.12. The main precursors are 4α-methylsterols, such as those in dinoflagellates, which yield Δ^4 4-methylsterenes upon dehydration. Reduction can then occur, yielding the 20R isomer of predominantly $5\alpha(H),4\beta$-methylsteranes. With increasing maturity the relative abundance of the thermally more stable 4α epimer increases, presumably via a hydrogen exchange isomerisation process. As an alternative to reduction the 4-methylsterenes may undergo rearrangement, producing mainly 4β-methyl-$10\alpha(H)$-diaster-13(17)-enes. Like their demethylated counterparts, these 4-methyldiasterenes isomerise at the C-20 position before undergoing reduction to 4-methyldiasteranes, which possess mainly the $13\beta(H),17\alpha(H)$ configuration. Small amounts of 4-methylspirosterenes and their rearranged counterparts are also produced, which can undergo further rearrangements analogous to their 4-desmethyl counterparts.

There is, however, a difference between the behaviours of 4-methyl and 4-desmethyl steroids during diagenesis. While sterenes are produced at an early stage of diagenesis, 4-methylsterenes are formed later. One possible explanation lies in the observation that dehydration occurs more readily for a 3α configuration of the hydroxy group than for a 3β configuration and that while both configurations are found among stanols, to date, only the 3β-OH isomers of 4-methylstanols have been detected in immature sediments. Consequently, 3α-stanols may provide an early diagenetic source of sterenes, while 3β-stanols undergo dehydration at a later stage; this is

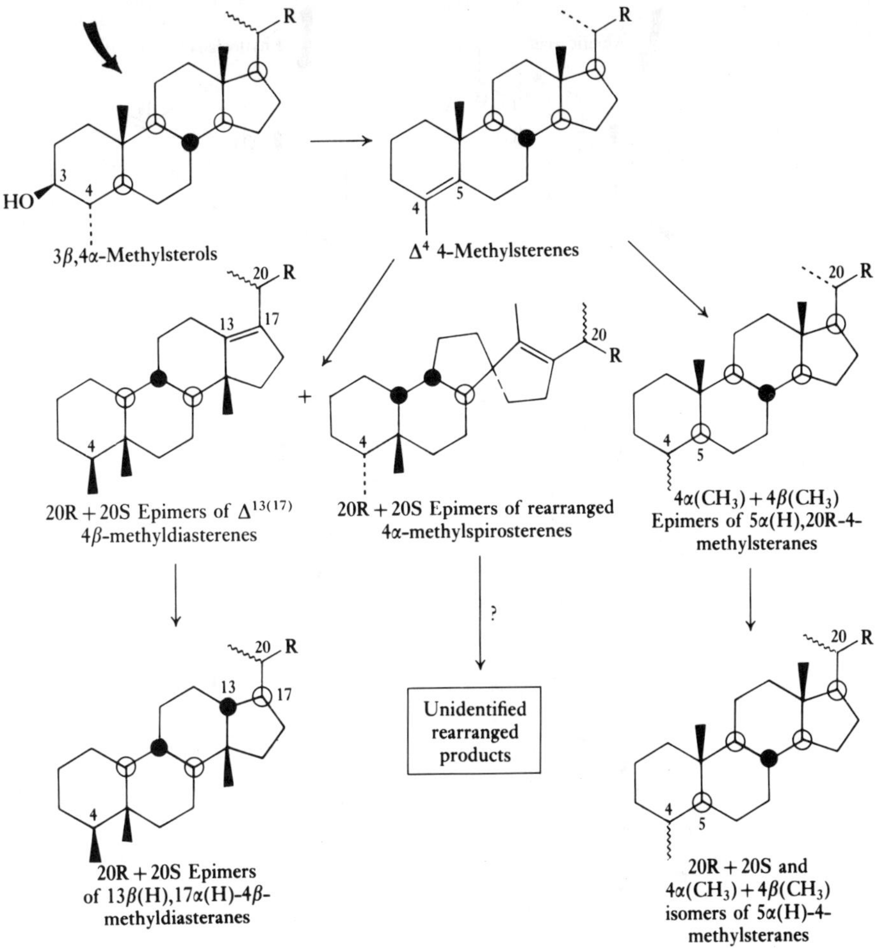

Figure 6.12   Summary of postulated diagenetic pathways for 4-methylsterols (large arrow indicates biogenic input). (After Wolff et al., 1986a, 1986b; Peakman and Maxwell, 1988.)

consistent with the observed more rapid disappearance of 3α-stanols from sediments than of 3β-stanols. In contrast, only a later stage formation of 4-methylsterenes is possible from 3β-4-methylstanols.

Terpenoids   Of the major terpenoidal classes the monoterpenoids are quite volatile and labile, and do not generally survive diagenesis in recognisable form. The sesquiterpenoids are less volatile and can survive diagenesis, undergoing defunctionalisation and either reduction (hydrogenation) or aromatisation to yield hydrocarbons. They may be the major source of the range of $C_{14}-C_{16}$ dicyclic sesquiterpanes often found in oils.

Diterpenoids can be an important source of the saturated and aromatic hydrocarbons remaining at the end of diagenesis, following the usual defunctionalisation and reduction/aromatisation reactions. Tricyclic diterpenoid products can be particularly abundant where there has been a significant contribution from higher plant resins, and they are commonly found in brown coals. Examples of the diagenetic transformations involving aromatisation of some higher plant diterpenoids are shown

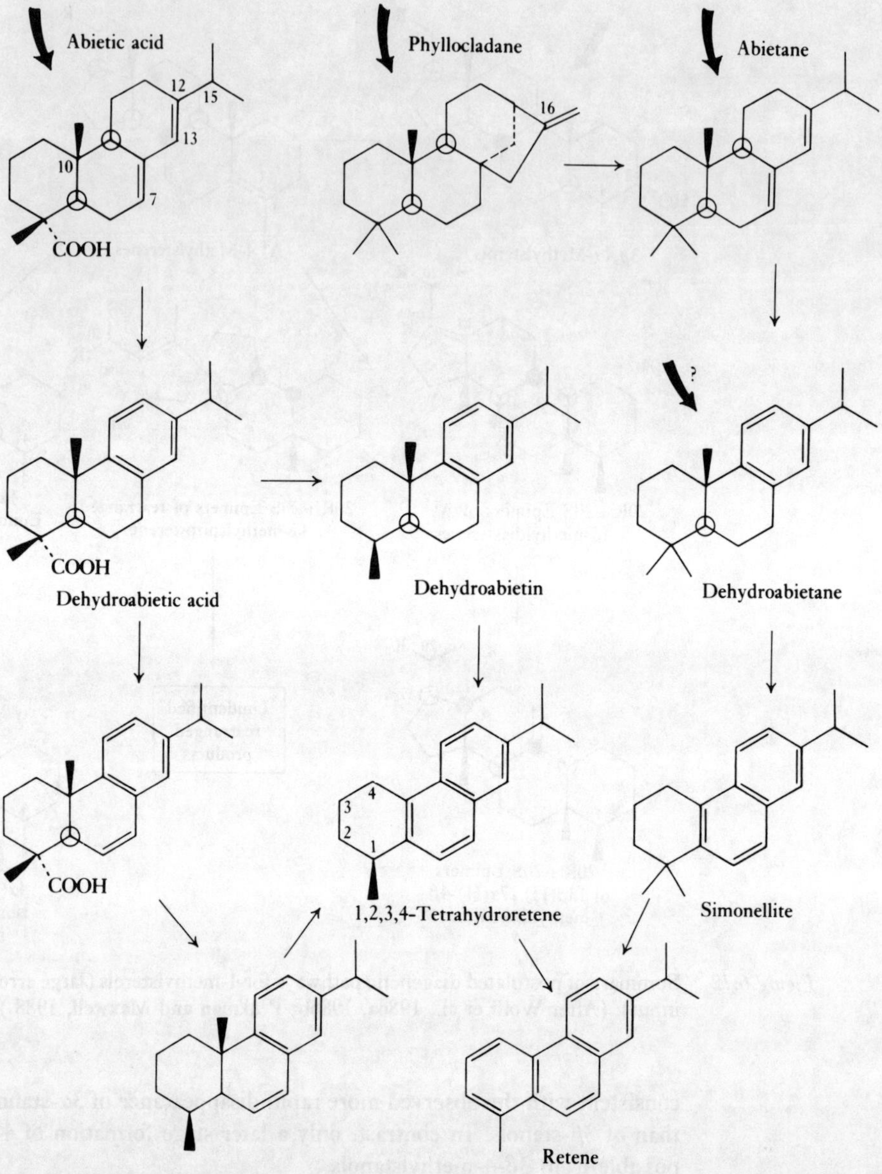

Figure 6.13   Proposed diagenetic pathways for the formation of retene from diterpenoidal resin components (large arrows indicate biogenic inputs). (After Simoneit, 1977; Simoneit, 1986; Alexander et al., 1987.)

in Fig. 6.13. The precursor diterpenoids often exist as several isomers, resulting from variation in configuration or C=C bond position; for example, abietic acid has a 10β-methyl and $\Delta^{7,12}$ unsaturation, epiabietic acid differs from abietic acid in having a 10α-methyl group and neoabietic acid differs from abietic acid in having $\Delta^{8(14),12(15)}$ unsaturation. However, upon aromatisation these differences vanish, leading to a simplified range of products; for example, retene is formed by all the above abietic acid isomers. Similarly, pimaric acid and its 13α-methyl isomer, sandaracopimaric acid, both yield 1,7-dimethylphenanthrene.

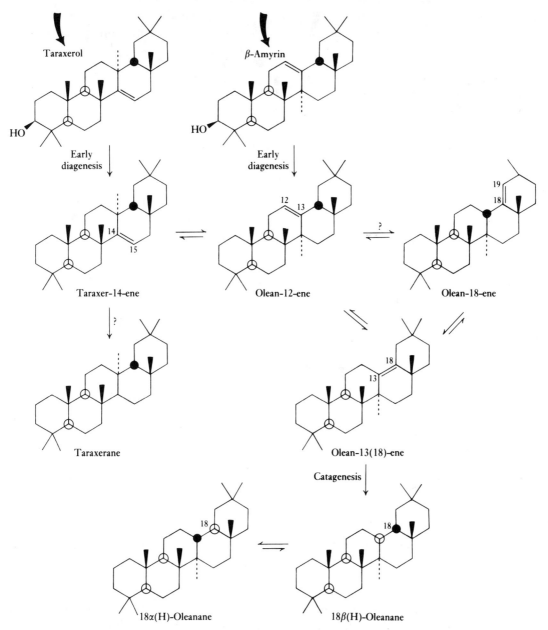

Figure 6.14   Formation of oleanane from higher plant triterpenoids (large arrows indicate biogenic inputs). (After ten Haven and Rullkötter, 1988; Ekweozor and Telnaes, 1990.)

In contrast to aromatisation, hydrogenation initially preserves the biologically conferred stereochemistry, as previously observed for steranes. New chiral centres can also be generated by the addition of hydrogen across a C=C bond, such as at the C-16 position in phyllocladane (see Fig. 6.13). Reduction of phyllocladane initially leads to a 16α(H) configuration but, with increasing maturity, this is largely converted to the more stable 16β(H) epimer (Alexander et al., 1987).

As for steroids, defunctionalisation of pentacyclic triterpenoids occurs during early

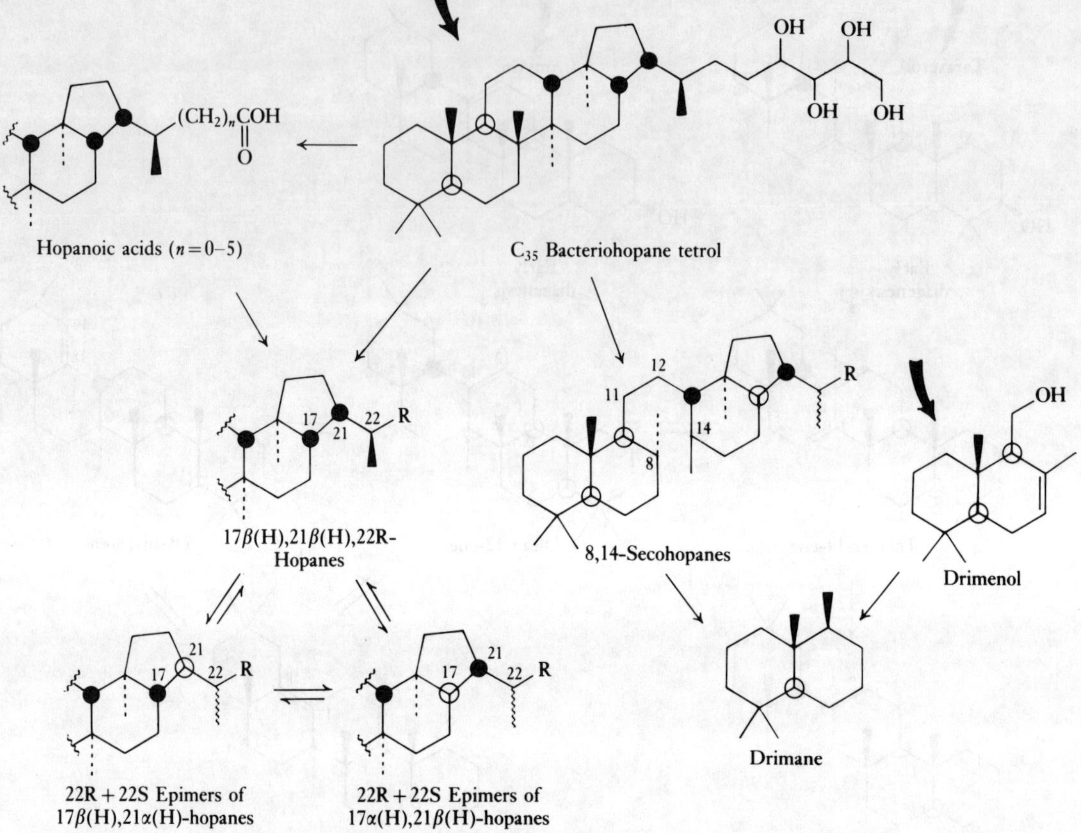

Figure 6.15   Examples of pathways to saturated hydrocarbons from bacteriohopanetetrol (large arrow indicates biogenic input). The A, B and C-rings of hopanes and hopanoic acids are omitted but are identical to those of the C_{35} tetrol precursor (R = H to C_6H_{13}). (After Schmitter et al., 1982; Mackenzie, 1984; Brassell, 1985.)

diagenesis and reduction or aromatisation later. Higher plant triterpenoids often contain a ring C=C bond, which can migrate around the pentacyclic skeleton. The most stable position for the C=C bond in oleanene (which is the diagenetic product of taraxerol and β-amyrin) is $\Delta^{13(18)}$, as shown in Fig. 6.14. Subsequent reduction yields 18β(H)-oleanane. Little taraxerane, the expected reduction product of taraxerol, is found in sediments. Instead, the intermediate taraxer-14-ene appears predominantly to undergo isomerisation to olean-12-ene, finally yielding oleanane (Fig. 6.14).

Hopanoids (alkenes, acids, ketones and alcohols) are ubiquitous components of sediments and soils, and are more widespread than higher plant-derived triterpenoids. Polyhydroxy-bacteriohopanes are thought to be the major precursors, yielding ketones and carboxylic acids during early diagenesis. The latter are the chief products, with C_{31}–C_{33} components occurring widely in Recent sediments. These acids are more resistant towards alteration than other oxygenated hopanoids during diagenesis. Often the C_{32} acid is dominant (31,32-bishomohopanoic acid; $n = 2$, Fig. 6.15), as in *Sphagnum* peats (Quirk et al., 1984) and soils (Ries-Kautt and Albrecht, 1989). Diploptene is also often a major hopanoidal component in soils. As shown in Fig.

6.15, reduction of the hopanoidal acids and polyols yields **hopanes** with a $17\beta(H),21\beta(H)$ configuration. Where the R group is C_2H_5 or larger (i.e. $C_{31}-C_{35}$ components) in Fig. 6.15 there is an additional chiral centre at C-22 which is biosynthesised with the R configuration. Commonly, $C_{29}-C_{35}$ hopanes are produced, arising from fragmentation of the n-alkyl chain attached at C-22 on the E-ring of the precursor polyols. A C_{27} component is also produced by complete loss of the side chain, but a C_{28} product is only rarely formed.

The 8(14) bond in the C-ring of pentacyclic triterpenoids generally appears to be susceptible to acid-catalysed rupture. The **8,14-secohopanes** produced from hopanoids by this bond cleavage are shown in Fig. 6.15. Similar **8,14-secotriterpanes** can be formed from higher plant and other triterpenoids, the reaction appearing to be favoured by the presence of an oxygenated functional group at C-3 in the precursors. Another group of triterpanes, the onoceranes, lack the 8(14) bond, but it is not due to cleavage during diagenesis. The plant-derived precursor compounds, onocerins (see Fig. 2.17), are biosynthesised without an 8(14) C—C bond and so simple reduction yields the onoceranes.

It is believed that the 11(12) bond in all the above C-ring-opened compounds is relatively weak and can undergo cleavage. Drimane, a sesquiterpane, can be produced from the AB-rings fragment of secohopanes and secotriterpanes upon breaking of the 11(12) bond. However, the presence of drimane in sediments is not conclusive of such fragmentation reactions because it may also be produced from reduction of drimenol (Fig. 6.15). Complete C-ring cleavage of higher plant triterpanes can also generate a range of bicyclic sesquiterpanes similar to drimane from the DE-rings fragment.

As with the steroids, triterpenoids may undergo aromatisation during diagenesis rather than reduction. Higher plant triterpenoids, such as α-amyrin (Fig. 6.16), ultimately form fully aromatised pentacyclic or tetracyclic hydrocarbons (polymethyl-picenes or polymethylchrysenes, respectively). The latter are formed by degradation of the A-ring, which appears to be a common process. Regardless of the initial position of any C=C bond, aromatisation of higher plant triterpenoids appears to begin in the A-ring upon loss of the oxygenated functional group at C-3, and progresses sequentially through the B, C, D and E rings. In contrast, hopanoidal precursors have no removable functional group at C-3 and aromatisation begins in the D-ring and progresses through the C, B and A rings. This sequence seems consistent with the C=C bond isomerisation that has been observed for diploptene (Fig. 6.16). Unsaturated hopanoidal hydrocarbons, or **hopenes**, are common in Recent sediments, particularly the $\Delta^{17(21)}$ and $\Delta^{13(18)}$ isomers. The latter is the most stable, and would be anticipated to yield a D-ring monoaromatic species, which may then undergo further aromatisation from the C-ring through to the A-ring. The major aromatic hopanoidal products have an ethyl (C_2H_5) group attached to the E-ring, rather than the iso-propyl (i-C_3H_7 or $CH(CH_3)_2$) that would be expected from diploptene, and it remains to be established whether these compounds are derived from diploptene by loss of a methyl group from the side chain as suggested in Fig. 6.16. Aromatic hopanoids with the same carbon number distributions as the hopanes are sometimes observed, although they are usually minor components. They must arise from the same precursors as the hopanes, presumably via hopenes formed from dehydration reactions of bacteriohopanepolyols.

Aromatisation can also occur in 8,14-secotriterpenoids during diagenesis and, as for the saturated counterparts, complete cleavage of the C-ring can occur. Examples

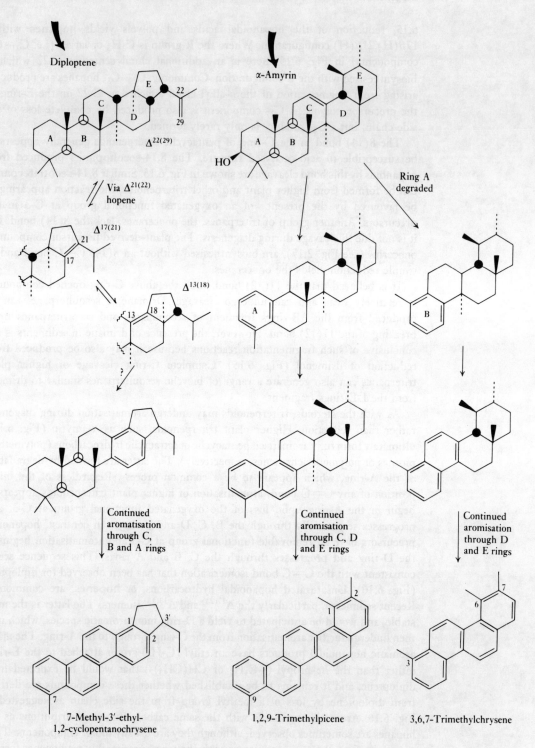

Figure 6.16 Examples of triterpenoid aromatisation (large arrows indicate biogenic inputs). For $\Delta^{17(21)}$ and $\Delta^{13(18)}$ hopenes only D and E-rings are shown (A, B and C-rings are identical to those in diploptene). (After Greiner et al., 1976; Tan and Heit, 1981; Chaffee and Johns, 1983; Simoneit, 1986; Hayatsu et al., 1987.)

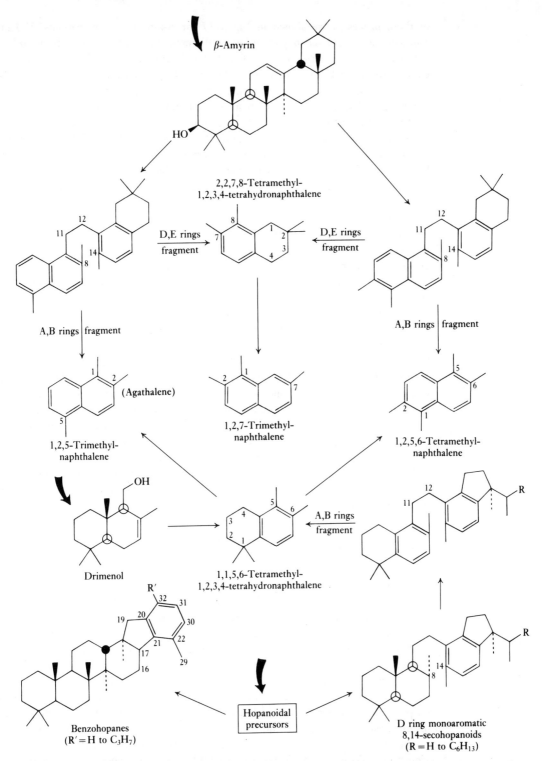

Figure 6.17   Examples of the formation of polymethylnaphthalenes from terpenoidal precursors (large arrows indicate biogenic inputs). (After Chaffee et al., 1984; Hayatsu et al., 1987; Püttmann and Villar, 1987; Strachan et al., 1988.)

of these processes are shown in Fig. 6.17 for β-amyrin and hopanoids. D-Ring monoaromatic 8,14-secohopanoids in sediments and oils are usually found in association with a series of $C_{32}-C_{35}$ benzohopanes (Fig. 6.17). Both series of compounds probably originate from the same C_{35} bacteriohopanoidal precursors, but the benzohopanes seem to be formed at an earlier stage of diagenesis by reactions involving cyclisation of the side chain in the E-ring (Hussler et al., 1984).

The initial product from the AB-rings fragment upon complete C-ring cleavage in aromatic secohopanoids is 1,1,5,6-tetramethyl-1,2,3,4-tetrahydronaphthalene, which can also be formed from the sesquiterpenoid drimenol. This hydrocarbon becomes fully aromatised with increasing maturity, yielding 1,2,5-trimethylnaphthalene (agathalene) and 1,2,5,6-tetramethylnaphthalene (Fig. 6.17). Formation of the latter requires a methyl group to move between adjacent carbon atoms (termed a **1,2-methyl shift**), not an uncommon process. During the aromatisation process a methyl group at C-1 must either be lost or migrate to an adjacent C atom (where it replaces a H atom), as the C-1 atom can only be bonded to three other atoms or groups upon aromatisation. Similarly, methyl groups cannot remain at ring junctions during aromatisation and in Fig. 6.17 it can be seen that they are generally lost (rather than undergoing migration), as in the formation of 1,1,5,6-tetramethyl-1,2,3,4-tetrahydronaphthalene from drimenol.

Aromatic secotriterpenoids of higher plant origin give the same two polymethyl-naphthalene products from the AB-rings fragment as their hopanoidal counterparts (Fig. 6.17). There is a slight difference in the mechanism of formation in that the 1,2-methyl shift is believed to occur at the same time as loss of the OH group at C-3 and while the 11(12) bond is still intact. A further polymethylnaphthalene can arise from the DE-rings fragment of aromatic triterpenoids with a six-membered E-ring, its methyl substitution pattern being characteristic of the structural type of triterpenoid from which it is derived. In Fig. 6.17 this compound is 1,2,7-trimethylnaphthalene, deriving from β-amyrin. Angiosperms are rich in β-amyrin and so large amounts of 1,2,7-trimethylnaphthalene relative to other polymethyl-naphthalenes can be a good indicator of an angiosperm input in immature sediments. At higher maturity levels, typical of catagenesis, this source indicator tends to be overwhelmed by larger amounts of non-specific polymethylnaphthalenes expelled from kerogen, and so the input signature is lost.

The greatest abundances of higher plant-derived aromatic triterpenoids and their degradation products are, not surprisingly, found in brown coals. However, aromatic hopanoids and related compounds can sometimes be abundant in coaly sediments, demonstrating the importance of bacterial activity in general during diagenesis.

The mechanisms for the diagenetic transformations of hopanoids and other triterpenoids have not been studied in the same depth as those of the steroids. The aliphatic and aromatic secohopanoids appear to be as stable as the hopanes during catagenesis and the ratio of the amounts of secohopanoids to their bicyclic degradation products remains fairly constant. This behaviour is consistent with the complete cleavage of the C-ring occurring in one concerted step, in a reaction that competes for precursors with a reaction involving cleavage of only the 8(14) bond. The bicyclic sesquiterpanes, secohopanes and hopanes may, therefore, represent the products of competitive reactions. These reactions are also competing with corresponding aromatisation reactions. Such a situation would parallel the transformations of steroids and is capable of providing valuable information on conditions during diagenesis.

6.4 Palaeotemperature measurement

The study of the sequence of glacial and interglacial episodes during the Quaternary (last 2 Ma) and their geographical extents benefits from an indication of temperature at given locations. Some molecular ratios exhibit the appropriate degree of temperature sensitivity and have been examined as potential palaeothermometers. Two examples, amino acid epimerisation and the degree of unsaturation of long-chain ketones, are considered below.

6.4.1 Amino acid epimerisation

The degree of epimerisation of amino acids in preserved proteinaceous materials such as mollusc shells has been used for determining the age of Quaternary sedimentary strata. This application is possible because living organisms synthesise proteins from the L form of amino acids (see Section 2.3), but after death isomerisation occurs, eventually resulting in a racemic mixture. The rate of this isomerisation for some amino acids can be dependent on organism genus and the age of the shell, although valine and isoleucine appear to provide relatively reliable data (Kimber and Griffin, 1987).

Temperature is an important factor controlling the rate of amino acid isomerisation. The isoleucine reaction in Eq. [6.1] involves epimerisation at one of the chiral carbons in the molecule (the β-carbon, yielding D-allo-isoleucine) and equilibrium is reached over ca. 2 Ma at 10°C. As the kinetics of this reaction are known, and if sedimentary horizons can be independently dated (e.g. using isotopic decay or luminescence methods) it is possible to determine differences in mean temperature between horizons (e.g. Miller et al., 1987), yielding information on Quaternary climatic changes.

$$H_3C\!-\!CH_2\!-\!\overset{\overset{\displaystyle H_3C}{\blacktriangledown}}{\underset{\underset{\displaystyle H}{\blacktriangle}}{C}}\!-\!\overset{\overset{\displaystyle NH_2}{\blacktriangledown}}{\underset{\underset{\displaystyle H}{\blacktriangle}}{C}}\!-\!COOH \rightleftharpoons H_3C\!-\!CH\!-\!\overset{\overset{\displaystyle H}{\blacktriangledown}}{\underset{\underset{\displaystyle H_3C}{\blacktriangle}}{C}}\!-\!\overset{\overset{\displaystyle H}{\blacktriangledown}}{\underset{\underset{\displaystyle NH_2}{\blacktriangle}}{C}}\!-\!COOH \qquad [6.1]$$

L-Isoleucine   D-Allo-isoleucine

6.4.2 Degree of unsaturation in long-chain ketones

It is possible to obtain information on ambient temperatures in past aquatic environments. While the ratio of $^{18}O : {}^{16}O$ in carbonate tests of marine invertebrates can be used to examine sea surface temperatures, its prime use is in determining the extent of glaciation and hence sea level variation during the Quaternary (see Box 6.2). There are also limitations to this stable isotope ratio technique in that carbonates are not preserved in marine sediments lying below the carbonate compensation depth (CCD, Box 6.2). Recently, however, an organic geochemical temperature parameter, U_{37}^K, has been proposed, based on distributions of long-chain (C_{37}–C_{39}) unsaturated ketones (Brassell et al., 1986). These compounds appear to be found only in a few living algae, particularly the ubiquitous coccolithophore *Emiliania huxleyi*, which first appeared during the late Pleistocene (ca. 250 ka ago).

They are also believed to have been characteristic components in morphologically related species of the family Gephyrocapsaceae dating back to the Eocene (ca. 45 Ma) (Marlowe et al., 1990). *Emiliania huxleyi* has been found to increase the degree of unsaturation of its long-chain ketones as temperature decreases, resulting in a lowering in the melting points of these lipids and so enabling the micro-organism to maintain cellular fluidity and function in colder climates (see Section 6.1.3).

Box 6.2

Oxygen isotope composition of carbonate tests and the CCD

Oxygen exists as three stable isotopes, ^{16}O, ^{17}O and ^{18}O, their relative abundances in the Earth as a whole being 99.759%, 0.037% and 0.204%, respectively. All are present in the water of the oceans and so take part in the equilibrium reaction involving carbonate shown in Eq. [1.1]. Organisms that secrete calcium carbonate tests incorporate the isotopes of oxygen broadly in the ratio in which they occur in ambient seawater (although there is some variation in uptake depending on temperature and species of organism). The ratio of the two most abundant isotopes can be expressed by:

$$\delta^{18}O\%_0 = \left(\frac{(^{18}O/^{16}O \text{ in sample})}{(^{18}O/^{16}O \text{ in standard})} - 1 \right) \times 1000 \qquad [6.2]$$

Water containing ^{16}O is lighter than that containing ^{18}O and so it evaporates quicker. This effect is more pronounced in cold than warm water because the partial pressure of $H_2^{18}O$ increases faster than that of $H_2^{16}O$ with increasing temperature. Consequently, the lower the ambient temperature, the lower is the ratio of $^{18}O : ^{16}O$ in rain and snow. For example, the $\delta^{18}O$ values for ice in Greenland, ice at the south pole and rain in the tropics are, respectively, -30, -50 and $0\%_0$.

During periods of glaciation the ^{18}O content of seawater increases as more ^{16}O is locked up in ice-caps. The $\delta^{18}O$ value of carbonate in the preserved shells of marine organisms can give a measure of the degree of water locked up in ice. The remains of benthic foraminiferans are usually chosen, as bottom water temperatures in the Quaternary oceans were fairly constant and so temperature-dependent variations are not expected. It has been determined that a change in $\delta^{18}O$ of 0.1$\%_0$ corresponded to a sea level change of ca. 100 m during the last glaciation, due to the change in amounts of seawater trapped as ice on continental areas. In contrast, data from pelagic foraminiferans can provide information on sea surface temperature. This information can be more difficult to interpret because it is superimposed on the ice volume effect and because sea surface temperature patterns and the factors that affect them can be quite complex.

Carbonate tests are not always preserved in marine sediments. Below the **carbonate compensation depth** (CCD) seawater is sufficiently undersaturated in carbonate for previously precipitated carbonate to redissolve at a significant rate. The CCD varies with crystalline form and ocean locality, e.g. for calcite it is ca. 4500 m in the Atlantic and ca. 3000 m in the Pacific.

The degree of unsaturation of C_{37} ketones in *Emiliania huxleyi* cultured at various temperatures can be evaluated by the expression (Brassell et al., 1986):

$$U_{37}^K = \frac{[C_{37:2}] - [C_{37:4}]}{[C_{37:2} + C_{37:3} + C_{37:4}]} \qquad [6.3]$$

where square brackets represent the concentrations of the di-, tri- and tetra-unsaturated C_{37} methylketones. The di- and triunsaturated components are the main C_{37} methylketones in sediments (see Fig. 6.1) and so a simplified expression that omits $C_{37:4}$ can be used:

$$U_{37}^K = \frac{[C_{37:2}]}{[C_{37:2} + C_{37:3}]} \qquad [6.4]$$

Values of U_{37}^K derived from Eq. [6.4] have been found to increase linearly with sea surface temperature (SST) according to the expression:

$$U_{37}^K = (0.033 \times \text{SST}) + 0.043 \qquad [6.5]$$

This temperature calibration has been confirmed by comparison with $\delta^{18}O$ values for planktonic species of the foraminiferan *Globigerina* (Prahl and Wakeham, 1987). The U_{37}^K parameter can be measured to an accuracy of 0.02 units, potentially allowing temperature to be determined within 0.5°C. Its application is not limited by CCD, and the long-chain ketones seem to be more stable than most unsaturated lipids and can survive diagenesis.

Chapter 7

Molecular assessment of ancient sediments and petroleum formation

7.1 Source indicators

7.1.1 Introduction

By the end of diagenesis the biogenic organic material in sediments either has been degraded and recycled by micro-organisms or has been largely converted into insoluble polymeric material (kerogen or brown coal). The most useful molecular source indicators at this stage of maturity, corresponding to ancient sediments, are biomarkers, which are mostly in the form of hydrocarbons. The biologically conferred configuration of biomarkers is often conserved at most chiral centres during diagenesis, although with increasing thermal maturity isomerisation occurs.

7.1.2 Hydrocarbons

Normal and methyl-branched alkanes

In Chapter 6 it was noted that some acyclic alkanes are produced directly by organisms (although usually at lower levels than fatty acids) and any usefulness they may have as source indicators in Recent sediments is largely retained in older sediments. For example, mid-chain methyl-branched alkanes, chiefly C_{18}, are characteristic of bacteria. Although normal (i.e. straight-chain), *iso* and *anteiso* alkanes are widely distributed in organisms, the presence of an odd-over-even predominance (OEP) in carbon number can provide source information. In particular, an OEP for *n*-alkanes $>C_{20}$ is characteristic of higher plant waxes, while a predominance of C_{15} and C_{17} components is characteristic of algae. During catagenesis, however, such *n*-alkane source indicators become less distinct with the generation of large amounts of non-specific *n*-alkanes.

After the defunctionalisation and reduction that occur in fatty acids and other functionalised acyclic lipids during diagenesis, there can be significant loss of information on source organisms that was provided by features such as the nature and position of functional groups (e.g. in hydroxy fatty acids) and the position and stereoisomerism of any C=C bonds. However, it may be possible to obtain some source-related information from the resulting acyclic alkanes from their carbon number and the position of any branch points in the alkyl chains. Prior to the onset of thermal cracking, when short-chain alkanes are produced from long-chain precursors, it is still possible to make a broad differentiation between short-chain ($<C_{20}$) microbial and long-chain ($>C_{20}$) acyclic components. The distributions of acyclic alkanes in organisms can often be seen to parallel those of the fatty acids, reflecting their biosynthetic relationship. In this way the OEP observed in higher plant wax *n*-alkanes is augmented by the decarboxylation of wax-derived fatty acids with an OEP.

Acyclic   Acyclic isoprenoidal alkanes occur widely in organisms but there are some that are
isoprenoids   source-specific. The biphytanyl ethers characteristic of some members of the
archaebacteria (methanogens and thermoacidophiles) can be preserved as a series of
long-chain (i.e. $> C_{20}$) isoprenoidal alkanes (e.g. C_{38} alkane in Fig. 7.1), which
retain the characteristic head-to-head linkage where the two phytanyl units join
together (Moldowan and Seifert, 1979; Albaigés, 1980). Lycopane (Fig. 6.9), a
saturated C_{40} isoprenoidal alkane, is also likely to be a bacterial marker, deriving
from reduction of lycopene (Fig. 2.21). However, squalane (Fig. 7.1), a saturated
C_{30} isoprenoidal alkane, may represent a direct archaebacterial input (e.g. Brassell
et al., 1981), or it may derive from diagenetic reduction of squalene (Fig. 2.19),
which occurs in a variety of organisms.

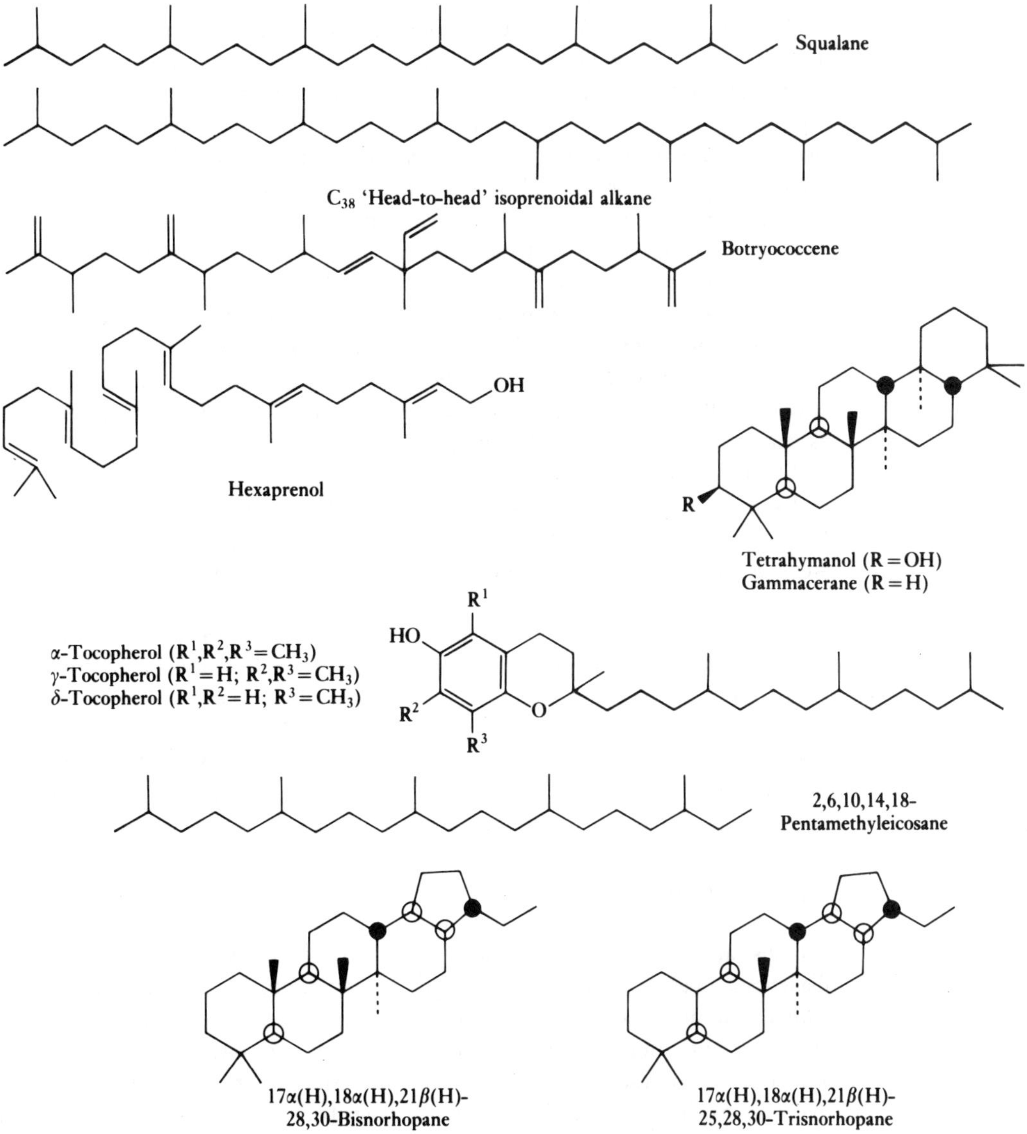

Figure 7.1   Some sedimentary biomarker hydrocarbons and precursors.

Botryococcane, a saturated C_{34} isoprenoidal alkane, is a particularly useful source- and environment-specific indicator. It appears to be derived only from the alga *Botryococcus braunii*, which is widely distributed in freshwater lakes and often forms large mats. It has two growth stages, a green and a brown, during which it produces different lipids. It is in the brown stage that characteristic alkenes (botryococcenes) are produced with six double bonds, of which the C_{34} component in Fig. 7.1 is the most abundant and undergoes complete hydrogenation during diagenesis to yield botryococcane. In some oils from Sumatra (Moldowan and Seifert, 1980) and South Australia (McKirdy et al., 1986) botryococcane is the dominant aliphatic component. These oils appear to be sourced largely by the remains of *Botryococcus* deposited in freshwater lakes (see Section 4.4.3).

Cycloalkanes   It was noted in Chapter 6 how monocyclic, dicyclic and tricyclic hydrocarbons, both aliphatic and aromatic, can derive from biomarkers such as carotenoids, sesquiterpenoids and diterpenoids. Again, some source information is lost with the loss of functional groups and C=C bonds, and it is possible for a number of biogenic precursor molecules to yield the same hydrocarbon at the end of diagenesis. For example, drimane can be formed from 8,14-secotriterpanes and drimenol (Fig. 6.15), and the same C_{29} sterane is produced by stigmasterol, sitosterol and fucosterol (Fig. 2.20). Extended tricyclic alkanes with isoprenoidal side chains (C_{19}–C_{30}, Fig. 5.10) are commonly found in oils and they may be microbial in origin. The hexaprenol in Fig. 7.1, a common cellular component, seems a likely precursor. Tetracyclic alkanes (C_{24}–C_{35}, Fig. 5.10) are also frequent oil constituents and they are clearly related to the hopanes, from which they may be formed by cleavage of the 17(21) C—C bond. These tricyclic and tetracyclic series are absent from Recent sediments, suggesting that their formation from kerogen requires higher temperatures than those associated with diagenesis (Aquino Neto et al., 1983).

Steranes and triterpanes can provide useful source information in ancient sediments and sedimentary rocks. The total carbon number of biogenic precursor sterols is largely preserved in steranes so that, with care, a triangular plot of C_{27} : C_{28} : C_{29} steranes can be used to evaluate sources of organic matter in a similar manner to that applied to the sterols in Fig. 6.1. From this approach it may be possible to gain an idea of the importance of vascular plant inputs relative to those from aquatic organisms, although the same reservations regarding this simplistic interpretation of carbon number distributions apply to steranes as they do to sterols (Section 6.2.2). Also, 4-methylsterols, which usually indicate a dinoflagellate input, are preserved as 4-methylsteranes, mostly occurring in the range C_{28}–C_{30}. In addition to C_{27}–C_{29} members, the regular (i.e. 4-desmethyl) steranes of some oils and source rocks have been found to contain C_{26} (Moldowan et al., 1991) and C_{30} (Moldowan, 1984) components. The C_{30} steranes are generally found in low relative abundance and appear to derive from marine organisms.

Hopanes are ubiquitous components of sediments and sedimentary rocks, demonstrating the importance of bacterial contributions to the sediment during diagenesis. High levels of hopanes relative to steranes in oils often appear to correlate with a predominance of C_{29} steranes over other steranes. The latter suggests a major vascular plant input to sedimentary organic matter, and so the high levels of hopanes are believed to result from pronounced bacterial reworking of this higher plant material. Such oils are probably sourced from bacterial remains and the lipid-rich structures of the vascular plants. Various other series of hopanes have been detected in which the number and/or position of ring-methyl groups differ from those in the regular hopanes. For example, there are series with one more ring-methyl group than usual (e.g. 2- and 3-methylhopanes), or with one less ring-methyl group than usual (e.g. 25-norhopanes). Components with differing methyl substitution patterns include a C_{29} homologue of Ts (Fig. 2.18c), i.e. $18\alpha(H)$, $21\beta(H)$-30-norneohopane, and a series of apparently rearranged hopanes, $17\alpha(H)$,15α-methyl-27-norhopanes, termed diahopanes (Killops and Howell, 1991; Moldowan et al., 1991). These components may represent inputs from different types of bacteria or they may derive from the same precursors as the regular hopanes but via different diagenetic pathways.

Higher plant triterpanes are particularly useful source indicators. For example, β-amyrin, an abundant constituent of angiosperms, is transformed into oleanane during diagenesis (Fig. 6.14). The presence of oleanane, therefore, suggests an input from vascular plants that only became widespread during the late Cretaceous–early Tertiary. A similar input is suggested by a high abundance of 1,2,7-trimethylnaphthalene relative to all other trimethylnaphthalenes, resulting from fragmentation of the C-ring of β-amyrin during diagenesis, as shown in Fig. 6.17. In contrast, high relative abundances of 1,2,5-trimethylnaphthalene and 1,2,5,6-tetramethylnaphthalene are likely to be accompanied by abundant aromatic 8,14-secohopanoids (from which they are derived) and hopanes from bacterial sources. However, with increasing thermal maturity during catagenesis (or coalification) the relative abundance of these specific polymethylnaphthalenes decreases until they are no longer conclusive source indicators. This occurs because of the generation of other polymethylnaphthalenes in significant quantities from kerogen (or coal), and because of the conversion of the source indicators into other isomers by methyl group migration (see Section 7.4.1).

7.1.3 Carbon isotopes

The carbon isotopic signature of both bulk organic matter and individual molecules can be helpful in determining the types of organisms contributing to ancient sediments in the same way as was noted for Recent sediments (Section 6.2.5). It has been found that molecular isotopic signatures can vary more widely for different compounds in the same sedimentary rock sample than between different rock samples. Some examples from the Eocene Messel Shale are presented in Table 7.1. The $\delta^{13}C$ value for the total kerogen in this shale is $-28.21\%_{oo}$, while that for the solvent-extractable organic matter (bitumen) is $-29.72\%_{oo}$. This relative lightness of lipid-derived organics is widely observed and is reflected in the values in Table 7.1. The differing values of $\delta^{13}C$ for pristane and phytane indicate that they do not share the same

Table 7.1 **Carbon isotope composition of some compounds in Messel Shale** (Data after Freeman et al., 1990)

Compound	$\delta^{13}C\%_{oo}$
Pristane	-25.4
Phytane	-31.8
$17\beta(H)$,$21\beta(H)$-30-Norhopane	-65.3
$17\beta(H)$,$21\beta(H)$-Hopane	-35.2

source. The value for pristane is compatible with an algal origin, but that of phytane probably reflects a phytanyl ether input from methanogenic bacteria. The value of $\delta^{13}C$ for $17\beta(H),21\beta(H)$-hopane is characteristic of the hopanoids in general in Messel Shale, which probably originate from chemosynthetic bacteria. The intense depletion in ^{13}C of $17\beta(H),21\beta(H)$-30-norhopane suggests that it does not share the same source as the other hopanoids but derives from a precursor already exhibiting a light isotopic signature. Such a value is expected for methylotrophic bacteria, as the methane they use for a substrate already bears the light isotopic signature of methanogenic bacteria. The hopanoids of methylotrophs, such as *Methylomonas*, are dominated by a C_{35} pentahydroxyamine which may degrade to the norhopane.

This molecular carbon isotope approach can be applied to other compound classes, such as porphyrins, allowing assessment of the relative contributions from algal and bacterial chlorophylls (e.g. Hayes et al., 1987). The resulting information cannot be obtained from bulk carbon isotopic data and biomarker distributions, and so molecular carbon isotope studies offer considerable scope for improving our understanding of the processes operating during the formation of organic-rich sediments. It can be seen that the complexity of fractionation processes and the information they can convey are hidden within the bulk isotopic measurements described in Section 5.6.1.

7.2   Indicators of depositional environment

To some extent, molecular source indicators provide information on depositional environments, as noted for botryococcane and C_{30} steranes in Section 7.1.2. Autochthonous inputs to sediments reflect the communities of organisms that flourish under the conditions in the overlying water column, while allochthonous inputs, particularly of vascular plant material, generally suggest relatively nearby sources and water transport of debris. For example, levels of vascular plant-derived oleanane approaching those of bacterial hopanes appear to be characteristic of nearshore deposition, often in marine or lacustrine deltas, as in oils sourced in the Niger Delta (Africa) and the Mahakam Delta (Indonesia). Freshwater lakes often exhibit characteristics of vascular plant and phytoplankton inputs, and significant inputs of long-chain acyclic isoprenoids from methanogenic bacteria may occur where sulphate levels are very low. In contrast, bacterial activity in marine and saline lake sediments is dominated by sulphate reducers. The absence of terrestrial indicators in marine sediments does not necessarily imply offshore deposition. Nearshore areas virtually devoid of vegetation (e.g. deserts) or with limited river drainage (e.g. west side of the Andes) will not exhibit a significant vascular plant fingerprint.

7.2.1   Hypersalinity

Hypersaline environments are interesting in that very few organisms are tolerant of such extreme conditions and relatively simple communities are usually found. Cyanobacteria can flourish in these environments because the mats they form are not subject to extensive grazing by herbivores, although the associated communities usually include some invertebrates together with some algae. One group of the archaebacteria, the halophilic bacteria, also thrives under these conditions and actually requires a minimum salt concentration. Bacterial inputs can, therefore, be significant in organic-rich sediments deposited in hypersaline lakes (see Section 3.4.1). This material is less subject to degradation in the sediment than in other environments because all microbial activity, including sulphate reduction, is suppressed by hypersalinity. Oil source rocks deposited in salt lakes occur widely in China (Jiamo et al., 1986; Jiamo and Guoying, 1989). Biomarker distributions are characterised by relatively large amounts of phytane, β-carotane, C_{34} and C_{35} hopanes and gammacerane (Fig. 7.1). In addition, C_{22} n-alkane is often relatively abundant (ten Haven et al., 1985). Gamacerane, which has been found in sediments dating back to the Proterozoic, is believed to be the diagenetic product of tetrahymanol (Fig. 7.1), a triterpenoidal alcohol that occurs widely in Recent marine sediments and apparently derives from some ciliates (protozoans). However, tetrahymanol may have occurred more widely among micro-organisms in ancient environments, possibly serving a similar function to the hopanoids (ten Haven et al., 1989; Venkatesan, 1989).

β-Carotane is found only in ancient lacustrine sediments and it has been suggested that it may be a by-product of methanogenesis (Repeta, 1989). Reduction of carotenoids to form saturated hydrocarbons does not appear to occur in Recent sediments and, as sulphate reduction causes degradation of carotenoids, the survival of potential precursors of β-carotane until reduction occurs requires the absence or limitation of sulphate reduction. The most likely precursors are those that do not contain a 5,6-epoxide group, such as β-carotene, and are, therefore, degraded less rapidly by anaerobic oxidation (see Section 6.3.3). The major phytoplankton carotenoids, such as fucoxanthin, contain a 5,6-epoxide grouping and are rapidly degraded at the beginning of diagenesis even in anoxic sediments. They are unlikely to be precursors of β-carotane. The exceptionally high levels of β-carotane in salt-lake sediments reflects the abundance of the above-mentioned suitable precursors in contributing organisms and the conditions necessary for their preservation during diagenesis (e.g. presence of methanogens but virtual absence of sulphate reducers).

7.2.2 Redox conditions

Phytol diagenesis While anoxic conditions are generally required for the formation of organic-rich sediments, oxygen is unlikely to be absent at all stages of diagenesis. Levels of oxygen in the water column can fluctuate over time and so can the position of the oxic/anoxic boundary. It is possible to gain some information on redox conditions (see Box 3.2) during diagenesis from molecular indicators. Phytol is an important part of phytoplankton chlorophylls, from which it appears to be released early on in diagenesis in the water column. Studies of the diagenetic fate of phytol (Didyk et al., 1978) have suggested that different products are formed under different redox conditions (Fig. 7.2). Under relatively oxidising conditions phytol, a C_{20} compound, appears to be oxidised mainly to phytenic acid, which then undergoes decarboxylation to pristene, a C_{19} compound, before finally being reduced to pristane. In contrast, under relatively anoxic conditions phytol is reduced to dihydrophytol (or phytanol), which is converted into phytane by dehydration and further reduction, with the preservation of all 20 carbon atoms in the product. The ratio pristane : phytane may, therefore, provide a measure of redox conditions during diagenesis. Values <1 are generally recorded for anoxic conditions that extend from the sediment up into the water column, while values >1 suggest a degree of oxicity.

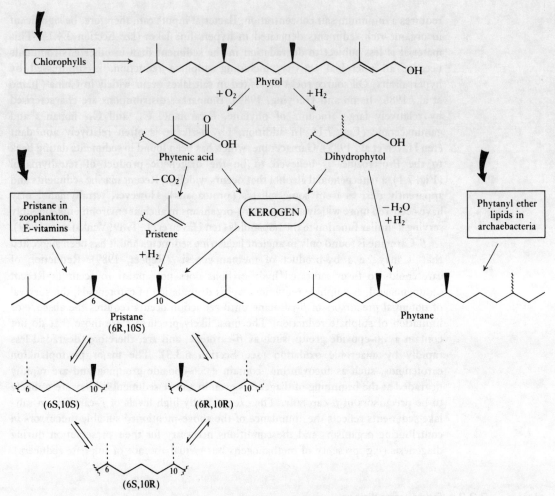

Figure 7.2   Summary of phytol sedimentary diagenesis, showing pristane isomerism (phytane undergoes similar isomerism).

Isoprenoidal hydrocarbons with < 20 carbon atoms (e.g. C_{14}–C_{16} and C_{18}) are common in sediments and may also derive from phytol. Degradation of the C_{20} chain appears to occur as part of the digestive process in zooplankton and during sedimentary diagenesis.

Phytol geochemistry is quite complex and the scheme described above is an oversimplification. For example, chlorophylls can reach the sediment intact (Johns et al., 1980) and so when hydrolysis liberates phytol in the anoxic zone only reduction to phytane is likely. In addition, the effects of geological constraints on redox conditions have been overlooked. These include the effect of variations in sedimentation rate on the thickness of the oxic sediment layer, the interbedding of turbidite and pelagic sequences, and the effect of micro-environments (the conditions within two adjacent micro-environments can vary significantly). Furthermore, there can be sources of pristane and phytane other than chlorophylls (ten Haven et al., 1987). These include methanogen or halophile phytanyl lipid sources for phytane and zooplankton or tocopherol (i.e. E vitamins, Fig. 7.1) sources for pristane

(Goossens et al., 1984). In hypersaline environments halophilic bacteria are very important and they contain more phytanyl lipids than methanogens. Consequently, a low pristane:phytane ratio is usually recorded, reflecting the salinity-dependent growth rate of the bacteria and not the degree of anoxicity. Despite all these possible restrictions on the use of pristane:phytane as a redox condition indicator, this ratio appears to work surprisingly well for a large number of oil source rocks. During catagenesis, however, the pristane:phytane ratio tends to increase as additional pristane appears to be generated from kerogen. This increase is paralleled by an increase in n-alkanes relative to phytane (usually measured by the C_{18}:phytane ratio). The generation of additional pristane during catagenesis reflects incorporation of phytol products into kerogen during diagenesis (see Section 7.4.1).

Nickel and vanadium distributions   The interaction of metal ions with organic species is an important aspect of diagenesis and is affected by redox potential (E_h) and acidity (pH) (see Box 3.2). These factors control the oxidation state of a metal ion and the availability of anions that may compete with organic species for the metal. The distribution of metals in oils can, therefore, provide information on depositional environments. Two metals that are particularly useful are nickel (Ni) and vanadium (V). The bulk of these two metals in oils appears to be strongly associated with high-molecular-weight organic components, especially asphaltenes, and the ratio of Ni to V appears not to be affected by post-diagenetic processes. Although Ni and V are also associated with porphyrins, which do undergo some changes with increasing maturity, the amounts in porphyrins are generally relatively minor (but there are exceptions, such as Venezuelan oils) and do not affect the overall interpretation of Ni:V ratios. Figure 7.3 shows the main field of E_h and pH conditions for marine organic sediments, which is divided into three areas (Lewan, 1984). While the average pH of seawater is 8.2, it can be seen that the pH of sedimentary pore water can vary significantly.

The chemical form of sulphur is important in determining the availability of metal ions in marine environments. In area I of Fig. 7.3 sulphur is generally in the form of sulphate (SO_4^{2-}) and nickel is available for bonding, in the form of Ni^{2+},

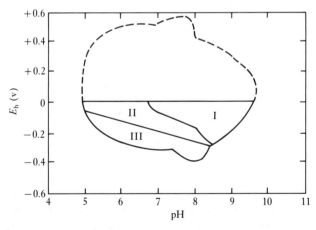

Figure 7.3 E_h and pH stability fields in sediments. Area bounded by outer solid line and broken line is the natural stability field for marginal and open marine sediments. Areas I, II, and III are discussed in the text, and the solid line encompassing them is the field for organic sediments. (After Lewan, 1984.)

but vanadium is unavailable. Oils from source rocks deposited under these conditions, such as those from Mesozoic and Tertiary reservoirs in the Uinta Basin (USA), have $V/(V + Ni)$ values <0.1 and S content $<1\%$. In area II sulphate is again the major form of sulphur, while Ni^{2+} and VO^{2+} (vanadyl) are available for bonding, but the vanadyl species may be partially hindered by sulphide complexation. Crude oils associated with these conditions include those from Tertiary reservoirs in eastern Nigeria and they contain low amounts of S (ca. $0.2–0.5\%$) with $V/(V + Ni)$ values in the range $0.1–0.9$. In area III sulphur exists mainly as H_2S at pH <7 and as HS^- at higher pH levels. Vanadium is available in the form of VO^{2+} and V^{3+}, but Ni^{2+} may be partially hindered by sulphide complexation. Associated oils, such as those from Jurassic reservoirs in Saudi Arabia, have high S content ($>1\%$) and $V/(V + Ni)$ values >0.5.

7.2.3   Recognition of different types of marine and lacustrine environments

It is apparent that different types of depositional environments are characterised by inputs of organic matter from particular groups of organisms, and the transformations that this organic matter undergoes during diagenesis are also influenced by the environment (e.g. E_h and pH conditions, inorganic mineral composition). Sufficient variety can exist in the composition of the sedimentary organic matter to permit the differentiation of a range of depositional environments. Unfortunately, there are few types of depositional environments that are exclusively characterised by the presence of just one or two specific biomarker hydrocarbons. However, differentiation can be achieved by detailed examination of the distributions of a range of biomarkers (involving determination of absolute as well as relative concentrations) combined with bulk organic matter properties (e.g. $\delta^{13}C$ values, kerogen typing, content of S, Ni and V).

Such an approach has enabled seven different types of depositional environment to be distinguished among Brazilian marginal offshore basins, formed during the opening of the South Atlantic by the rifting of the African and South American continental plates (Mello et al., 1988). The basins resulted from the progressive subsidence of the crust that occurred with the onset of rifting in the early Cretaceous. Initially, freshwater and then saline lacustrine deposits were formed, followed in the mid Cretaceous by marine evaporite deposition as rifting proceeded, resulting from the combined effects of intermittent marine transgressions and evaporation of the waters trapped by topographical barriers in a hot dry climate. These three depositional environments were relatively readily distinguished from each other and from the range of open marine environments that succeeded them during the late Cretaceous and Tertiary. For example, the freshwater lacustrine deposits of mainly type I kerogen typically exhibited high alkane concentrations with an OEP among n-alkanes, more abundant pristane than phytane, low sulphur content (ca. 0.3%) and abundant hopanes relative to steranes. In contrast, the type II kerogen in the marine evaporite deposits contained very high concentrations of gammacerane, β-carotane, squalane and a related C_{25} isoprenoidal alkane (2,6,10,14,18-pentamethyleicosane, Fig. 7.1), hopanes (especially hopane itself) and steranes in general. In addition, n-alkanes exhibited an EOP, phytane was more abundant than pristane and sulphur content was high (up to 2.5%). As noted in Section 7.2.1, all these features appear to be related to high salinity in general and were also exhibited, but to a lesser extent, in the predominantly type II kerogens of the saline-lake deposits in the Brazilian marginal basins. However, the saline-lake deposits and marine evaporites could be differentiated by a number of characteristics, such as the presence of C_{30} steranes in the marine deposits only and the moderate sulphur content of the saline-lake deposits (0.3–0.6%).

Nickel and vanadyl porphyrins were abundant in both marine and lacustrine saline environments in the Brazilian basins, with nickel porphyrins predominating. In contrast, these compounds, together with β-carotane, were virtually absent in the freshwater lacustrine deposits. This demonstrates that the absence of some biomarkers can be as informative as the presence of others. Among the readily identified characteristics of the open marine depositional environments were the presence of 18α(H)-oleanane in Tertiary deltaic sequences, the dominance of 17α(H),18α(H),21β(H)-28,30-bisnorhopane and 17α(H),18α(H),21β(H)-25,28,30-trisnorhopane (Fig. 7.1) among the hopanes in highly anoxic marine sediments, and the dominance of vanadyl porphyrins over their nickel counterparts in marine carbonates.

7.3   Thermal maturity and molecular transformations

7.3.1   Configurational isomerism

Acyclic isoprenoidal alkanes

As temperature increases with increasing burial depth of sediments, thermally mediated changes supersede biologically controlled transformations. Configurational isomerism is one such process, which begins during diagenesis for some components. For example, in pristane the biologically conferred 6R,10S isomer is converted into equal amounts of the four possible isomers (6R,10S; 6R,10R; 6S,10S and 6S,10R) prior to the onset of catagenesis (Fig. 7.2). The 6S,10S and 6R,10R forms of pristane are, in fact, identical and can be interconverted simply by rotating the molecule about a vertical axis through the central C atom. This isomer is known as meso pristane (rather than 6S,10S or 6R,10R).

Steranes

At the end of diagenesis the configuration of chiral carbon atoms in the sterane skeleton is largely that inherited from the biogenic precursor, which provides the degree of flatness required in the cell membrane rigidifying molecules. At certain of the chiral centres the stereochemistry already corresponds to the thermodynamically most stable configuration (e.g. at C-8 and C-9, see Fig. 2.18 for carbon numbering scheme). At C-10 and C-13 hydrogen exchange cannot alter the configuration because there is no hydrogen directly bonded to the chiral carbon atom to allow the necessary process to occur (see Section 6.3.1 and Box 6.1). Isomerisation is, therefore, limited to the C-14, C-17 and C-20 positions. A mixture of isomers is usually present at C-5 and C-24, although the 5α(H) configuration is normally more abundant.

Only the 20R isomers of the regular steranes and 4-methylsteranes exist initially, but they undergo isomerisation to form an equilibrium mixture containing approximately equal amounts of 20R and 20S isomers (Figs 6.11 and 6.12). This process appears to require higher temperatures than the isomerisation of pristane. Similar temperatures to those at which the 20S/R isomerisation occurs are required for another isomerisation process of the regular steranes that affects the C-14 and C-17 cyclic positions in concert. The $14\alpha(H),17\alpha(H)$ isomer is converted into an equilibrium mixture of $14\alpha(H),17\alpha(H)$ and $14\beta(H),17\beta(H)$ isomers (Fig. 6.11) that contains rather more of the latter isomer.

A variable amount of $5\alpha(H),14\beta(H),17\beta(H)$-steranes may be present at the end of diagenesis. A possible source of these components at a relatively early stage of maturity is the reduction of Δ^{14} sterenes produced from Δ^7 stenols, a much simplified scheme for which is shown in Fig. 7.4 and involves spirosterene intermediates (Peakman and Maxwell, 1988; Peakman et al., 1989). Spirosterenes may be important intermediates in many sterene rearrangements, even though they are not themselves preserved in substantial amounts at the end of diagenesis (e.g. as spirosteranes).

Triterpanes   Isomerisation processes also affect terpanes. The diagenetic product of triterpenoids with the β-amyrin skeleton is $18\beta(H)$-oleanane; however, subsequent isomerisation at the C-18 position during catagenesis yields a final equilibrium mixture favouring the $18\alpha(H)$ epimer (Fig. 6.14).

In C_{31}–C_{35} hopanes the biologically conferred 22R configuration is preserved during the initial stages of diagenesis. Subsequent isomerisation results in a final equilibrium mixture containing approximately equal amounts of 22R and 22S isomers.

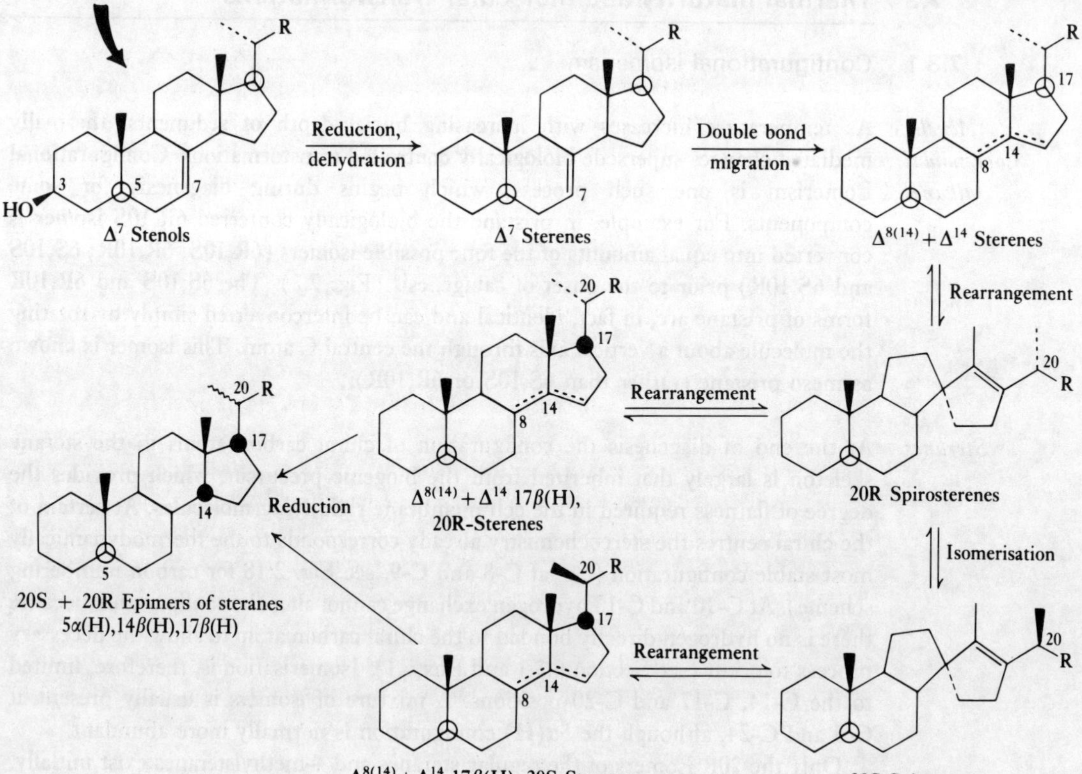

Figure 7.4   Possible origin of 20S and 20R epimers of $14\beta(H),17\beta(H)$-steranes in immature sediments from Δ^7 stenols. (N.B. The Δ^{14} $17\alpha(H)$,20R-sterene intermediates can give rise to $14\alpha(H)$, $17\alpha(H)$,20R-steranes upon reduction, as an alternative to undergoing rearrangement to spirosterenes.)

At the C-17 and C-21 positions (see Fig. 2.18 for carbon numbering scheme) the configurations are initially mainly $17\beta(H),21\beta(H)$. The $17\beta(H),21\beta(H)$ isomer is much less thermally stable than either the $17\beta(H),21\alpha(H)$ or $17\alpha(H),21\beta(H)$ isomer and is rapidly converted into a mixture of the latter two isomers. The other possible isomeric configuration, $17\alpha(H),21\alpha(H)$, does not appear to be formed at any stage. With increasing temperature the $17\beta(H),21\alpha(H)$ isomer is converted almost completely into the more stable $17\alpha(H),21\beta(H)$ isomer, so that the final equilibrium mixture is dominated by the latter (Fig. 6.15). It should be remembered that these compounds are usually collectively referred to as hopanes, regardless of configuration at the various chiral centres. Strictly speaking, however, only those with a $21\beta(H)$ configuration are hopanes, while those with a $21\alpha(H)$ configuration are called **moretanes**.

7.3.2 Aromatisation

Another important, thermally mediated, transformation is aromatisation, which we have already examined for diterpenoids (Fig. 6.13) and triterpenoids (Fig. 6.16). One of the major classes of diagenetic products of steroids is the C-ring monoaromatics (Fig. 6.11). As temperature rises with increasing burial, complete aromatisation of the three hexacyclic rings occurs, generally with the loss of the C-19 methyl group at the AB-ring junction. Only small amounts of BC-ring diaromatics have been detected, suggesting that these intermediates in the aromatisation process are short-lived. Aromatisation is not a reversible reaction, unlike isomerisation. Initially there are no triaromatic steroidal hydrocarbons, but during catagenesis C-ring monoaromatics tend to be converted into triaromatics. Other series of triaromatics are often observed in oils, in which there is an additional methyl group attached to the ring system at C-1 (possibly from migration rather than loss of the methyl group on C-10) or C-4 (possibly from 4-methylsteroids).

In humic coals aromatisation of cycloalkanes in general increases throughout coalification and is virtually complete towards the end of the bituminous coal stage.

7.3.3 Enrichment of short-chain hydrocarbons and cracking processes

Steroids   As maturity progresses through the oil window there is an apparent increase in the amounts of short-chain steroidal hydrocarbons (e.g. C_{21} and C_{22} alkanes, and C_{20} and C_{21} triaromatics). These do not appear to be formed from thermal cracking of longer-chain components, as was originally thought, and so their increase in abundance relative to the longer-chain homologues probably reflects a greater resistance towards thermal degradation. Although there is not a direct product–precursor relationship, the ratio of short-chain to long-chain components appears to be a useful indicator of maturity in closely related samples. However, the ratio in immature sediments of the same maturity but representing different sources/depositional environments can vary significantly.

Porphyrins   Porphyrin distributions in oils are dominated by C_{27}–C_{33} DPEP and etio species complexed with Ni^{2+} and VO^{2+}, and with increasing temperature the ratio DPEP : etio for vanadyl porphyrins decreases. This may in part be due to a cracking process involving cleavage of the isocyclic ring (i.e. the non-pyrrole ring) of the DPEPs to yield etio porphyrins (Fig. 6.8c), but more important may be the generally faster degradation rate for DPEP than etio porphyrins with increasing temperature. Some studies have suggested that the abundance of DPEP porphyrins decreases from the onset of oil generation, while that of etio porphyrins appears to parallel the amount of oil generated, exhibiting a maximum in the middle of the oil window (Barwise and Roberts, 1984). The latter is consistent with release of etio porphyrins from kerogen rather than an origin from cracking of previously released DPEP porphyrins.

Further studies on related oils derived from source rocks of differing maturities suggest that, as catagenesis progresses, the vanadyl porphyrins released from kerogen tend to have higher degrees of alkylation (i.e. $>C_{32}$ species) and the proportion of etio to DPEP increases. When porphyrin generation ceases or when migrated oil has become isolated from its source kerogen, terminating the supply of porphyrins, continued maturation results in a further decrease in the DPEP : etio ratio and also a decrease in average carbon number because some dealkylation occurs (Baker and Louda, 1986).

7.4   Molecular maturity and source parameters in petroleum exploration

The distribution of individual hydrocarbons, and particularly biomarkers, in oils and in bitumens isolated from source rocks can provide information on the level of maturity of the source kerogen. Biomarkers are especially useful in the correlation of oils with other oils and with their source rocks.

7.4.1   Molecular maturity parameters

Light hydrocarbons   Composition rather than amount of bitumen is a more widely applicable maturity parameter for oil source rocks. Useful maturity indications can be obtained from the light hydrocarbons up to C_8, particularly those in the C_3-C_8 range, which are virtually absent in Recent sediments but are formed in increasing amounts during catagenesis. These hydrocarbons can be divided into gaseous (C_1-C_4) and gasoline (C_4-C_8) fractions. During diagenesis the gasoline fraction is lean and methane is the main gas. During catagenesis abundant C_2-C_4 gases are generated together with a rich gasoline fraction. During metagenesis the composition of the light hydrocarbons generated is similar to that during diagenesis, with methane being the main product.

One method of assessing maturity is to construct a triangular plot from the relative amounts of C_7 components according to the three groupings: paraffins (i.e. acyclic alkanes), naphthenes (i.e. cycloalkanes) and aromatics. With increasing maturity the relative paraffinic content increases (Phillippi, 1975). This approach requires the gas chromatographic separation of all C_7 hydrocarbons from each other and from other components. It is possible to simplify the method to an examination of the *n*-hexane : methylcyclopentane ratio. However, care must be taken in applying such simplifications because the ubiquitous methyl-branched acyclic alkanes in oils may be formed by catalysed rearrangements of previously generated hydrocarbons (e.g. formation of methylhexanes and dimethylpentanes from *n*-heptane; Kissin, 1987; Mango, 1990).

Carbon preference index   The carbon preference index (CPI) is a numerical means of representing odd-over-even predominance in *n*-alkanes in a particular carbon number range. It can be used as a maturity measurement when there is an obvious OEP in C_{25}–C_{33} *n*-alkanes resulting from higher plant waxes (Bray and Evans, 1961). In this application it is calculated from the amounts of *n*-alkanes according to Eq. [7.1]:

$$CPI = \frac{1}{2}\left(\frac{C_{25} + C_{27} + C_{29} + C_{31} + C_{33}}{C_{24} + C_{26} + C_{28} + C_{30} + C_{32}} + \frac{C_{25} + C_{27} + C_{29} + C_{31} + C_{33}}{C_{26} + C_{28} + C_{30} + C_{32} + C_{34}}\right) \quad [7.1]$$

Values of CPI will initially be >1.0 but will tend towards a final value of 1.0 with increasing maturity, as the *n*-alkanes in this carbon number range become diluted by the generation of large amounts of additional *n*-alkanes in the same range without any preference for odd or even carbon numbers. However, this method is not universally applicable and a CPI of 1.0 may equally represent a sample of advanced maturity (within catagenesis) or an immature sample that did not receive a significant higher plant input.

Pristane formation index   Pristane, like other biomarker hydrocarbons, is evolved during catagenesis from suitable precursor molecules (e.g. phytyl chains on various chlorophylls) that have been incorporated into the kerogen structure. The ratio of the amount of pristane already generated to the amount of remaining precursor compounds can, therefore, provide information on the maturity of a kerogen sample. During pyrolysis of a sample, precursors are released as alkenes (prist-1-ene and prist-2-ene, Fig. 7.5) because hydrogen is not available during this rapid simulation of catagenesis for the formation of the saturated alkane (pristane) upon rupture of the bonds to the kerogen macromolecules. The pristane formation index (PFI) is calculated from the relative amounts of pristane and the two pristenes:

$$PFI = \frac{[\text{pristane}]}{[\text{pristane}] + [\text{prist-1-ene}] + [\text{prist-2-ene}]} \quad [7.2]$$

Recorded values of PFI lie in the range 0.09–0.95 and correlation with other maturity parameters has been noted (Goossens et al., 1988a, 1988b). For example, in the Mahakam Delta (Indonesia) the approximate relationship between PFI and vitrinite reflectance is:

$$PFI = 2.86(\%R_o) - 1 \quad [7.3]$$

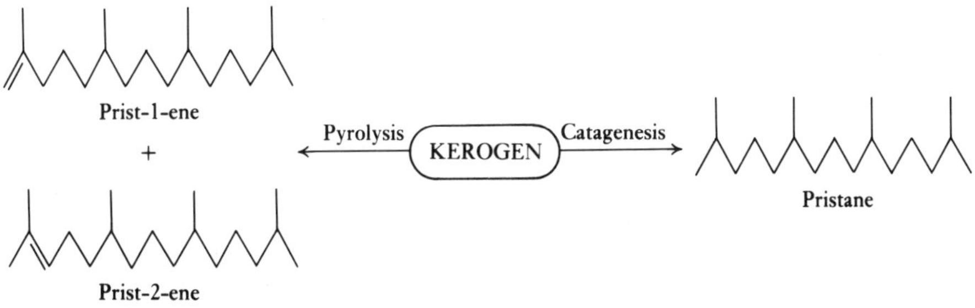

Figure 7.5   Relationship between pristane generated during catagenesis and pristenes formed during laboratory pyrolysis of kerogen.

Biomarker transformations　　Maturity measurements can be based on biomarker distributions, such as the isomerisation at cyclic and acyclic chiral centres in steranes and triterpanes, and the aromatisation of C-ring monoaromatic steroidal hydrocarbons discussed in Section 7.3. The general operational maturity ranges for some of the more commonly used molecular maturity indicators in petroleum geochemistry are given in Fig. 7.6, in comparison with vitrinite reflectance (R_o), thermal alteration index (TAI) and time–temperature index (TTI).

The most useful reactions are those in which only one component is present in the beginning in immature sediments. These include isomerisation of pristane at C-6 and C-10, of steranes at C-20 and of hopanes at C-22, and also the aromatisation of C-ring monoaromatic steroidal hydrocarbons. The components usually used for sterane and hopane maturity parameters are shown in Fig. 7.7 and are chosen for

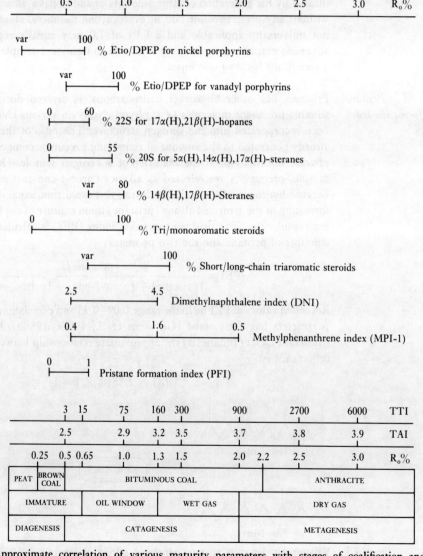

Figure 7.6　　Approximate correlation of various maturity parameters with stages of coalification and petroleum generation.

chromatographic reasons. It is not possible to separate all steranes and hopanes by the gas chromatographic methods used; some components **co-elute** (i.e. they emerge from the chromatographic column and enter the detector at the same time; see Section 7.5.2), making it impossible to determine the amount of each component accurately. The steranes and hopanes in Fig. 7.7 are generally free from such co-elution problems and so provide reliable maturity information. The analysis of steranes and hopanes by gas chromatography–mass spectrometry (GC–MS) is summarised in Section 7.5.

Methyl group isomerism in aromatic hydrocarbons

It can be seen from Fig. 7.6 that most of the biomarker hydrocarbon maturity indicators operate at relatively low maturities and have reached their end points before the end of the oil window, except for the ratio of short-chain to long-chain triaromatic steroids. This limits their use in petroleum exploration mainly to the assessment of whether potential source rocks have reached the onset of catagenesis. However, another molecular maturity parameter, not based on biomarkers, has been developed from studies on coals and also seems to be applicable to petroleum source rocks. It is the methylphenanthrene index, which is based on the relative abundances of phenanthrene and its methyl homologues. There are five possible isomers of methylphenanthrene, the 1, 2, 3, 4 and 9 isomers; all other substitution positions merely repeat these five isomers. 4-Methylphenanthrene (4-MP) is a relatively minor component in oils when present ($<1\%$ total MPs; Garrigues and Ewald, 1983), and can be discounted. Of the other four isomers, 1-MP and 9-MP have methyl groups in positions where there is some steric interaction with hydrogen atoms on adjacent carbon atoms (i.e. close proximity resulting in repulsive forces), while in 2-MP and 3-MP the methyl groups are further from these hydrogens and so their steric interactions are reduced. These steric interactions are represented in Fig. 7.8a and result in the 1 and 9 isomers being slightly less thermodynamically stable than their 2 and 3 counterparts. With increasing temperature it appears that methyl groups become mobile and consequently, with increasing maturity, 1-MP and 9-MP are gradually converted into the more stable 2-MP and 3-MP. In addition, methylation of phenanthrene appears to occur. A measure of the progress of these processes with increasing maturity can be represented by the amounts of the various components in the ratio designated MPI-1 (Radke and Welte, 1983):

$$\text{MPI-1} = 1.5 \times \left(\frac{(2\text{-MP}) + (3\text{-MP})}{(\text{P}) + (1\text{-MP}) + (9\text{-MP})} \right) \qquad [7.4]$$

MPI-1 increases up to the end of the oil window and then decreases. This decrease is due to the loss of methyl groups at higher temperatures, apparently regenerating phenanthrene. Various other ratios of methylphenanthrenes have also been used as maturity indicators but MPI-1 is probably the most commonly used. It has been calibrated against vitrinite reflectance for various types of coals (Fig. 7.8b) and it is possible to obtain a calculated value of vitrinite reflectance (R_c) from MPI-1:

$$\%R_\text{c} = 0.60(\text{MPI-1}) + 0.40 \qquad \text{(for } R_\text{o} < 1.35\%) \qquad [7.5\text{a}]$$

$$\%R_\text{c} = -0.60(\text{MPI-1}) + 2.30 \qquad \text{(for } R_\text{o} > 1.35\%) \qquad [7.5\text{b}]$$

It can be seen that it is possible for MPI-1 values to correspond to two different levels of maturity, but it is usually obvious from other maturity indicators whether

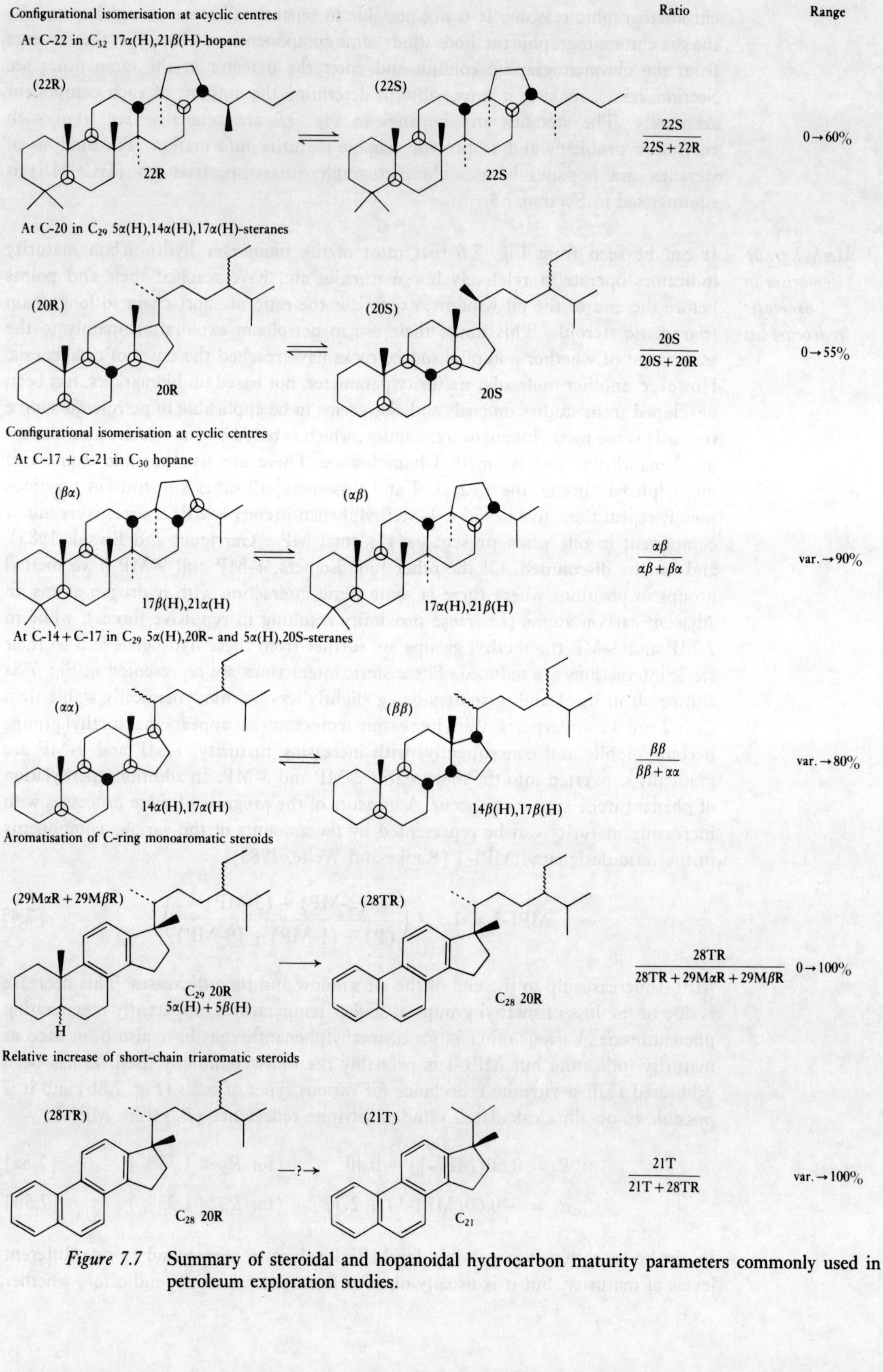
Configurational isomerisation at acyclic centres
Ratio
Range
At C-22 in C32 17α(H),21β(H)-hopane
(22R)
(22S)
22R
22S
22S/(22S+22R)
0→60%
At C-20 in C29 5α(H),14α(H),17α(H)-steranes
(20R)
(20S)
20R
20S
20S/(20S+20R)
0→55%
Configurational isomerisation at cyclic centres
At C-17 + C-21 in C30 hopane
(βα)
(αβ)
17β(H),21α(H)
17α(H),21β(H)
αβ/(αβ+βα)
var.→90%
At C-14 + C-17 in C29 5α(H),20R- and 5α(H),20S-steranes
(αα)
(ββ)
14α(H),17α(H)
14β(H),17β(H)
ββ/(ββ+αα)
var.→80%
Aromatisation of C-ring monoaromatic steroids
(29MαR + 29MβR)
(28TR)
C29 20R
5α(H) + 5β(H)
H
C28 20R
28TR/(28TR+29MαR+29MβR)
0→100%
Relative increase of short-chain triaromatic steroids
(28TR)
(21T)
—?→
C28 20R
C21
21T/(21T+28TR)
var.→100%

Figure 7.7 **Summary of steroidal and hopanoidal hydrocarbon maturity parameters commonly used in petroleum exploration studies.**

the lower or higher maturity level is correct. It should be noted that facies changes (see Box 3.3) can adversely affect the MPI-1 parameter.

Other maturity indicators have been based on the relative stabilities of methyl-substituted aromatic hydrocarbons. An example is the dimethylnaphthalene index (DNI), which is based on the decrease in the ratio of the least thermodynamically stable isomer, 1,8-dimethylnaphthalene (1,8-DMN; Fig. 7.8a), to the total amount of all dimethylnaphthalenes with increasing maturity (Alexander et al., 1984). Amounts of 1,8-DMN are usually very small in relation to other dimethylnaphthalenes (DMNs), and DMNs can be lost during the isolation procedure because they are very volatile. However, these drawbacks are ameliorated

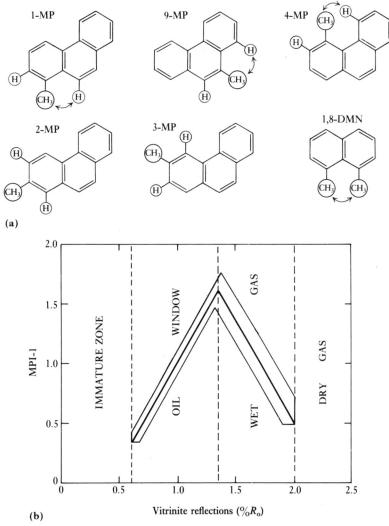

(a)

(b)

Figure 7.8 (a) Representation of major steric interactions (double-headed arrows) in methylphenanthrenes and 1,8-dimethylnaphthalene; (b) correlation of methylphenanthrene index (MPI-1) with vitrinite reflectance (R_o) for type III kerogens (shown as bold line with error spread indicated by surrounding envelope). (After Radke and Welte, 1983.)

by the use of a logarithmic scale, such that:

$$DNI = -\log\left(\frac{(1,8\text{-DMN})}{(\Sigma DMNs)}\right)$$ [7.6]

Comparison with other molecular maturity indicators suggests that DNI probably operates across the oil window, as shown in Fig. 7.6.

In evaluating maturity it is advisable to use a range of molecular, optical and pyrolytic parameters because any maturity indicator may be adversely affected by one factor or another (e.g. migration or source-related effects) and so provide misleading information when considered in isolation.

7.4.2   Effect of geothermal gradient on molecular maturity parameters

The relationships between various maturity indicators in Fig. 7.6 is intended as a guide only, because the greater temperature dependence of some processes will alter their relative operational range in basins with higher than average heat flows, as shown for sterane isomerisation at C-20 and steroid aromatisation in Fig. 7.9. Similarly, vitrinite reflectance values may underestimate maturity, in terms of the onset of catagenesis, in rapidly subsiding basins with high geothermal gradients such as the Mahakam Delta, Indonesia (Radke, 1987).

The kinetics of some of the reactions described above have been examined in laboratory simulation experiments and the results obtained compared with data from geological samples of known thermal histories. The extent of a reaction depends on both time and temperature, and so its kinetics can be described by the Arrhenius

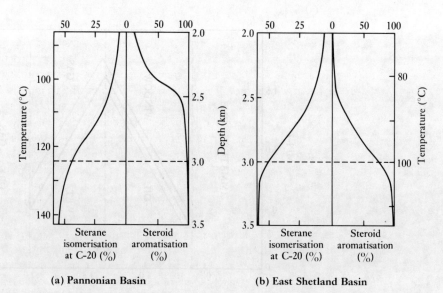

(a) Pannonian Basin   (b) East Shetland Basin

Figure 7.9   Plots of sterane isomerisation at C-20 and steroid aromatisation (calculated as in Fig. 7.7) vs. burial depth in (a) the Pannonian Basin (Pliocene (late Tertiary) deposits, geothermal gradient $\geqslant 50°C/km$) and (b) the East Shetland Basin, North Sea (Jurassic, geothermal gradient ca. $30°C/km$). Broken line represents approximate depth of onset of oil generation. (Data after Mackenzie and McKenzie, 1983; Mackenzie, 1984.)

Table 7.2 **Activation energy (E_{act}) and frequency factor (A) for some biomarker reactions** (After Mackenzie et al., 1984; Abbott et al., 1985; Goossens et al., 1988b)

Reaction	E_{act} (kJ/mol)	A (/s)
Isomerisation of 20S to 20R in steranes	91	7×10^{-3}
Isomerisation of 22S to 22R in hopanes	91	2.5×10^{-2}
Isomerisation of 6R,10S to 6R,10R + 6S,10S in pristane	120	2.1×10^{7}
Aromatisation of C-ring monoaromatic steroids	200	1.8×10^{14}
Pristane generation from kerogen (PFI)	59	2.2×10^{-7}

equation (see Box 5.2), which is characterised by the value of two constants, the activation energy (E_{act}) and the Arrhenius constant (a frequency factor, A). These constants are presented in Table 7.2 for isomerisation of hopanes at C-22 and of steranes at C-20, for aromatisation of C-ring monoaromatic steroidal hydrocarbons and for the pristane formation index.

The data in Table 7.2 allow the rate of each reaction to be calculated at a particular temperature and show that steroidal aromatisation is much more temperature-sensitive than the isomerisation reactions (i.e. it has larger E_{act} and A values). This is more obvious from the plot of $\log_e k$ vs. $1/T$ (a log plot of the Arrhenius equation) for these sterane and hopane reactions in Fig. 7.10a. Hopane isomerisation is faster than sterane isomerisation at all temperatures. In most oils the level of maturity is such that equilibrium has been reached in the isomerisation at C-22 in hopanes, while isomerisation at C-20 in steranes may not have reached equilibrium.

At temperatures above about 100°C steroidal aromatisation occurs at an increasingly faster rate than sterane isomerisation. As a result, in young extensional basins in which there is a high heat flow, such as the Pannonian Basin in Hungary, aromatisation can be virtually complete while sterane isomerisation is at an early stage (Fig. 7.9a). In older basins with more moderate heat flows, such as the North Sea (Fig. 7.9b), the steroid isomerisation and aromatisation reactions may proceed at approximately equivalent rates (Mackenzie and McKenzie, 1983). This behaviour is useful in assessing the maximum temperatures experienced by sedimentary material in different types of basins and the time periods over which these temperatures persisted (e.g. Beaumont et al., 1985). This approach is shown in Fig. 7.10b for extensional basins in which the degree of extension is 1.5, and is based on the assumption that crustal stretching typically leads to higher temperatures and is followed by uplift, usually resulting in a decrease in temperature. It can be considered that the uplift effectively freezes the steroid isomerisation and aromatisation reactions. The position at which the steroid isomerisation and aromatisation values for a particular sedimentary rock sample plots in Fig. 7.10b, therefore, gives the maximum temperature experienced prior to uplift and the time spent at this temperature (i.e. the time between extension and uplift). For example, a sterane isomerisation value of 30% and an aromatisation value of 50% suggest that a maximum temperature of 90°C was experienced for ca. 130 Ma.

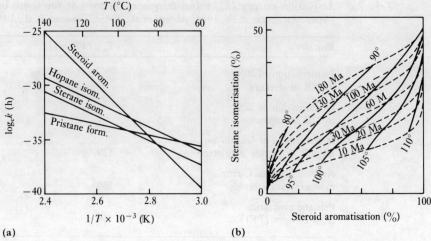

Figure 7.10   (a) Arrhenius plots for isomerisation of steranes at C-20 and of hopanes at C-22, for aromatisation of C-ring monoaromatic steroidal hydrocarbons and for the pristane formation index, showing the greater temperature dependence of the aromatisation reaction. (After Mackenzie et al., 1984; Goossens et al., 1988b.) (b) Plot of the extent of sterane isomerisation at C-20 vs. steroid aromatisation (as given by the ratios in Fig. 7.7) for a basin undergoing 50% extension. The plot point for a particular sedimentary rock sample gives the elapsed time between stretching and subsequent uplift (broken lines), and the maximum temperature prior to uplift (solid lines), based on the assumption that the cooling resulting from uplift freezes both reactions. (After Mackenzie et al., 1984.)

The pristane formation index is less sensitive to variations in temperature than the other reactions in Fig. 7.10a. In addition, it proceeds at a faster rate than the other reactions at lower temperatures (i.e. ca. $<65°C$) and consequently the cooling associated with uplift in a basin does not appear to freeze this reaction effectively. Maturity indications from PFI values can, therefore, give useful additional information when compared with steroidal isomerisation and aromatisation data, providing the basin is not so old that PFI has reached its end value (Goossens et al., 1988a, 1988b).

7.4.3   Correlation of oils and source rocks

It was noted in Section 5.6.3 that the identification of the source rock of a particular oil can be quite complicated, due to the possibility of large secondary migration distances and the existence of several potential source rocks and a number of reservoired oils. The distributions of the major hydrocarbons can be used to an extent (e.g. Phillippi, 1981), but they are often not sufficiently distinctive to permit definite correlation, although sulphur compounds can be more useful in this respect. However, biomarker hydrocarbons are particularly useful in the correlation of oils and source rocks. Correlation is achieved by comparison of the relative distributions of components within a class of biomarkers, and the presence of one or more of the less commonly occurring biomarkers (often specific to source and/or depositional environment) in a particular source rock (and hence in its related oil) is advantageous. A triangular plot of the relative amounts of $C_{27}:C_{28}:C_{29}$ steranes is a commonly employed correlation parameter, with related oils and source rocks plotting close to one another.

The variety in component distributions that is possible within the biomarker classes of steranes and triterpanes alone often permits a degree of certainty in correlation that was not possible before advances in GC–MS technology enabled the development of this type of biomarker analysis in the 1970s. Correlation is usually possible even if there are maturity differences between the bitumen in a potential source rock sample and an oil generated from the source rock succession in a deeper location. Allowance can also be made for the effects of migration on biomarker distributions. Biomarkers can be used in correlation even where some biodegradation has occurred because many are relatively resistant to this process. Under more advanced biodegradation the changes that can be expected in biomarkers are quite well understood (Section 5.4.4). However, under severe biodegradation there may be little molecular information remaining in the hydrocarbons fraction of an oil, but it may be possible to examine the biomarkers released from asphaltenes by pyrolysis. Biomarkers in asphaltenes appear to be protected from the effects of biodegradation and have similar, but not identical, distributions to those in the hydrocarbons fraction of the oil (Cassani and Eglinton, 1986).

7.5   Biomarker hydrocarbon analysis

7.5.1   Introduction

A detailed description of all the analytical techniques applied in organic geochemistry is beyond the scope of this book. However, much of the information gained from molecular distributions that has been discussed in this chapter and in Chapter 6 is obtained by the use of the separation technique of gas chromatography (GC). The choice of detector system coupled to the gas chromatograph depends on the relative concentrations of the components in a mixture. For the more abundant and readily identifiable hydrocarbon components in oil (at typical, individual component concentrations of around the per mil level up to the per cent level for n-alkanes) a flame ionisation detector (FID) is suitable. In this detector organic compounds are burnt in a hydrogen–air flame, producing ions which result in a measurable increase in current between two electrodes from as little as 100 pg (10^{-10} g) of material. Using this detector, components such as n-alkanes are identified from their recognisable elution patterns. Terpenoidal and steroidal hydrocarbons are generally present only in trace quantities, at concentrations of around two orders of magnitude lower than the more abundant components, and require a more specific and sensitive detector, a mass spectrometer. The combined gas chromatography–mass spectrometry method (GC–MS) allows terpenoids and steroids to be analysed effectively free of interference from the more abundant components.

7.5.2   Gas chromatography–mass spectrometry

Gas chromatography is a typical chromatographic technique in that a mixture of components is passed over the surface of an immobile material (or stationary phase, a solid or liquid) which has varying affinity for the different compound types in the mixture. The mixture is moved over the surface of the stationary phase in a suitable fluid (a gas in the case of GC), which is termed the mobile phase, and separation of components results from their differing degrees of retention by the stationary phase. In the application of GC to liquid hydrocarbons analysis, the stationary phase is in the form of a thin film lining the interior wall of a long, open tubular, capillary column. Columns are typically made of vitreous silica, ca. 25 m long with an internal diameter of ca. 0.25 mm, and are coiled up for convenience. The stationary phase is often a methylsilicone liquid film of ca. 0.25 μm thickness, which can be immobilised by direct chemical bonding to the glass wall of the column. The mobile phase is an inert gas, usually helium, passed through the column under pressure. This separation technique is strictly called gas–liquid chromatography because it involves a gaseous mobile phase (also termed the carrier gas) and a liquid stationary phase.

Ideally, the individual compounds present in a mixture emerge from the end of the column at varying intervals, carried by the mobile phase (the process of **elution**), with no two compounds eluting at the same time. Components are separated as a result of spending different lengths of time in the stationary phase. As a compound can only move along the column when in the gaseous phase, the time it takes to elute (its retention time) depends on its vapour pressure (i.e. boiling point) and its chemical affinity for the stationary phase (i.e. solubility). The lower the vapour pressure (i.e. the higher the boiling point) and/or the higher the solubility of a component, the longer it will take to elute because the proportion of time it spends in the gaseous phase will be lower. For a homologous series like the n-alkanes, the chemical nature of individual members is similar and so their interaction with the stationary phase is similar. Their order of elution is, therefore, effectively governed by their volatility, which decreases with increasing carbon number. The structure of a hydrocarbon can influence its interaction with the stationary phase and so diastereomers can potentially be resolved.

A mixture is introduced in solution in a small amount of a suitable solvent (e.g. ca. 1 μl hexane) at the head of the column. A common introduction system is that of syringe injection through a septum (which causes minimum disruption of carrier gas flow) into an injector port which is heated (ca. 300°C) so that all the sample vaporises rapidly and enters the column in a short period. Gas–liquid chromatography requires components to spend time in the gaseous phase, so they must exhibit sufficiently high vapour pressures. Involatile compounds cannot be analysed by this technique as they would never emerge from the end of the column. The elution of the less volatile components in a mixture is aided by increasing the column temperature uniformly throughout the analysis. The time between the emergence of the first and last molecules of a single compound from the end of the column is very short (ca. 5 s).

Upon leaving the GC column, molecules enter the ionisation region of a mass spectrometer. Here they are bombarded with fast electrons which expel an electron from each molecule, producing positively charged molecular ions which are accelerated in the direction of a detector system. The mass spectrometer operates at very low pressure (i.e. near perfect vacuum) to allow the unimpeded passage of ions to the detector. The molecular ions carry a lot of energy and, therefore, are usually quite unstable, undergoing fragmentation in various ways into smaller, relatively stable ions. By the application of various electromagnetic fields the trajectories of these ions can be deflected, the amount of deflection depending on

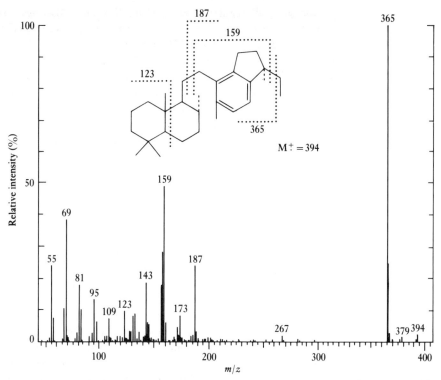

Figure 7.11   Mass spectrum of a C_{29} D-ring monoaromatic 8,14-secohopanoidal hydrocarbon. The origins of some major ions from fragmentation of the molecular ion (M^+, m/z 394) are shown. (After Killops, 1991.)

the mass : charge ratio (represented by m/z, or sometimes by m/e) of an individual ion. At any instant these fields cause ions of just one m/z value to impinge on a detector, usually an electron multiplier, which produces a measurable current. By varying the fields a sweep over a range of m/z values can be achieved over a specific time period (usually ca. 1 s). The ion current, which is proportional to the number of ions produced, is recorded for each m/z value during the sweep, allowing a mass spectrum to be constructed like that for the C_{29} D-ring monoaromatic 8,14-secohopanoid in Fig. 7.11. A plot (or **chromatogram**) of total ion current vs. time is equivalent to the chromatogram obtained from a FID.

7.5.3   Evaluation of biomarker distributions

The fragment ions produced from a molecular ion are characteristic of the structure of the compound and can aid the structural identification of a previously unidentified compound. As can be seen from Fig. 7.11, some fragment ions are particularly abundant and some of them may be shared by related compounds that have structural units in common. For example, in Fig. 7.11 the most intense ion (base peak) at m/z 365 and the second most abundant ion at m/z 159 are found in all the C_{29}–C_{35} D-ring monoaromatic 8,14-secohopanoids of this type (the series being formed by elongation of the ethyl side chain on the E-ring).

Similarly, all regular steranes share the same AB-rings structure and methyl substitution pattern and produce an abundant and characteristic fragment ion of m/z 217, resulting from the AB-rings fragment (see Fig. 7.12). The corresponding ion for 4-methylsteranes is of m/z 231, but it can readily lose a methyl group to generate the m/z 217 fragment ion. Diasteranes also produce an abundant m/z 217 fragment ion. Most tricyclic, tetracyclic and pentacyclic terpanes yield a common and abundant AB-rings fragment ion of m/z 191 (see Fig. 7.12). Few other compounds produce these fragment ions in any comparable abundance and so the response of the mass spectrometer to m/z 191 plotted against time (known as a single ion chromatogram or mass chromatogram) reveals the distribution of most diterpanes and triterpanes, while the corresponding chromatogram of m/z 217 gives the distribution of steranes. This is a useful feature as it allows distributions of a particular biomarker hydrocarbon class to be seen without the interference of other, more abundant components that cannot form an ion of suitable m/z value, even if such components co-elute with any of the steranes or terpanes. In addition to providing this selectivity of detection, the mass spectrometer can provide improved sensitivity of detection if it is set to monitor just a few, abundant fragment ions, such as m/z 191 and m/z 217, rather than a wide range of m/z values.

The identity of many steranes and terpanes has been established and so they can be recognised by their characteristic elution patterns on methylsilicone stationary phase. An example of these patterns is shown in Fig. 7.12 for a Lower Liassic paper shale (from Kilve, North Somerset, UK). The relative immaturity of the shale is indicated by the dominance of the $5\alpha(H),14\alpha(H),17\alpha(H)$ configuration for the C_{27}–C_{29} steranes (Fig. 7.12c). The elution ranges of the terpanes and steranes are indicated on the FID chromatogram for the aliphatic hydrocarbons (Fig. 7.12a), which emphasises the trace levels of biomarker hydrocarbons in this sample relative to the major components, n-alkanes.

Other biomarker hydrocarbon classes can be monitored similarly according to their characteristic fragment ion m/z values, some of which are given in Table 7.3.

Table 7.3   **Characteristic fragment ions of some biomarker hydrocarbons**

Biomarker hydrocarbon class	Fragment ion m/z
Acyclic isoprenoids	113, 183
Dicyclic sesquiterpanes	123
Tricyclic terpanes	191
Tetracyclic terpanes	191
Pentacyclic triterpanes (including hopanes)	191
25-Norhopanes	177
2-/3-Methylhopanes	205
8,14-Secotriterpanes	123
Benzohopanes	191
D-Ring monoaromatic 8,14-secohopanoids	159, 365
$14\alpha(H),17\alpha(H)$-Steranes	217
$14\beta(H),17\beta(H)$-Steranes	217, 218
Diasteranes	217, 259
4-Methylsteranes	217, 231
C-Ring monoaromatic steroids	253
Triaromatic steroids	231
Triaromatic 1-/4-methylsteroids	245

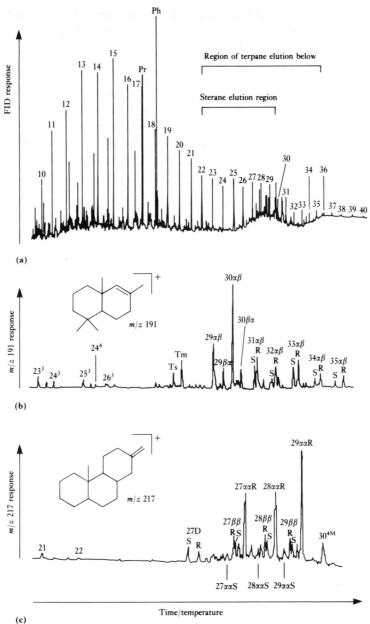

Figure 7.12   Aliphatic hydrocarbon distributions in a Liassic shale (Bridgwater Bay, Severn Estuary, UK). (a) Gas chromatogram of aliphatic hydrocarbons, with *n*-alkanes indicated by carbon number and pristane and phytane labelled Pr and Ph, respectively. (b) Terpane distributions shown by *m/z* 191 ion chromatogram. Tricyclic terpanes are indicated by carbon number with superscript 3, while C_{24} tetracyclic terpane is labelled 24^4. Ts and Tm are C_{27} pentacyclic triterpanes with hopanoidal skeletons. Other major components are C_{29}–C_{35} hopanes, labelled according to configuration at C-17 and C-21, and also at C-22 where appropriate. (c) Sterane distributions shown by *m/z* 217 ion chromatogram, with components labelled according to carbon number. Configuration at C-14, C-17 and C-20 is given for the C_{27}–C_{29} 5α(H)-steranes. The 20S and 20R isomers of C_{27} 13β(H),17α(H)-diasterane are labelled 27DS and 27DR, respectively. A C_{30} 4-methylsterane is denoted by 30^{4M}. The elution regions covered by (b) and (c) are indicated in (a).

It should be noted that the quantitative data obtained from peak area response in single ion chromatograms, which are used in evaluating the maturity parameters in Fig. 7.7, do not represent the absolute abundance of the components concerned. These fragment ions are not necessarily produced in the same total abundance during the ionisation of an equal amount of each component in the mass spectrometer (i.e. response factors differ). An example is provided by $17\alpha(H),21\beta(H)$-hopane in Fig. 7.12b (i.e. $30\alpha\beta$), which exhibits the greatest m/z 191 response. This is because, upon fragmentation, a DE-rings fragment ion is produced with the same m/z value of 191 as the AB-rings fragment ion, but in other hopanes the corresponding DE-rings fragment ion has different m/z values (e.g. 177 for C_{29} and 205 for C_{31} components) due to variation in length of the alkyl chain on the E-ring. Hence, for the same absolute amount, the C_{30} hopane gives a relatively enhanced m/z 191 response. This behaviour does not matter, fortunately, as long as the starting and end point values are known for the ratios based on the relative single ion responses.

Chapter 8

Environmental behaviour of anthropogenic organic compounds

8.1 Introduction

The preceding chapters have been concerned with the behaviour of biogenic compounds in the geosphere, and it is fitting to conclude this book by considering the environmental impact of organic compounds resulting from human activities. Inputs of man-made components to the environment are termed **anthropogenic**. Many anthropogenic compounds, such as CO_2 from fossil fuel combustion, also occur naturally. We shall begin by considering how anthropogenic inputs of such compounds influence the carbon cycle on global and localised scales. Some anthropogenic organic compounds, such as the insecticide DDT, do not have a natural source; they owe their existence entirely to humans and are, therefore, sometimes termed **xenobiotic**. We shall examine the environmental behaviour of some of these compounds which, although quantitatively minor anthropogenic inputs, nevertheless have significant effects on the biosphere.

8.2 Human influence on the carbon cycle

8.2.1 Carbon dioxide and the greenhouse effect

In Chapter 1 the global carbon cycle was treated as a steady state system. However, human activity, particularly since the industrial revolution in the mid-1700s, has had a significant influence on the size of some of the fluxes between various carbon reservoirs. The most obvious effect is the imbalance between fluxes of CO_2 into and out of the atmosphere (Fig. 1.1). There appears to be a net input of CO_2 into the atmosphere of 3 Gt C/a, chiefly resulting from fossil fuel burning, which has thrown the carbon cycle out of its steady state. The level of CO_2 in the atmosphere prior to the industrial revolution has been estimated at ca. 290 ppm (Freyer, 1979a). However, earlier agricultural practice and deforestation dating back to Neolithic times (ca. 10 ka ago) have also affected the carbon cycle. Burning timber during deforestation returns the carbon stored in woody tissue to the atmosphere as CO_2, while the crops that replace the forests do not provide any long-term storage capacity. When these effects are taken into consideration the level of atmospheric CO_2 before humans made a significant impact may have been ca. 265 ppm. By 1976 ca. 140 Gt of CO_2 (ca. 38 Gt C) had accumulated in the atmosphere from human activities.

In absolute terms, anthropogenic additions of CO_2 to the atmosphere are relatively small, and levels of the gas have varied widely throughout the Earth's history. For example, compared with present levels, CO_2 concentration was approximately double at the start of the Cambrian (570 Ma ago), rose to ca. 18 times greater by the end of the Cambrian (500 Ma ago), had fallen to less than double during the Carboniferous (300 Ma ago) and by the mid Cretaceous had risen again to about five times current levels. However, even the relatively small increases resulting from anthropogenic inputs of CO_2 may have significant climatic repercussions for humanity.

The atmosphere presently contains ca. 2200 Gt of CO_2, corresponding to ca. 345 ppm, and there is an annual input of ca. 1–1.5 ppm, ca. 80% of which is from fossil fuel burning (ca. 5.2 Gt/a) and about 20% from deforestation and agricultural land clearance (ca. 1 Gt/a). However, the observed increase in atmospheric levels is only about half this value. Half the anthropogenic input appears to be absorbed by the oceans, to take part in the carbonate–bicarbonate equilibrium (Eq. [1.1]). Increased phytoplankton productivity may also aid the uptake of CO_2 by the oceans. Unfortunately, it is difficult to estimate global atmospheric CO_2 levels accurately, as superimposed on spatial (e.g. latitudinal) variations there are temporal (diurnal and seasonal) variations. In addition, there have also been variations in natural fluxes of CO_2 over the last 150 ka which may have caused climatic changes. Predicting the likely effects of continued inputs of anthropogenic CO_2 at current rates is, therefore, difficult. It is not known whether the increased oceanic uptake represents a long-term sink, or whether this CO_2 may be released if global warming occurs (as the solubility of gases in water decreases with increasing water temperature).

It is difficult to think of carbon dioxide as a pollutant because it occurs naturally and is a vital part of the photosynthesis and respiration cycle. However, a major effect of increasing atmospheric levels of this gas is an increase in surface temperature on a global scale. This 'greenhouse effect' arises from the balance between the amount of incident solar energy (insolation) adsorbed by the Earth and the amount of energy that it emits back into space. The high surface temperature of the sun (ca. 6000 K) means that solar energy is confined to the shorter wavelength region of the electromagnetic spectrum, with a maximum in the visible region (Fig. 8.1a). Some of the incident solar energy (totalling 342 W/m^2 a) is reflected back into space by clouds, atmospheric dust and also the ground (a total of 103 W/m^2 a), but the remainder heats up the Earth's surface (239 W/m^2 a). Because the Earth is warmer than surrounding space it radiates energy back into space, but at longer wavelengths than the incoming radiation, in the infrared region of the electromagnetic spectrum (like heat emission from an electric fire), due to its relatively low surface temperature (ca. 288 K). An exact balance between this back radiation and adsorbed insolation would maintain a constant mean temperature at the Earth's surface.

While the gases in the atmosphere cause only minor attenuation of the incoming energy at short wavelength, they can adsorb significant amounts of the longer wavelength, outgoing energy. This is because infrared wavelengths correspond to the resonant energies associated with vibration of chemical bonds. The energy adsorbed by gas molecules is re-emitted, but equally in all directions, and so roughly half of this energy (ca. 148 W/m^2 a) is trapped and travels back to the Earth's surface. The trapping of this back radiation by molecules of various gases is so effective that only a narrow window (ca. 8–12 nm; Fig. 8.1b) exists by which radiation can escape to space. Increase in atmospheric CO_2 results in an increase in trapped radiation, causing an increase in the Earth's surface temperature, an effect analogous to that caused by the glass in a greenhouse. As temperature rises so does the level of back radiation from the Earth's surface so that, if atmospheric CO_2 levels were allowed to stabilise at a constant higher value, an equilibrium would once again be achieved between incident and emitted energy fluxes, but at a higher surface temperature.

It is predicted that, with continued emission of anthropogenic gases (CO_2 and others, see Table 8.2), the mean global temperature will probably have risen by $1-5°C$ by the year 2050 (Dickinson and Cicerone, 1986). However, the effects of rising mean temperature on climate patterns, cloud cover and ocean current systems make it extremely difficult to predict local temperature changes, which may actually decrease in some areas while increasing significantly in others. Nevertheless, temperature changes and increased CO_2 levels are likely to have marked effects on agriculture in terms of the types of crops that can be grown in a particular area (which will also be affected by changes in rainfall patterns) and in controlling weeds (which may compete more effectively than at present against currently cultivated crops). Increased mean global temperature will also result in thermal expansion of the water in the oceans, causing a rise in sea level on top of any due to melting of the polar ice-caps.

In order to halt the increase in atmospheric CO_2 levels either the rate of output from the atmosphere into sinks must be increased or the rate of input must be

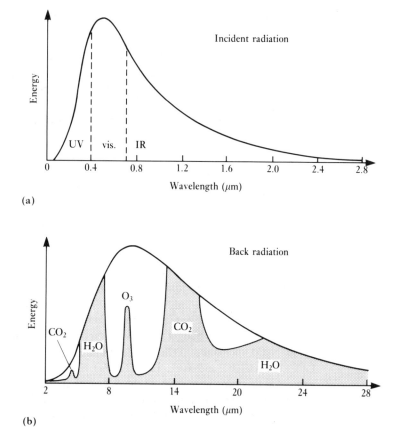

Figure 8.1   Variation of radiation intensity with wavelength for (**a**) incident solar radiation on and (**b**) back radiation from the Earth's surface. The latter shows adsorption (stippled areas) by atmospheric gases.

Table 8.1   **Anthropogenic sources of carbon dioxide emissions in the UK** (After Aldhous, 1990)

Source	Percentage of total emissions
Power stations	31
Other industries	25
Road transport	17.5
Domestic	14
Commercial/public services	6.5
Refineries	4
Shipping	2

decreased. As has been mentioned, the former appears to have occurred to an extent in the form of increased oceanic uptake. Whether this represents a long-term sink in organic-rich sediments via increased primary production is not clear. Interestingly, this uptake of anthropogenic CO_2 may be aided by the increased supplies of biolimiting nutrients entering the seas from anthropogenic sources. Decreasing anthropogenic emissions of CO_2 would appear to be the most reliable way of halting global warming, and analysis of major sources of CO_2 is required to determine the most effective remedial action. Major anthropogenic sources for the UK are shown in Table 8.1 and suggest that the following steps are likely to bring about the greatest reduction in emissions:

(a) a lowering of demands for power by increased efficiency of domestic and industrial heating, insulation and lighting systems;

(b) increased exploitation of alternative, 'clean' power sources to fossil fuels (e.g. winds and tides);

(c) more efficient industrial machinery and less wasteful processes (i.e. greater recycling of materials and increased life expectancy of consumer items);

(d) a switch to more energy-efficient transportation systems.

There is also some scope for improvement in the way that fossil fuels are used. Combustion of methane provides more energy and less carbon dioxide than burning oil and coal. However, out of a world fuel consumption equivalent to 7811 Mt oil in 1987, only ca. 20% was provided by methane, while approximately equal amounts of oil and coal together provided ca. 68%.

8.2.2   Effects of other trace gases on global warming

Other gases that are present in the atmosphere at sufficient concentrations to influence the Earth's thermal budget are water vapour, ozone, nitrogen(I) oxide (nitrous oxide, N_2O), methane and the most commonly used **chlorofluorocarbons** (CFCs), CFC-11 (CCl_3F) and CFC-12 (CCl_2F_2). In terms of radiation trapping, water vapour is the most important of these gases but the other gases, despite their low relative concentrations, can have a significant effect. Table 8.2 compares the radiation trapping capacity of these gases and carbon dioxide at 1985 concentrations with estimated values for the pre-industrial period and projected values for the year 2050. It can be seen from Table 8.2 that an increase in concentration for each of the trace gases has a disproportionately greater effect on radiation trapping than an equivalent increase in CO_2 levels. This is because radiation trapping increases approximately

Table 8.2 **Estimated changes in trace tropospheric gas concentrations and their contributions to radiation trapping in pre-industrial times and in 2050 compared with 1985 values** (After Dickinson and Cicerone, 1986)

	1985		Pre-industrial		2050	
	Concn. (ppb)	Trapped heat (W/m²)	Concn. (ppb)	Decrease in trapped heat cf. 1985 (W/m²)	Concn. (ppb)	Increase in trapped heat cf. 1985 (W/m²)
Carbon dioxide	345 000	~50	275 000	1.3	$(4-6) \times 10^5$	0.9–3.2
Methane	1 700	1.7	700	0.6	2100–4000	0.2–0.9
Ozone	10–100	1.3	0–25% less	0–0.2	15–50% more	0.2–0.6
Nitrous oxide	304	1.3	285	0.05	350–450	0.1–0.3
CFC-11	0.22	0.06	0	0.06	0.7–3.0	0.2–0.7
CFC-12	0.38	0.12	0	0.12	2.0–4.8	0.6–1.4

uniformly with concentration for CFC-11, CFC-12 and ozone, with the square root of concentration for methane and nitrous oxide, but only with the logarithm of concentration for CO_2. As methane and chlorofluorocarbon levels are projected to rise more rapidly than CO_2, by 2050 the increase in trapped radiation due to methane and CFC-11 could be 25–70% of that due to carbon dioxide. The concentrations of gases in Table 8.2 for 2050 are based on rough estimates of projected consumption levels of fossil fuels and CFCs derived from previous increases, and do not take into account possible legislatively enforced reduction in emissions (Dickinson and Cicerone, 1986).

The data in Table 8.2 indicate the warming (i.e. trapped radiation) attributable to a given concentration of trace gas at a particular time. However, the overall effect of a given amount of a particular gas on global warming also depends on its residence time in the atmosphere. Table 8.3 compares the **global warming potential** (GWP) of various trace gases to carbon dioxide (arbitrarily assigned a GWP value of 1.0) for the same number of molecules of each gas. Atmospheric residence time is taken into account by GWP, from which the importance of CFC emissions becomes more apparent.

Methane   After carbon dioxide, methane is the next most significant contributor to global warming at present. There are a variety of sources for methane entering the atmosphere, both natural and anthropogenic. It is essential to quantify all inputs as

Table 8.3 **Global warming potential (GWP) of various tropospheric gases, molecule for molecule, with CO_2 as a reference** (After Lashof and Ahuja, 1990; Rodhe, 1990)

Gas	Residence time (a)	1990 Concn. (ppb by vol.)	Increase (% per a)	1990 Contribution to warming (%)	GWP (mol : mol cf. CO_2)
Carbon dioxide	230	353 000	0.5	60	1
Methane	14	1 700	1.0	15	4
Nitrous oxide	160	310	0.2	5	180
Ozone	0.1	10–50	0.5	8	4
CFC-11	60	0.28	4.0	4	4 000
CFC-12	120	0.48	4.0	8	10 000

accurately as possible to determine the importance of individual anthropogenic inputs and so to allow evaluation of the best approach to limiting emissions. Unfortunately, estimation of fluxes is very difficult and the values given in Table 8.4 may prove to be dramatically inaccurate for some sources.

Some biogenic sources of methane can be considered primarily anthropogenic because of the significant influence of human activities. For example, methanogenesis occurs in paddy fields under anaerobic conditions during flooding. Much of the methane produced appears to be oxidised by bacteria (methylotrophs) and the amount reaching the atmosphere is seasonally variable. Estimates of methane emissions reaching the atmosphere from this source are in the range $5-25$ mg/m^2 h, but taking a mean value gives a total annual flux of 110 Mt (i.e. 110×10^{12} g). Other important anthropogenic sources of methane arise from methanogenesis in land-fill sites and from vegetation burning during deforestation/land clearance (which involves incomplete oxidation, so that CO_2 is not the only product). There is a significant production of enteric methane in ruminants, which can also be considered as anthropogenic because it arises mainly from domestic animals (chiefly cattle). This input accounts for ca. $15-20\%$ of the total methane flux. Probably ca. $2-2.5\%$ of natural gas production is lost to the atmosphere, and methane is also emitted during incomplete combustion of fossil fuels.

A number of methane sources can be classified as predominantly natural. For example, oceanic surface waters are oversaturated with respect to methane, due to bacterial (methanogenic) activity in localised anaerobic environments, such as the digestive systems of zooplankton. This results in a net flux of methane to the air. Methane is similarly produced in freshwater environments and is voided into the atmosphere. Deep ocean waters contain much lower methane concentrations than surface waters and the methane generated within anaerobic sediments and released to the water column is largely oxidised by methylotrophic bacteria. In the oceans there are localised hydrothermal sources of methane but they do not make a significant contribution to atmospheric methane inputs; nor do deep mantle (i.e. abiogenic) sources. Marine and lacustrine environments as a whole do not make a large contribution to the methane flux, but natural wetlands do. The bacterial gases produced by these and other anaerobic soils are dominated by methane but contain other hydrocarbons (e.g. ethane and propane).

Table 8.4　**Major sources of methane, their mean carbon isotope signatures and estimated contributions to total methane flux entering the atmosphere** (After Whiticar, 1990)

Source	$\delta^{13}C^o/_{oo}$	Flux into atmosphere (Mt/a)	Flux into atmosphere $(\% \ of \ total)$
Termites	−70.0	40	7.4
Rice paddies	−63.0	110	20.4
Ruminants	−60.0	80	14.8
Oceans	−60.0	10	1.9
Methane hydrates	−60.0	5	0.9
Wetlands	−58.3	115	21.3
Fresh waters	−58.0	5	0.9
Land-fill	−55.0	40	7.4
Natural gas	−44.0	45	8.3
Coal	−37.0	35	6.5
Vegetation burning	−25.0	55	10.2

The methane produced in some deep-sea sediments can be trapped within the crystalline structure of an ice layer (**methane hydrate**) which is stable under certain depth-related pressure and temperature conditions. A significant amount of methane is similarly trapped by tundra permafrost. While these methane hydrates do not make a major contribution to atmospheric methane at present, their potential for releasing significant amounts of methane as a result of ice melting during global warming is important.

The quantity of methane generated by enteric bacteria in termites is extremely difficult to estimate. Production figures as high as $75-310$ Mt/a have been postulated, which mean that termites could account for $> 50\%$ of the total flux. The total annual input of methane from all sources to the atmosphere from Table 8.4 is 540 Mt, while the estimated output from atmosphere to sinks is 500 Mt. A major sink for atmospheric methane is oxidation by highly reactive hydroxyl radicals (the unpaired electron in which can be represented by $OH\cdot$), formed by photodissociation of water molecules in the atmosphere. The potential inaccuracies in flux data can be seen by comparing the observed carbon isotopic signature of atmospheric methane of $-47\%_{00}$ with that calculated from the data in Table 8.4 of ca. $-54\%_{00}$ (the latter is actually equivalent to $-58\%_{00}$ upon correcting for the kinetic isotope effect (see Box 1.5) that operates during the hydroxyl abstraction reaction). There are clearly major gaps in our understanding of the pathways of methane into and out of the atmosphere and the fluxes involved. Unfortunately, models for the environmental behaviour of many anthropogenic substances suffer from similar inadequacies in information on pathways and fluxes.

As well as influencing global temperatures by the trapping of thermal radiation, increasing methane levels in the atmosphere will influence atmospheric chemistry in two further ways: firstly by the increase in its oxidation products, such as H_2O, CO, and CO_2, and secondly by the rapid abstraction of hydroxyl radicals, which are important in a variety of oxidation processes and in ozone chemistry.

Carbon monoxide Carbon monoxide has a GWP value of 2.2 (Lashof and Ahuja, 1990) and its chief anthropogenic source is the incomplete combustion of fossil fuels, amounting to $\geqslant 450$ Mt/a. Natural sources, however, outweigh anthropogenic sources. For example, atmospheric oxidation of methane to carbon monoxide (mainly by photolysis involving hydroxyl radicals) probably contributes $\geqslant 1500$ Mt/a. Microbial activity in ocean surface waters leads to a net transfer of CO from sea to air of a further ca. 100 Mt/a. Despite these large inputs, atmospheric CO levels do not appear to be increasing, which implies that various sinks are removing CO as fast as it enters the atmospheric reservoir. Among these sinks are oxidation to CO_2 by soil bacteria and uptake of CO by plants, but the major sink appears to be hydroxyl radical oxidation to CO_2 (Freyer, 1979b). In terms of its contribution to global warming, therefore, it is convenient to consider anthropogenic CO as simply an additional source of CO_2.

Dimethyl sulphide Dimethyl sulphide (CH_3—S—CH_3 or DMS), like carbon monoxide, is produced by micro-organisms in ocean surface waters and exhibits a net flux from ocean to atmosphere. The precursor of DMS (dimethylsulphoniopropionate, DMSP) is produced by certain phytoplankton and DMS is liberated during the decay of these organisms and as a by-product of their ingestion by zooplankton. Surface waters become enriched in DMS, leading to a net flux to the atmosphere. An important factor determining the rate of this gaseous transfer is the concentration of DMS in surface water but processes that control it are, as yet, poorly understood (Kiene and Bates, 1990).

DMS is quantitatively the most important volatile sulphur compound involved in air–sea transfer processes, its flux amounting to ca. 30–50 Mt/a (Liss, 1983). This corresponds to ca. 15–25 Mt/a of S, estimated to account for nearly half the total biogenic input of sulphur to the atmosphere. In comparison, volcanic emissions of sulphur to the atmosphere are of a similar size, while total anthropogenic emissions (particularly SO_2 from coal and oil burning) amount to ca. 100 Mt/a (Bolin et al., 1983). In the atmosphere DMS undergoes oxidation and some of the products (e.g. methanesulphonate, $CH_3SO_3^-$, or MSA) are believed to act as cloud condensation nuclei (CCN; Ayers and Gras, 1991). It is probable, then, that higher productivity among DMS-producing phytoplankton results in more CCN and hence increased cloud cover, which in turn causes a decrease in incident radiation levels, possibly reducing heating of the Earth. However, lower light levels would provide a negative feedback on phytoplankton production, reducing both CO_2 uptake and amelioration of the greenhouse effect. It is not clear, at present, whether these opposing effects are likely to offset or reinforce global warming, or whether they have any significant influence at all. Any effects are likely to be most strongly felt in the tropics, where DMS transfer to the atmosphere is highest, partly due to the latitudinal distribution of phytoplankton species (such as coccolithophores) that produce most DMS (Charlson et al., 1987).

8.2.3   Eutrophication

On a more localised basis, human activity can affect the carbon cycle indirectly by increasing the supply of mineral nutrients to aquatic environments. The high nutrient supply supports a high level of primary productivity by phytoplankton (algae and cyanobacteria), a process described as **eutrophication**. Under natural conditions, eutrophication is usually seasonal, being related to the stratification of the water column and the succession of phytoplankton (see Section 3.2.4). When nutrients are exhausted in the upper layers of water in thermally stratified lakes the phytoplankton bloom, which can form a substantial mat of material on the water surface, comes to an end. The cycle can only begin again when light levels are adequate and there are sufficient nutrients in the surface waters, resulting from mixing of the water column when strong winds and cooling destroy the stratification. However, when a constant anthropogenic supply of nutrients is available a high level of productivity can be sustained over longer periods, with seasonal variation in light levels probably becoming the limiting growth factor.

Eutrophication of still, freshwater bodies has become more of a problem in recent years with increased inputs of mineral nutrients via rivers and streams. Important anthropogenic sources of these nutrients include run-off of agricultural applications (nitrates and phosphates) and domestic waste water (which contains phosphates from cleaning agents). Degrading organic detritus, such as sewage, can release large amounts of water-soluble nutrients such as nitrate and phosphate. These nutrients are present in discharges from plants treating domestic sewage and in run-off of rain water that has leached agriculturally derived sewage. The dependence of algal growth on nutrients can readily be seen by the lush green areas of growth around outfalls of untreated sewage in coastal areas.

Aerobic decomposition of large amounts of detritus resulting from phytoplankton blooms in still bodies of fresh water can lead to severe depletion of oxygen in the water column, causing high mortality among fish. Under the right conditions, the anoxicity that may be associated with eutrophication can lead to increased preservation of organic material in underlying sediments. Eutrophication can even affect stretches of rivers, usually in the summer in temperate regions, when light intensity is high and low water levels result in a sluggish flow. Oxygen enters the water from the air relatively slowly in warm calm weather when the water surface is still. Consequently, degradation of organic matter in the water column may result in oxygen depletion, but artificial oxygenation of the water can be undertaken to preserve fish stocks. The marine environment has also been affected by eutrophication; areas with restricted flow where nutrients can accumulate are particularly vulnerable. Densely inhabited and industrialised coastal areas in the Mediterranean, such as the Adriatic, have suffered from sporadic eutrophication.

Some of the species involved in phytoplankton blooms are toxic and even pose a danger to humans. Cyanobacterial blooms (of e.g. *Anabaena*) resulting from eutrophication in lakes and other still waters in late summer can release toxins into the water. Cyanobacteria occur naturally in the succession of phytoplankton species during a typical growth season in the UK, but eutrophication can cause a population explosion. The development of suitable control methods requires an understanding of the factors that affect the bloom. It is in the late summer that cyanobacteria grow vigorously, when competition with other species for light as at its lowest. The demise of other species in the phytoplankton succession (e.g. diatoms) is generally the result of depletion of silicate and nitrate within the epilimnion. Cyanobacteria contain gas vacuoles to prevent their sinking out of the photic zone and they are not affected by low nitrate levels as they can fix nitrogen directly from the atmosphere. Phosphate is, therefore, usually the biolimiting nutrient for cyanobacteria and controlling anthropogenic inputs of phosphate will reduce the likelihood of a bloom. Chemical treatments can be used to remove phosphate from effluents and to prevent remobilization of sedimentary phosphate. Alternatively, if the dominance of cyanobacteria can be sufficiently delayed their maximum density may be decreased, so reducing potential toxicity problems. This may be possible if early summer conditions can be maintained, permitting early species in the succession which can compete effectively against cyanobacteria to persist for a longer period. For example, maintaining a supply of nutrients to the epilimnion, particularly silicate, allows diatom growth to continue (Reynolds et al., 1983).

8.3   Halocarbons and ozone depletion

Ozone is a vital component of the stratosphere (see Box 8.1) because it adsorbs most of the UV radiation that is harmful to life. In contrast, it is an undesirable gas in the troposphere, the layer of the atmosphere underlying the stratosphere and in immediate contact with the Earth's surface and the biosphere (see Box 8.1), because it is toxic to organisms. Unfortunately, the burning of fossil fuels is a major source of tropospheric ozone. In addition, little of the ozone produced in the troposphere enters the stratosphere, where it could have a beneficial action. This is mainly because it is produced near the ground and, being highly reactive, it has a short life and is destroyed by photochemical reactions (some of which may play a part in the oxidative degradation of other gaseous pollutants) before it can reach the stratosphere. Also, the thermal stratification of the atmosphere acts against rapid transport of ozone and other gases between the troposphere and stratosphere.

Box 8.1

Atmospheric stratification and the ozone layer

The atmosphere is composed of a number of layers in which temperature either decreases or increases with increasing altitude. This thermal stratification into discrete layers can be likened to that of the ocean (Fig. 3.1). The lowest two layers are the **troposphere** (ca. 0–15 km altitude, where weather occurs) and the **stratosphere** (ca. 15–50 km altitude), the boundary between the two being the tropopause (Fig. 8.2). With increasing altitude in the troposphere air density and temperature both decrease until the tropopause is reached; then temperature increases again in the stratosphere, until the stratopause.

The increase in temperature in the stratosphere is due to the adsorption of UV radiation by the ozone present in this layer. The ozone is itself formed by the action of UV radiation on oxygen, which splits the molecule into two extremely reactive atoms. These atoms then react with further oxygen molecules to produce ozone:

$$O_2 \xrightarrow{\text{UV}} O + O \qquad\qquad [8.1]$$

$$O + O_2 \longrightarrow O_3 \qquad\qquad [8.2]$$

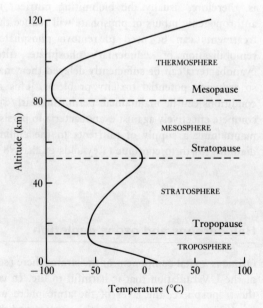

Figure 8.2   Typical variation of temperature with altitude in the atmosphere and the resulting thermal layers.

While ozone levels are increasing in the troposphere, they are being depleted in the stratosphere by the action of anthropogenic gases, particularly chlorofluorocarbons (CFCs). CFCs are still widely used as refrigerants, blowing/propulsion agents and cleaning agents for electronic components. They have long residence times in the atmosphere (ca. 60 and 120 years, respectively, for CFC-11 and CFC-12) and so are able to reach the stratosphere. However, once in the stratosphere they undergo photolytic degradation to provide chemical species that are extremely effective scavengers of ozone in reactions that take place on the surfaces of ice crystals. The main reactive species is the chlorine atom, which behaves like a catalyst in its reaction with ozone because it is regenerated and can react with many ozone molecules before finally being scavenged by other chemical species (Solomon et al., 1986). The range of reactions that can occur is complex, but the consumption of ozone can be simplified to:

$$Cl + O_3 \longrightarrow ClO + O_2 \qquad\qquad [8.3]$$

$$ClO + O \longrightarrow Cl + O_2 \qquad\qquad [8.4]$$

In effect, the reactions in Eqs [8.3] and [8.4] bring about the reverse of Eqs [8.1] and [8.2]. Another anthropogenic gas, nitric oxide (NO), scavenges ozone in a similar way (to form nitrogen dioxide, NO_2) and is produced by the dissociation of N_2O, which can be generated from fossil fuel combustion and from the decomposition of nitrate fertilisers.

CFCs are just one class of **halocarbons**, which are compounds that contain carbon and at least one halogen atom (i.e. fluorine, chlorine, bromine or iodine). Halocarbons containing bromine have also been found to destroy ozone. Because of the long-term effects of the build-up of stratospheric halocarbons and other gases on ozone levels, various measures are being taken to reduce emissions of these compounds; one measure being the introduction of less harmful alternatives in industrial applications (e.g. Prather and Watson, 1990). Alternatives to the most commonly used CFCs (CFC-11 and CFC-12) are HCFCs (hydrochlorofluorocarbons) and HFCs (hydrofluorocarbons). These compounds degrade more rapidly in the atmosphere than CFCs because of the presence of hydrogen. In particular, HFCs contain no chlorine and the hydrogen fluoride that is formed upon decomposition is not believed to be detrimental to ozone. However, the effects of these compounds on global warming may not be insignificant. For example, HCFC-22 ($CHClF_2$) has an estimated residence time in the atmosphere of 15 years and a GWP of 810 (see Section 8.2.2).

8.4 Hydrocarbon pollution in aquatic environments

8.4.1 Fossil fuel combustion

Polycyclic aromatic hydrocarbons in Recent sediments Carbon dioxide and water are the major products from combustion of fossil fuels. However, combustion is rarely totally efficient and among the products from oil and coal are usually various aromatic hydrocarbons which contain several fused benzenoid rings and are hence generally referred to as **polycyclic aromatic hydrocarbons** (PAHs). They are formed by the action of the heat generated during the combustion process, but under conditions of local oxygen deficiency, so they can be considered as pyrolysis products.

Distributions of **pyrolytic PAHs** are characterised by the dominance of the parent, non-alkylated species shown in Fig. 8.3a. Particularly abundant are the highly peri-condensed compounds, such as pyrene, benzopyrene, benzo[*ghi*]perylene and coronene, which result from extensive angular fusion of benzenoid systems. The presence of such PAH distributions in the aromatic hydrocarbon fractions of Recent sediments is generally considered to reflect inputs from the combustion of organic matter, such as wood and fossil fuels. Pyrolytic PAHs are widespread in sediments post-dating the industrial revolution and demonstrate the extent of human influence on the environment.

Highly peri-condensed PAHs are more reactive than PAHs exhibiting lower degrees of angular fusion and so their high concentrations among pyrolysis products are attributed to rapid quenching by adsorption (through hydrogen bonding) on to particles of soot, which is itself a polycondensed PAH material. The observed distributions of pyrolytic PAHs in Recent sediments from a variety of environments are remarkably constant (Youngblood and Blumer, 1975; Laflamme and Hites, 1978; Wakeham et al., 1980; Gschwend and Hites, 1981); an example is shown in Fig. 8.3b. This uniformity suggests that the PAH distributions finally preserved in the sediment are largely protected by their association with soot particles against the actions of environmental agents during transportation and sedimentation (Prahl and Carpenter, 1983). For example, although all the PAHs are virtually insoluble in water, slight differences in solubility exist, which would be expected to lead to preferential leaching of some components, but this does not appear to occur. Also, some PAHs (e.g. anthracene) are more susceptible to photo-oxidation than others and might be expected to be preferentially degraded during aeolian transport and while subaerially exposed in sediments, but again such losses appear to be minimal.

The immobilisation of PAHs within the sediment matrix is important as many of these compounds are proven carcinogens and it would not be desirable for them to be readily released for incorporation into the aquatic food chain. Benzo[*a*]pyrene is a particularly potent carcinogen and its concentration in sediments in the Severn Estuary (UK) has been estimated at 9 ppm (based on sediment dry weight; Thomson and Eglinton, 1978). A significant proportion of pyrolytic PAHs associated with soot particles is likely to be distributed by wind, the influence of which can extend a great distance out into the oceans when the Trade Winds are involved. Fluvial transport of PAHs is also important for lacustrine and nearshore marine environments. Street dust and asphalt have been found to contain pyrolytic PAHs and so the disposal pattern for urban drainage water can influence local sedimentary concentrations of PAHs. However, in a high-energy environment like the Severn Estuary, much surface sediment is periodically resuspended and redistributed by the complex current systems.

Polycyclic aromatic hydrocarbons in ancient sediments

It appears that a recognisable pyrolytic PAH fingerprint can survive over geological time-scales. Wildfires, primarily initiated by lightning strikes, are the most likely source of these PAHs in ancient sediments. Such fires have probably been a feature of terrestrial ecosystems from at least the late Devonian, as suggested by the occurrence in sediments of fusinites and semi-fusinites (see Section 4.3.1), the proposed products of vegetation fires (Chaloner, 1989). Observed PAH distributions in ancient sediments are often like those seen in Recent sediments and are consistent

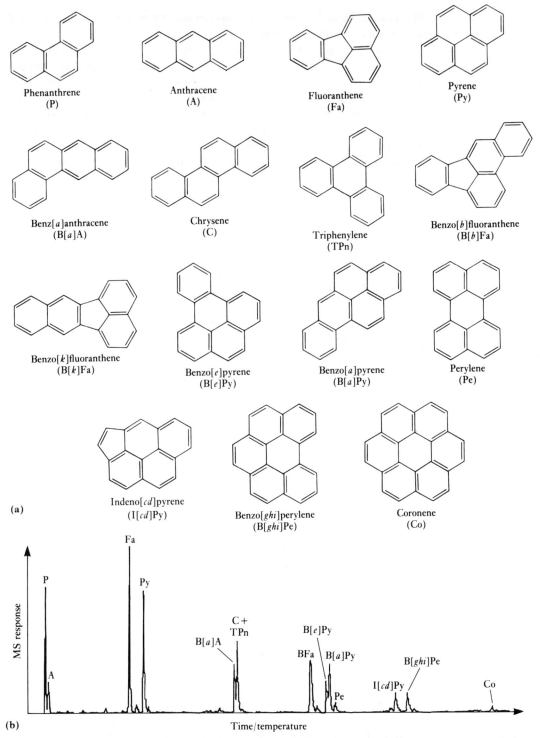

Phenanthrene
(P)

Anthracene
(A)

Fluoranthene
(Fa)

Pyrene
(Py)

Benz[a]anthracene
(B[a]A)

Chrysene
(C)

Triphenylene
(TPn)

Benzo[b]fluoranthene
(B[b]Fa)

Benzo[k]fluoranthene
(B[k]Fa)

Benzo[e]pyrene
(B[e]Py)

Benzo[a]pyrene
(B[a]Py)

Perylene
(Pe)

Indeno[cd]pyrene
(I[cd]Py)

Benzo[ghi]perylene
(B[ghi]Pe)

Coronene
(Co)

(a)

(b)

Figure 8.3   (a) Major pyrolytic polycyclic aromatic hydrocarbons (PAHs) and (b) their typical gas chromatographic distributions in Recent sediments. The mass spectrometer (MS) response corresponds to selective monitoring of PAH molecular ions only. In (b) the peak labelled BFa contains both B[b]Fa and B[k]Fa.

with moderate to high combustion temperatures of 400–800°C, as established for contemporary vegetation combustion (Youngblood and Blumer, 1975).

Suitably high temperatures for the formation of PAHs can occasionally arise in sedimentary environments as a result of igneous activity. For example, pyrolytic PAHs associated with ash, coal and wood fragments and apparently linked with igneous activity have been noted in sedimentary material from the Midland Valley of Scotland (Murchison and Raymond, 1989). Pyrolytic PAHs can, therefore, be adsorbed by charred woody remains as well as by soot. An interesting example of PAH generation by igneous activity is the pyrolytic PAHs detected in oil seeps in the Guaymas Basin, California (Kawka and Simoneit, 1990). Their formation has been attributed to the *in situ* pyrolysis of contemporary organic matter (with a significant terrestrial component) at temperatures in excess of 300°C. The heat is supplied by hydrothermal activity associated with the spreading centre in the basin.

Pyrolytic PAH distributions in ancient sediments sometimes differ from those typical of Recent sediments in exhibiting enhanced levels of the more highly peri-condensed structures, especially benzo[*e*]pyrene, benzo[*ghi*]perylene and coronene (Killops and Massoud, 1992). The reasons are as yet unknown but may reflect the effects of either different formation conditions or varying geochemical processes over geological time periods.

It can be seen that there are natural inputs of pyrolytic PAHs to the environment as there are for atmospheric carbon dioxide. However, unlike CO_2, anthropogenic PAH contributions in contemporary sediments exceed natural inputs. This increase in environmental burden may, therefore, be significant in its effects on organisms.

8.4.2   Oil spills

Effects of oil pollution

Oil spills in coastal areas cause immediate and obvious problems to animals and plants from external fouling; ingestion is an added problem for animals. Unseen effects include the suppression of primary production in phytoplankton due to reduction in light levels, the suppression of oxygen transfer from the atmosphere and the damage caused to benthic communities when the oil sinks to the sea bed. There are also long-term effects on ecosystems related to the release of toxic components (**toxicants**) over a prolonged period as the oil breaks up and the concentration of toxicants in organisms towards the top of the food chain increases.

Once again, however, oil is a naturally produced material and there are examples of natural oil seeps. It seems likely, therefore, that ecosystems can cope to an extent with the presence of oil, providing the amount is not overwhelming. Anthropogenic inputs are highest along coasts which, being among the most productive areas, are important in fish production. Recovery of spilt oil is the most effective response in terms of minimising ecological effects. However, weathering occurs rapidly, making recovery difficult. For example, approximately half the oil released into the Arabian Gulf during the 1991 hostilities evaporated within 24 hours. Loss of the more volatile, liquid hydrocarbons left a tarry material, rich in asphaltenes, that broke up with wave and current action and drifted to the sea floor as its density increased. The asphaltenes fraction contains more polar components, including organometallics, which gradually leach into the water. These components exhibit varying toxicities, and once in solution they become readily available for ingestion by organisms. Emulsification also occurs soon after an oil spill, as wind and wave action breaks up the oil into fine droplets, mixing them with water. Recovery of the dispersed oil in an emulsion is very difficult.

Beached oil is also difficult to deal with. Spraying with detergent-based dispersing agents drives the oil deeper into the substrate on sandy and muddy shores and, until the recent development of new agents, the toxicity of these agents to sediment-dwelling communities was similar to that of oils. The use of dispersants is most effective while the oil spill is still at sea. Oil will degrade naturally with time under the effects of weathering and bacterial action. As well as leaching of soluble components, weathering processes include photo-oxidation, which fragments molecules and introduces polar oxygen groups. The products are generally more water-soluble and may also be more readily metabolised by organisms. In contrast, apolar components are difficult to metabolise and can build up in certain body tissues, with possible long-term toxicity implications. Some types of bacteria, however, are able to use oil components as a carbon source (e.g. *Nocardia* species; Chosson et al., 1991), as was noted when considering in-reservoir biodegradation (Section 5.4.3). They may be present in relatively large numbers in areas where oil pollution is a chronic problem, as in the Arabian Gulf, and provide a degree of buffering to the ecosystem against small spills. Cultures of these bacteria are now being exploited in combatting major spills.

Oil pollution monitoring

Monitoring levels of oil contamination can be difficult, as only the hydrocarbons fraction is readily amenable to analytical techniques. Unfortunately, the dominant components, *n*-alkanes, are biodegraded more rapidly than other components and there are also natural sources of *n*-alkanes, such as higher plants (mainly odd carbon numbers in the range C_{23}–C_{33}) and benthic algae (in which C_{15} and C_{17} often dominate). When the obvious signs of oil fouling have disappeared, oil contamination is usually confirmed by the presence of certain molecular indicators of petrogenic inputs, particularly characteristic sterane and terpane distributions. These biomarkers can also be useful in quantifying pollution levels, providing the source of the oil is known so that concentrations of biomarkers can be related to absolute amounts of a specific oil.

Biodegradation proceeds quite rapidly in subaerial environments, eventually leading to the removal of all the major components that can be resolved by gas chromatography (see Fig. 7.11a). However, another apparent characteristic of petrogenic contamination becomes visible in gas chromatograms with increasing biodegradation: a 'hump' or unresolved complex mixture (UCM), extending over much of the range formerly occupied by *n*-alkanes. A gas chromatogram of the total hydrocarbons fraction of a heavily biodegraded oil is shown in Fig. 8.4 as an example. The chemical characteristics of the UCM suggest that it is generated from kerogen along with other oil components, but only becomes noticeable during gas chromatography when the levels of major resolved components have been reduced by bacterial action (Killops and Al-Juboori, 1990). The area of the UCM in gas chromatograms can be used to determine levels of environmental contamination and is particularly useful for assessing chronic pollution.

Chronic, low-level inputs of oil into aquatic environments can be conveniently studied using suitable sessile, filter-feeding organisms, such as mussels (e.g. *Mytilus edulis*). Mussels have been found to accumulate hydrocarbons in their tissues and accumulation appears to continue with increasing exposure. These organisms naturally colonise oil production platform legs in the North Sea, providing a useful

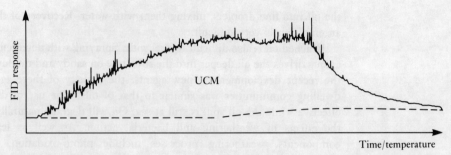

Figure 8.4   Gas chromatogram of the total hydrocarbons fraction from a heavily biodegraded oil, showing the unresolved complex mixture (UCM). Elution range of the UCM corresponds to $C_{10}-C_{35}$ *n*-alkanes.

means of monitoring pollution during oil production. Acyclic alkanes are also absorbed by the mussels from algal food sources and so monitoring the UCM may be the most accurate way of measuring petrogenic contamination levels.

In an examination of the association of petrogenic inputs with various grades of sedimentary material it was found that significantly more of the UCM was associated with the clay fraction than with the sand and silt fractions (Brassell and Eglinton, 1980). This behaviour is not surprising because the approximately spherical silt would be expected to exhibit a smaller available adsorption surface area than the plate-like clay particles. In addition, the aluminosilicates in clay may exhibit a greater adsorption affinity for UCM components. It is important to remember that **adsorption** is a surface effect which can involve strong chemical interactions (chemisorption) whereas **absorption** is a weaker, physical interaction in which a fluid enters pores within a solid (e.g. water absorption by a sponge). The occurrence of particulate association in the depositional environment for oil components contrasts with that of pyrolytic PAHs, which remain associated with the relatively low-density particulates by which they are adsorbed at their time of formation (Prahl and Carpenter, 1983; Readman et al., 1984). These variations in particulate associations may result in significant differences in the behaviour of petrogenic and pyrolytic hydrocarbons in sedimentary environments.

8.5   Some xenobiotic organic substances

With the exception of the chlorofluorocarbons discussed in Sections 8.2.2 and 8.3, which have no natural source, most of the anthropogenic inputs that we have considered so far have involved substances that also occur naturally. Since the Second World War there has been a dramatic increase in the number, amounts and applications of substances that are entirely man-made. Because they have no natural source their presence in the environment is unambiguously indicative of pollution. It can be argued that, although absolute amounts are small, the environmental impact of these xenobiotic substances is greater than that for naturally occurring materials because organisms have not had the opportunity to evolve in their presence. Some of these compounds are highly toxic, teratogenic (i.e. causing fetal deformities) and carcinogenic (i.e. causing cancer). In particular, the more stable, hydrophobic compounds tend to accumulate in certain tissues in higher organisms and cannot be metabolised, and because they are fat-soluble they can even be passed on to the offspring of animals (e.g. via milk for mammals and via egg yolk for birds and fish). Examples of this type of substance are the chloro-aromatic group of compounds, two of which are considered below.

8.5.1 DDT and related compounds

The insecticide DDT (1,1,1-trichloro-2,2-di(p-chlorophenyl)ethane, previously known as p,p'-dichlorodiphenyltrichloroethane; Fig. 8.5) was widely used after the Second World War. It appeared to offer the ultimate means of controlling a range of insects, including those attacking crops and those carrying disease. It was believed that excess DDT would be washed away, finally being diluted to negligible levels in the oceans. However, as the harmful effects of the accumulation of the compound in vertebrates became apparent, its use was banned in many countries in 1975. DDT is hydrophobic and so its application by aqueous spraying is achieved by the use of a wettable powder formulation. DDT is efficiently adsorbed by detritus and colloidal humic material (mainly by hydrogen bonding). It is concentrated in the tissue of detritus feeders and is passed up the food chain to the top consumers, generally large vertebrates, including humans. Adult vertebrates may not be killed directly by the poisonous compound, but they may not be able to produce the next generation. The deleterious effects of DDT have been found to extend to the plankton. At each step in the food chain accumulation levels increase and this is shown by the data in Table 8.5, which represent the effects of the careful, restricted use of DDT to control mosquitoes on Long Island, USA.

Measurements of DDT residues in soils and sediments often include two related compounds DDD (1,1-dichloro-2,2-di(p-chlorophenyl)ethane, Fig. 8.5) and DDE

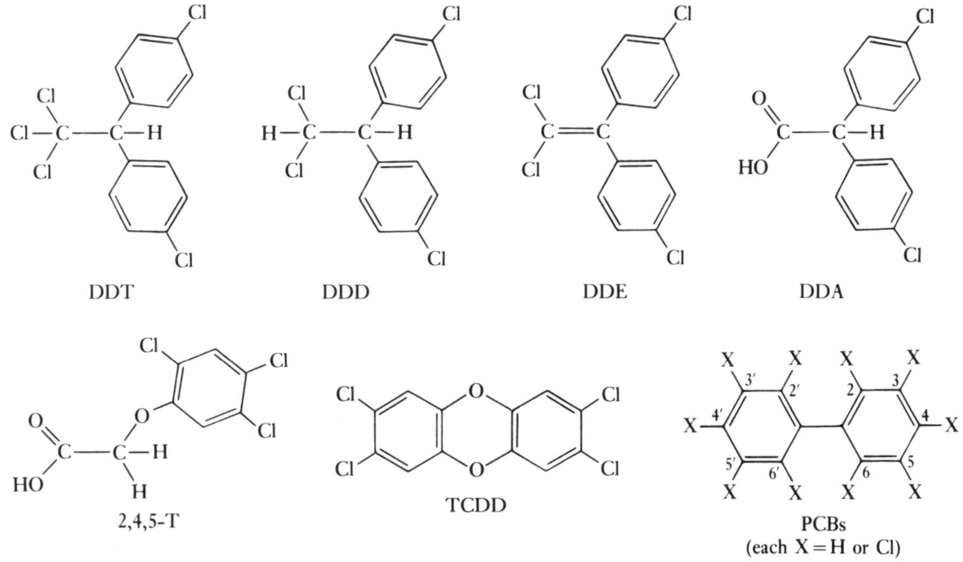

Figure 8.5 Some xenobiotic chloro-aromatic compounds.

Table 8.5   **Example of DDT concentration in an aquatic food chain** (After Woodwell et al., 1967)

	DDT residues* (ppm)
Water	0.00005
Plankton	0.04
Silverside minnow	0.23
Pickerel (predatory fish)	1.33
Heron (feeds on small animals)	3.57
Herring gull (scavenger)	6.00
Merganser (fish-eating duck)	22.8
Cormorant (feeds on larger fish)	26.4

* Total residues of DDT and related toxic compounds DDE and DDD as ppm of organism wet weight.

(1,1-dichloro-2,2-di(*p*-chlorophenyl)ethylene, Fig. 8.5), which are degradation products of DDT. This degradation is probably mainly microbially mediated, with reductive dechlorination being a major pathway. Further degradation can lead to the conversion of the CCl_3 group of DDT into a COOH group, the product being called DDA (2,2-di(*p*-chlorophenyl)acetic acid, Fig. 8.5). DDA is polar and will dissolve in water but DDD and DDE, like DDT, are insoluble and can persist for a long time in soils and sediments. This is partly attributable to the relative chemical stability of the compounds. In addition, humic material is able to trap these chloro-aromatic compounds, like many other small organic compounds and also metal ions, but has the potential to release them as changes in pH and redox conditions affect the structural and chelating characteristics of humics (see Sections 4.2.2 and 8.6.2).

DDD has been used in its own right as an insecticide but it is less toxic than DDT. Other chlorinated pesticides, such as lindane (γ-hexachlorocyclohexane), have presented similar toxicological problems in the environment. There are also parallels among herbicides, such as 2,4,5-T (2,4,5-trichlorophenoxyacetic acid, Fig. 8.5), which is a defoliant and acts systemically. It persists for long periods in soil and, although its toxicity to animals is apparently low, an associated and extremely carcinogenic compound, TCDD (2,3,7,8-tetrachlorodibenzo-*p*-dioxin, Fig. 8.5), may be present which is formed as an impurity during the manufacturing process.

8.5.2   Polychlorinated biphenyls

Polychlorinated biphenyls (PCBs) have a variety of industrial applications related to their high chemical and thermal stabilities, electrical resistance and low volatilities. Some of their uses are as dielectric fluids in capacitors and transformers, as lubricants and hydraulic fluids, as heat exchangers and fire retardants and as plasticisers. Commercial production began in the USA in 1929 and the total amounts produced probably exceed 1 Mt (ca. 37 kt were produced in 1970; Goldberg, 1976). Manufacture involves the chlorination of the parent hydrocarbon, biphenyl, resulting in a mixture of individual PCBs (Fig. 8.5). For any degree of chlorination there are usually several isomers, or congeners (Table 8.6), and there is a total of 209 possible congeners. The properties of the congeners depend largely upon the

Table 8.6   **Number of PCB congeners possible according to degree of chlorination**

Formula of PCB	Number of congeners
$C_{12}H_9Cl$	3
$C_{12}H_8Cl_2$	12
$C_{12}H_7Cl_3$	24
$C_{12}H_6Cl_4$	42
$C_{12}H_5Cl_5$	46
$C_{12}H_4Cl_6$	42
$C_{12}H_3Cl_7$	24
$C_{12}H_2Cl_8$	12
$C_{12}HCl_9$	3
$C_{12}Cl_{10}$	1

degree of chlorination and so different commercial mixtures are manufactured by controlling the ratio of chlorine to biphenyl. Examples of commercial PCB mixtures are Aroclor 1242, which comprises mainly di-, tri- and tetrachlorobiphenyls, and Aroclor 1254, which contains tetra-, penta-, hexa- and heptachlorobiphenyls. Aroclor is the trade name used by Monsanto and the last two digits give the percentage of chlorine (by weight) incorporated in the PCBs.

PCBs are now widespread in the environment and can be detected in air, water, soils, sediments and organisms. They are persistent due to their chemical inertness but bacterial degradation, at least of the less chlorinated congeners, does occur slowly. Typical environmental concentrations are $0.05-5.0$ ng/m^3 in air, $2-50$ μg/kg in surface soil, 2 ng/l in water, 10 μg/kg in vegetation and $5-50$ mg/kg in the blubber of marine mammals (1 ng $= 10^{-9}$ g, 1 μg $= 10^{-6}$ g, 1 mg $= 10^{-3}$ g). Like DDT they have extremely low solubilities in water but they are lipophilic, which results in progressive enrichment on passing along a food chain. Significant accumulation can occur in the fatty tissues of animals, particularly marine mammals containing a high proportion of blubber. PCBs do not appear to be carcinogenic, but chronic exposure of organisms to low levels is probably important. A major difficulty in assessing the toxic effects of such exposure is that higher levels of PCBs in tissue are invariably associated with higher levels of other chloro-aromatics, such as DDT. Reproductive cycles appear to be affected and there may also be immune system damage. A further problem is the passing on of PCBs to the offspring of mammals in milk. Toxicity appears to be related to molecular structure and, in particular, the positions of chlorination (3,3',4,4' seems especially toxic).

As concern grew about the biological effects of PCBs, restrictions were imposed on their uses in the early 1970s and production virtually ceased by the end of that decade. Various studies have suggested that there was a corresponding decrease in environmental levels, in both substrates and organisms, over this period. However, levels in organisms appear to have remained fairly constant subsequently, suggesting that PCBs have remained biologically available. In contrast, levels of other organochlorine compounds, like DDT, continued to decline.

Individual PCB congeners exhibit different physical and chemical properties, such as the degree of solubility in water, volatility and susceptibility to bio-accumulation. The more highly chlorinated species tend to exhibit lower solubilities and volatilities. For example, the solubility of monochlorinated congeners is about 1 mg/l, while that of decachlorobiphenyl is around five orders of magnitude lower. In addition, the more highly chlorinated species are more resistant towards microbial degradation and are more lipophilic. However, the PCB distributions found in sediments often parallel those of particular commercial mixtures, suggesting that the congeners in the mixtures tend to behave as a single phase. This can be explained by the strong association with colloidal organic (humic) material, which has the effect of reducing the availability of PCBs for microbial degradation.

In the past, land-fill has been an important means for the disposal of PCBs. However, degradation occurs only slowly and the partial volatility of PCBs at ambient temperatures leads to their remobilisation and release into the atmosphere from both soil and water. Transfer into the air can occur by direct vaporisation or by association of PCBs with aerosols. Redeposition of PCBs is brought about by wash-out in rain water (**wet deposition**) or by fall-out of particulates (**dry deposition**). Atmospheric transport has resulted in the widespread occurrence of PCBs in surface sediments and soils (at ppb levels, i.e. one part in 10^9). In addition to waste burial, PCBs are most likely to be introduced into the environment by leaks or accidental spills, which is different from the mode of introduction of DDT. More recently, incineration has been used for disposal of PCBs, but unless temperatures are sufficiently high for efficient combustion even more toxic compounds are produced, collectively called dioxins, the most potent of which is TCDD (Fig. 8.5). Aeolian transport is, therefore, an important route for the introduction of dioxins into the environment.

Aeolian transport is also important for both DDT and PCBs. Although our detailed knowledge of the pathways and fluxes of these substances into the marine environment is limited, information has been obtained for the North American Basin, an area of ca. 10^7 km^2 centred on Bermuda in the North Atlantic, which has been studied since 1970. Estimates of annual atmospheric inputs of DDT and PCBs to this area from the eastern USA in the early 1970s are about 14 t and 140 t, respectively (Bidleman et al., 1976). These values represent the sum of the estimates of three contributions to the air-to-sea transfer: direct transfer of the gaseous phase, wet deposition and dry deposition. Factors controlling each of these transfer processes are poorly understood, restricting the reliability of the final flux estimates, but to a first approximation it appears that the three air–sea transfer mechanisms make broadly equivalent contributions for both DDT and PCBs.

Data for PCB concentrations in the North American Basin in the 1970s (Harvey and Steinhauer, 1976) reflect the physical processes operating in the water column and the association of PCBs with particulate organic matter. The isolation of the **surface mixed layer** (SML, ca. 100 m deep here) from the underlying water by the thermocline resulted in higher concentrations of PCBs in the SML due to the trapping of the aeolian input. Concentrations in the SML in 1972 were estimated at ca. 30 ng/l, but by 1975 they had fallen to ca. 1 ng/l, reflecting the reduction in industrial sources of PCBs from the eastern USA that occurred during the early 1970s. The rate of loss of PCBs from the SML during this period corresponded to a flux of ca. 2 mg/m^2 a.

The bulk of the PCBs was associated with particulate matter, at concentrations of ca. 2 mg PCBs per kilogram of particulate matter in the SML in 1972 (a concentration factor of ca. 2×10^5 compared with levels of dissolved PCBs). The sinking of particulates, to a first approximation, was found to account for the levels of PCBs detected in the underlying sediments of the Basin in the mid-1970s. Based on an estimated 270 t of accumulated PCBs in the sediments and the fact that major production of PCBs started around 1955, an average sedimentary accumulation rate of ca. 1.4 $\mu g/m^2$ a is obtained. This value agrees quite well with a measured PCB deposition rate of 4.9 $\mu g/m^2$ a from sediment trap experiments, which would require a particulate deposition rate of ca. 2–3 g/m^2 a. As most of the particulate matter in the SML is recycled (ca. 96%), a particulate loss rate from the SML of 2–3 g/m^2 a is reasonable. Even taking what is probably a maximum value of 13 g/m^2 a for particulate fall-out from the SML (corresponding to the sedimentary deposition rate under the Sargasso Sea), the loss of PCBs by this mechanism would be only ca. 26 $\mu g/m^2$ a, which is ca. 100 times less than the total loss from the SML.

Another process must have been responsible for most of the PCB flux out of the SML. Advection and diffusion were not considered to make significant contributions over the time-scale involved. However, it was found that PCB concentrations were relatively low immediately above and below the air/sea interface, suggesting that rapid loss of PCBs from the surface microlayer of the sea was occurring. This may be attributed to a large proportion of the aeolian input of PCBs being entrained back into the atmosphere during periods when water evaporation was high, at low and mid latitudes, and then being transported to areas of greater precipitation or condensation at higher latitudes. This behaviour is consistent with general models predicting that there would be higher PCB losses to the air from both soil and water in warmer mid latitudes, and that these PCBs would be carried by the dominant air mass movements to higher latitudes where redeposition would occur. At higher latitudes the lower temperatures result in greater persistence of PCBs because volatilisation and degradation processes are slower. Further support for such models has been obtained in the form of increasing PCB levels in Arctic biota in recent years. PCBs are one of the most intensely studied groups of anthropogenic components, and this example shows how difficult it can be to predict the areas that will act as the final reservoirs for anthropogenic inputs without detailed knowledge of the environmental behaviour of the compounds in question.

8.6   Factors affecting the fate of anthropogenic inputs

8.6.1   General considerations

In the above sections a number of important factors have been examined which influence the environmental impact of anthropogenic substances. These include the sources, pathways and associated fluxes of inputs into the environment, the residence times and reactions of individual components and the influence of possible associations with particular types of sedimentary material on transport, biological uptake and long-term sedimentary fate. All these factors, in addition to toxicity, require consideration in assessing the environmental impact of an anthropogenic component, but often there is insufficient information about a number of them. This limits our ability to predict the effects of anthropogenic inputs, which weakens the case for the introduction of effective legislation to control such inputs on both the national and international scales. When an anthropogenic compound also occurs naturally, there can be considerable problems in determining both the level and the environmental impact of the anthropogenic input relative to the natural background at a particular location.

It is worth noting that humans are not the only organisms to affect ecosystems by the input of potentially toxic materials. Phytoplankton succession is affected by various chemicals exuded by different species. Some of these chemicals have been found to be essential for the growth of the next group of organisms in the succession, while others may suppress the growth of competing organisms or inhibit predation. These compounds are generally called **allelochemicals**. Bromophenols are an example of toxins with bactericidal properties and which also offer some protection against predation. They are found in marine algae and invertebrates, particularly annelids, phoronids and hemichordates.

8.6.2   Humic substances and pollutants

Humic and fulvic acids play an important role in the complexation (or chelation) and release of metals. For example, lowering pH to a value of ca. 2 can release most of the humic-bound iron, and the binding power of $Fe(II)$ ions with humic substances appears to be greater than that of $Fe(III)$ ions. Carboxyl groups appear particularly important in metal complexation. In general, retention of metallic cations by humic acids and brown coals is favoured by increasing pH. However, the interaction of humic material with metals is complex and for the series of metals $Hg(II)$, $Fe(III)$, Pb, Cu, Al, Ni, $Cr(III)$, Cd, Zn, Co and Mn it has been found that Hg and Fe are adsorbed most efficiently, and Co and Mn least (Kerndorf and Schnitzer, 1980). Within this series, the order of the relative stabilities of metal complexes depends on pH. For example, $Cu(II)$ complexes are more stable than those of $Ni(II)$ at pH 3, but less stable at pH 5. These laboratory results have been found to parallel those in natural aquatic systems, resulting in metals such as Cu, Pb, Ni and Cr being enriched in the humic substances in the water column compared with the underlying sediment. There may be competition between metals for active sites on humic material and so the concentration of humic material and individual metal ions is an important factor.

The potential of humic substances for forming water-soluble complexes with toxic metals and organics is an important pollution problem. In addition, chlorination of humic-rich potable waters may release previously complexed toxic components and may also result in the formation of additional toxic compounds like chloroform from the humic material. One solution is to remove the humic substances by granular activated charcoal, but this is expensive. It is possible that the addition of humic acids may be useful in removing pollutants, upon subsequent precipitation, during waste water treatment. Humic acids can also be used to monitor certain types of pollution which often result in higher N and S levels but low COOH group content.

References

Abbott G.D., Lewis C.A. and Maxwell J.R. (1985) Laboratory models for aromatization and isomerization of hydrocarbons in sedimentary basins. *Nature* **318**, 651–653.

Abraham H. (1945) *Asphalts and Allied Substances.* Van Nostrand, New York.

Albaigés J. (1980) Identification and geochemical significance of long chain acyclic isoprenoid hydrocarbons in crude oils. In *Advances in Organic Geochemistry 1979* (eds A.G. Douglas and J.R. Maxwell) pp. 19–28. Pergamon, Oxford.

Aldhous P. (1990) Taxation or regulation? *Nature* **347**, 412.

Alexander G., Hazai I., Grimalt J. and Albaigés J. (1987) Occurrence and transformation of phyllocladanes in brown coals from Nograd Basin, Hungary. *Geochim. Cosmochim. Acta* **51**, 2065–2073.

Alexander R., Kagi R. and Sheppard P. (1984) 1,8-Dimethylnaphthalene as an indicator of petroleum maturity. *Nature* **308**, 442–443.

Allen P.A. and Allen J.R. (1990) *Basin Analysis – Principles and Applications.* Blackwell Scientific, Oxford.

Allen P.A. and Collinson J.D. (1986) Lakes. In *Sedimentary Environments and Facies* (ed. H.G. Reading) pp. 63–94. Blackwell Scientific, Oxford.

Alvarez L.W., Alvarez W., Asaro F. and Michel H.V. (1980) Extraterrestrial cause for the Cretaceous–Tertiary extinction. *Science* **208**, 1095–1108.

Aquino Neto F.R., Trendel J.M., Restle A., Connan J. and Albrecht P.A. (1983) Occurrence and formation of tricyclic and tetracyclic terpanes in sediments and petroleums. In *Advances in Organic Geochemistry 1981* (eds M. Bjorøy et al.) pp. 659–667. Wiley, Chichester.

Ayers G.P. and Gras J.L. (1991) Seasonal relationship between cloud condensation nuclei and aerosol methanesulphonate in marine air. *Nature* **353**, 834–835.

Baker E.W. and Louda J.W. (1986) Porphyrins in the geological record. In *Biological Markers in the Sedimentary Record* (ed. R.B. Johns) pp. 125–225. Elsevier, Amsterdam.

Barwise A.J.G. and Roberts I. (1984) Diagenetic and catagenetic pathways for porphyrins in sediments. *Org. Geochem.* **6**, 167–176.

Beaumont C., Boutilier R., Mackenzie A.S. and Rullkötter J. (1985) Isomerization and aromatization of hydrocarbons and the paleothermometry and burial history of Alberta Foreland Basin. *Am. Assoc. Pet. Geol. Bull.* **69**, 546–566.

Behar F. and Vandenbroucke M. (1987) Chemical modelling of kerogen. *Org. Geochem.* **11**, 15–24.

Berner R.A., Lasaga A.C. and Garrels R.M. (1983) The carbonate–silicate geochemical cycle and its effects on atmospheric carbon dioxide over the past 100 million years. *Am. J. Sci.* **283**, 641–683.

Bidleman T.F., Ritt C.P. and Olney C.E. (1976) High molecular weight chlorinated hydrocarbons in the air and sea: rules and mechanisms of air–sea transfer. In *Marine Pollutant Transfer* (eds H.L. Windom and R.A. Duce) pp. 323–351. D.C. Heath, Boston.

Blom L., Edelhausen L. and van Krevelen D.W. (1957) Chemical structure and properties of coal. XVIII – Oxygen groups in coal and related products. *Fuel* **36**, 135–153.

Bois C., Bouche P. and Pelet R. (1982) Global geologic history and distribution of hydrocarbon reserves. *Am. Assoc. Pet. Geol. Bull.* **66**, 1248–1270.

Bolin B., Degens E.T., Duvigneaud P. and Kempe S. (1979) The global biogeochemical carbon cycle. In *The Global Carbon Cycle* (eds B. Bolin, E.T. Degens, S. Kempe and P. Ketner) SCOPE Rep. No. 13, pp. 1–56. Wiley, Chichester.

Bolin B., Rosswall T., Richey J.E., Freney J.R., Ivanov M.V. and Rodhe H. (1983) C, N, P, and S cycles: major reservoirs and fluxes. In *The Major Biogeochemical Cycles and their Interactions* (eds B. Bolin and R.B. Cook) SCOPE Rep. No. 21, pp. 41–65. Wiley, Chichester.

Bonnett R., Burke P.J., Czechowski F. and Reszka A. (1984) Porphyrins and metalloporphyrins in coal. *Org. Geochem.* **6**, 177–182.

Bordovskiy O.K. (1965) Sources of organic matter in marine basins. *Mar. Geol.* **3**, 5–31.

Brassell S.C. (1985) Molecular changes in sediment lipids as indicators of systematic early diagenesis. *Phil. Trans. R. Soc. Lond. A* **315**, 57–75.

Brassell S.C. and Eglinton G. (1980) Environmental chemistry – an interdisciplinary subject. Natural and pollutant organic compounds in contemporary aquatic environments. In *Environmental Chemistry* (ed. J. Albaigés) pp. 1–22. Pergamon Press, Oxford.

Brassell S.C., Eglinton G., Marlowe I.T., Pflaumann U. and Sarnthein M. (1986) Molecular stratigraphy: a new tool for climatic assessment. *Nature* **320**, 129–133.

Brassell S.C., McEvoy J., Hoffmann C.F., Lamb N.A., Peakman T.M. and Maxwell J.R. (1984) Isomerisation, rearrangement and aromatisation of steroids in distinguishing early stages of diagenesis. *Org. Geochem.* **6**, 11–23.

Brassell S.C., Wardroper A.M.K., Thomson I.D., Maxwell J.R. and Eglinton G. (1981) Specific acyclic isoprenoids as biological markers of methanogenic bacteria in marine sediments. *Nature* **290**, 693–696.

Bray E.E. and Evans E.D. (1961) Distribution of *n*-paraffins as a clue to recognition of source beds. *Geochim. Cosmochim. Acta* **22**, 2–15.

Brooks J., Cornford C. and Archer R. (1987) The role of hydrocarbon source rocks in petroleum exploration. In *Marine Petroleum Source Rocks* (eds J. Brooks and A.J. Fleet) Geol. Soc. Spec. Publn. No. 26, pp. 17–46. Blackwell Scientific, Oxford.

Cardoso J.N. and Eglinton G. (1983) The use of hydroxyacids as geochemical indicators. *Geochim. Cosmochim. Acta* **47**, 723–730.

Cassani F. and Eglinton G. (1986) Organic geochemistry of Venezuelan extra-heavy oils. I. Pyrolysis of asphaltenes: a technique for the correlation and maturity evaluation of crude oils. *Chem. Geol.* **56**, 167–183.

Chaffee A.L. and Johns R.B. (1983) Polycyclic aromatic hydrocarbons in Australian coals. I. Angularly fused pentacyclic tri- and tetra-aromatic components of Victorian brown coal. *Geochim. Cosmochim. Acta* **47**, 2141–2155.

Chaffee A.L., Strachan M.G. and Johns R.B. (1984) Polycyclic aromatic hydrocarbons in Australian coals. II. Novel tetracyclic components from Victorian brown coal. *Geochim. Cosmochim. Acta* **48**, 2037–2043.

Chalansonnet S., Largeau C., Casadevall E., Berkaloff C., Peniguel G. and Couderc R. (1988) Cyanobacterial resistant biopolymers. Geochemical implications of the properties of *Schizothrix* sp. resistant material. *Org. Geochem.* **13**, 1003–1010.

Chaloner W.G. (1989) Fossil charcoal as an indicator of palaeoatmospheric oxygen level. *J. Geol. Soc., Lond.* **146**, 171–174.

Charlson R.J., Lovelock J.E., Andreae M.O. and Warren S.G. (1987) Oceanic phytoplankton, atmospheric sulphur, cloud albedo and climate. *Nature* **326**, 655–661.

Chosson P., Lannau C., Connan J. and Dessort D. (1991) Biodegradation of refractory hydrocarbon biomarkers from petroleum under laboratory conditions. *Nature* **351**, 640–642.

Chyba C.F., Thomas P.J., Brookshaw L. and Sagan C. (1990) Cometary delivery of organic molecules to the early Earth. *Science* **249**, 366–373.

Clayton C. (1991) Carbon isotope fractionation during natural gas generation from kerogen. *Mar. Pet. Geol.* **8**, 232–240.

Connan J. (1984) Biodegradation of crude oils in reservoirs. In *Advances in Petroleum Geochemistry, Vol. 1* (eds J. Brooks and D. Welte) pp. 299–335. Academic Press, New York.

Cowie G.L. and Hedges J.I. (1984) Carbohydrate sources in a coastal marine environment. *Geochim. Cosmochim. Acta* **48**, 2075–2087.

de Leeuw J.W. and Baas M. (1986) Early-stage diagenesis of steroids. In *Biological Markers in the Sedimentary Record* (ed. R.B. Johns) pp. 101–123. Elsevier, Amsterdam.

de Leeuw J.W., Cox H.C., van Grass G., van de Meer F.W., Peakman T.M., Baas J.M.A. and van de Graaf B. (1989) Limited double bond isomerisation and selective hydrogenation of sterenes during early diagenesis. *Geochim. Cosmochim. Acta* **53**, 903–909.

Demaison G.J. and Moore G.T. (1980) Anoxic environments and oil source bed genesis. *Am. Assoc. Pet. Geol. Bull.* **64**, 1179–1209.

Dereppe J.-M., Moreaux C. and Debyser Y. (1980) Investigation of marine and terrestrial humic substances by ^1H and ^{13}C nuclear, magnetic resonance and infrared spectroscopy. *Org. Geochem.* **2**, 117–124.

De Vooys C.G.N. (1979) Primary production in aquatic environments. In *The Global Carbon Cycle* (eds B. Bolin, E.T. Degens, S. Kempe and P. Ketner) SCOPE Rep. No. 13, pp. 259–292. Wiley, Chichester.

Dickinson R.E. and Cicerone R.J. (1986) Future global warming from atmospheric trace gases. *Nature* **319**, 109–115.

Didyk B.M. and Simoneit B.R.T. (1989) Hydrothermal oil of Guaymas Basin and implications for petroleum formation mechanisms. *Nature* **342**, 65–69.

Didyk B.M., Simoneit B.R.T., Brassell S.C. and Eglinton G. (1978) Organic geochemical indicators of palaeoenvironmental conditions of sedimentation. *Nature* **272**, 216–222.

Durand B. (1980) Elemental analysis of kerogens. In *Kerogen – Insoluble Organic Matter from Sedimentary Rocks* (ed. B. Durand). Editions Technip, Paris.

Ekweozor C.M. and Telnaes N. (1990) Oleanane parameter: verification by quantitative study of the biomarker occurrence in sediments of the Niger delta. *Org. Geochem.* **16**, 401–413.

Elliott T. (1986) Deltas. In *Sedimentary Environments and Facies* (ed. H.G. Reading) pp. 113–154. Blackwell Scientific, Oxford.

Espitalié J., Laporte J.L., Madec M., Marquis F., Leplat P., Paulet J. and Boutefeu A. (1977) Méthode rapide de caractérisation des roches mères, de leur potential pétrolier et de leur degré d'évolution. *Rev. Inst. Fr. Pét.* **32**, 23–42.

Eugster H.P. and Kelts K. (1983) Lacustrine chemical sediments. In *Chemical Sediments and Geomorphology* (eds A.S. Goudie and K. Pye) pp. 321–368. Academic Press, London.

FAO (1972) *Atlas of the Living Resources of the Seas.* Food and Agriculture Organisation of the United Nations, Dept. Fisheries, Rome.

Fenchel T.M. and Jørgensen B.B. (1977) Detritus food chains of aquatic ecosystems: the role of bacteria. In *Advances in Microbial Ecology, Vol. 1* (ed. M. Alexander) pp. 1–57. Plenum Press, New York.

Finar I.L. (1975) *Organic Chemistry, Vol. 2. Stereochemistry and the Chemistry of Natural Products.* Longman, London.

Freeman K.H., Hayes J.M., Trendel J.-M. and Albrecht P. (1990) Evidence from carbon isotope measurements for diverse origins of sedimentary hydrocarbons. *Nature* **343**, 254–256.

Freyer H.-D. (1979a) Variations in the atmospheric CO_2 content. In *The Global Carbon Cycle* (eds B. Bolin, E.T. Degens, S. Kempe and P. Ketner) SCOPE Rep. No. 13, pp. 79–96. Wiley, Chichester.

Freyer H.-D. (1979b) Atmospheric cycles of trace gases containing carbon. In *The Global Carbon Cycle* (eds B. Bolin, E.T. Degens, S. Kempe and P. Ketner) SCOPE Rep. No. 13, pp. 101–128. Wiley, Chichester.

Galimov E.M. (1976) Variations of the carbon cycle at present and in the geological past. In *Environmental Geochemistry, Vol. 1* (ed. J.O. Nriagu) pp. 3–11. Ann Arbor Science, Ann Arbor, Mich.

Garrigues P. and Ewald M. (1983) Natural occurrence of 4-methyl-phenanthrene in petroleums and recent marine sediments. *Org. Geochem.* **5**, 53–56.

Goad L.J., Lenton J.R., Knapp F.F. and Goodwin T.W. (1974) Phytosterol side chain biosynthesis. *Lipids* **9**, 582–595.

Goldberg E.D. (1976) *The Health of the Oceans.* UNESCO Press, Paris.

Goossens H., de Lange F., de Leeuw J.W. and Schenk P.A. (1988a) The Pristane Formation Index, a molecular maturity parameter. Confirmation in samples from the Paris Basin. *Geochim. Cosmochim. Acta* **52**, 2439–2444.

Goossens H., de Leeuw J.W., Schenck P.A. and Brassell S.C. (1984) Tocopherols as likely precursors of pristane in ancient sediments and crude oils. *Nature* **312**, 440–442.

Goossens H., Due A., de Leeuw J.W., van de Graaf B. and Schenck P.A. (1988b) The Pristane Formation Index, a molecular maturity parameter. A simple method to assess maturity by pyrolysis/evaporation–gas chromatography of unextracted samples. *Geochim. Cosmochim. Acta* **52**, 1189–1193.

Goth K., de Leeuw J.W., Püttmann W. and Tegelaar E.W. (1988) Origin of Messel Oil Shale kerogen. *Nature* **336**, 759–761.

Gould S.J. (1991) *Wonderful Life. The Burgess Shale and the Nature of History*. Penguin Books, London.

Greiner A.C., Spyckerelle C. and Albrecht P. (1976) Aromatic hydrocarbons from geological sources – I. New naturally occurring phenanthrene and chrysene derivatives. *Tetrahedron* **32**, 257–260.

Gschwend P.M. and Hites R.A. (1981) Fluxes of polycyclic aromatic hydrocarbons to marine and lacustrine sediments in the northern United States. *Gochim. Cosmochim. Acta* **45**, 2359–2367.

Harvey G.R. and Steinhauer W.G. (1976) Biogeochemistry of PCB and DDT in the North Atlantic. In *Environmental Biogeochemistry, Vol. 1* (ed. J.O. Nriagu) pp. 203–221. Ann Arbor Science, Ann Arbor, Mich.

Harvey H.R., Eglinton G., O'Hara S.C.M. and Corner E.D.S. (1987) Biotransformation and assimilation of dietary lipids by *Calanus* feeding on a dinoflagellate. *Geochim. Cosmochim. Acta* **51**, 3031–3040.

Hatcher P.G., Maciel G.E. and Dennis L.W. (1981) Aliphatic structure of humic acids; a clue to their origin. *Org. Geochem.* **3**, 43–48.

Hatcher P.G., Rowan R. and Mattingly M.A. (1980) ^{1}H and ^{13}C NMR of marine humic acids. *Org. Geochem.* **2**, 77–85.

Hayatsu R., Botto R.E., Scott R.G., McBeth R.L. and Winans R.E. (1987) Thermal catalytic transformation of pentacyclic triterpenoids: alteration of geochemical fossils during coalification. *Org. Geochem.* **11**, 245–250.

Hayes J.M., Takigiku R., Ocampo R., Callot H.J. and Albrecht P. (1987) Isotopic compositions and probable origins of organic molecules in the Eocene Messel shale. *Nature* **329**, 48–53.

Hayes M.H.B., MacCarthy P., Malcolm R.L. and Swift R.S. (eds) (1989) *Humic Substances II. In Search of Structure*. Wiley, Chichester.

Hedges J.I. and Ertel J.R. (1982) Characterisation of lignin by gas capillary chromatography of cupric oxide oxidation products. *Anal. Chem.* **54**, 174–178.

Hedges J.I. and Mann D.C. (1979) The characterisation of plant tissues by their lignin oxidation products. *Geochim. Cosmochim. Acta* **43**, 1803–1807.

Heredy L.A. and Wender I. (1980) Model structure for a bituminous coal. *Am. Chem. Soc. Div. Fuel Chem. Prepr.* **25**, 38–45.

Holloway P.J. (1982) The chemical constitution of plant cutins. In *The Plant Cuticle* (eds D.F. Cutler, K.L. Alvin and C.E. Price) 45–85. Linnean Soc. Lond. Academic Press, London.

Holmes S. (1983) *Outline of Plant Classification*. Longman, London.

Huang W.-Y. and Meinschein W.G. (1979) Sterols as ecological indicators. *Geochim. Cosmochim. Acta* **43**, 739–745.

Hunt J.M. (1991) Generation of gas and oil from coal and other terrestrial organic matter. *Org. Geochem.* **17**, 673–680.

Hussler G., Connan J. and Albrecht P. (1984) Novel families of tetra- and hexacyclic aromatic hopanoids predominant in carbonate rocks and crude oils. *Org. Geochem.* **6**, 39–49.

Jenkyns H.C. (1980) Cretaceous anoxic events: from continents to oceans. *J. Geol. Soc., Lond.* **137**, 171–188.

Jiamo F. and Guoying S. (1989) Biological marker composition of typical source rocks and related crude oils of terrestrial origin in The People's Republic of China: a review. *Appl. Geochem.* **4**, 13–22.

Jiamo F., Guoying S., Pingan P., Brassell S.C., Eglinton G. and Jiyang J. (1986) Peculiarities of salt lake sediments as potential source rocks in China. *Org. Geochem.* **10**, 119–126.

Jiang Z.S. and Fowler M.G. (1986) Carotenoid-derived alkanes in oils from northwestern China. *Org. Geochem.* **10**, 831–839.

Johns R.B., Gillan F.T. and Volkman J.K. (1980) Early diagenesis of phytyl esters in a contemporary temperate intertidal sediment. *Geochim. Cosmochim. Acta* **44**, 183–188.

Johns R.B., Nichols P.D. and Perry G.J. (1979) Fatty acid composition of ten marine algae from Australian waters. *Phytochemistry* **18**, 799–802.

Jones D.M., Douglas A.G., Parkes R.J., Taylor J., Giger W. and **Schaffner C.** (1983) The recognition of biodegraded petroleum-derived aromatic hydrocarbons in recent marine sediments. *Mar. Pollut. Bull.* **14**, 103–108.

Jørgensen B.B. (1983a) The microbial sulphur cycle. In *Microbial Geochemistry* (ed. W.E. Krumbein) pp. 91–124. Blackwell Scientific, Oxford.

Jørgensen B.B. (1983b) Processes at the sediment–water interface. In *The Major Biogeochemical Cycles and their Interactions* (eds B. Bolin and R.B. Cook) SCOPE Rep. No. 21, pp. 477–515. Wiley, Chichester.

Kawka O.E. and **Simoneit B.R.T.** (1990) Polycyclic aromatic hydrocarbons in hydrothermal petroleums from the Guaymas Basin spreading center. *Appl. Geochem.* **5**, 17–27.

Keely B.J., Eckardt C.B. and **Maxwell J.R.** (1992) A novel chlorophyll transformation pathway in the aquatic environment. *Org. Geochem.* **19**, 217–27.

Kempe S. (1979) Carbon in the rock cycle. In *The Global Carbon Cycle* (eds B. Bolin, E.T. Degens, S. Kempe and P. Ketner) SCOPE Rep. No. 13, pp. 343–377. Wiley, Chichester.

Kerndorf H. and **Schnitzer M.** (1980) Sorption of metals on humic acid. *Geochim. Cosmochim. Acta* **44**, 1701–1708.

Kiene R.P. and **Bates T.S.** (1990) Biological removal of dimethyl sulphide from sea water. *Nature* **345**, 702–705.

Killops S.D. (1991) Novel aromatic hydrocarbons of probable bacterial origin in a Jurassic lacustrine sequence. *Org. Geochem.* **17**, 25–36.

Killops S.D. and **Al-Juboori M.A.H.A.** (1990) Characterisation of the unresolved complex mixture (UCM) in the gas chromatograms of biodegraded petroleums. *Org. Geochem.* **15**, 147–160.

Killops S.D. and **Howell V.J.** (1991) Complex series of pentacyclic triterpanes in a lacustrine sourced oil from Korea Bay Basin. *Chem. Geol.* **91**, 65–79.

Killops S.D. and **Massoud M.S.** (1992) Polycyclic aromatic hydrocarbons of pyrolytic origin in ancient sediments – evidence for Jurassic vegetation fires. *Org. Geochem.* **18**, 1–7.

Kimber R.W.L. and **Griffin C.V.** (1987) Further evidence of the complexity of the racemization process in fossil shells with implications for amino acid racemization dating. *Geochim. Cosmochim. Acta* **51**, 839–846.

Kissin Y.V. (1987) Catagenesis and composition of petroleum: origin of *n*-alkanes and isoalkanes in petroleum crudes. *Geochim. Cosmochim. Acta* **51**, 2445–2457.

Klemme H.D. (1980) Petroleum basins – classification and characteristics. *J. Pet. Geol.* **3**, 187–207.

Koblenz-Mishke O.I.V., Volkonsky V.V. and **Kabanova J.G.** (1970) Planktonic primary production of the world oceans. In *Symposium on Scientific Exploration of the South Pacific* (ed. W.S. Wooster) pp. 183–193. Natl. Acad. Sci., Washington.

Laflamme R.E. and **Hites R.A.** (1978) The global distribution of polycyclic aromatic hydrocarbons in recent sediments. *Geochim. Cosmochim. Acta* **42**, 289–303.

Largeau C., Derenne S., Casadevall E., Kadouri A. and **Sellier N.** (1986) Pyrolysis of immature Torbanite and of the resistant biopolymer (PRB A) isolated from extant alga *Botryococcus braunii*. Mechanism of formation and structure of Torbanite. *Org. Geochem.* **10**, 1023–1032.

Larter S.R. and **Senftle J.T.** (1985) Improved kerogen typing for petroleum source rock analysis. *Nature* **318**, 277–280.

Lashof D.A. and **Ahuja D.R.** (1990) Relative contributions of greenhouse gas emissions to global warming. *Nature* **344**, 529–531.

Lee R.F., Hirota J. and **Barnett A.M.** (1971) Distribution and importance of wax esters in marine copepods and other zooplankton. *Deep-Sea Res.* **18**, 1147–1165.

Levorsen A.I. (1967) *Geology of Petroleum.* Freeman, San Francisco.

Lewan M.D. (1984) Factors controlling the proportionality of vanadium to nickel in crude oils. *Geochim. Cosmochim. Acta* **48**, 2231–2238.

Liss P.S. (1983) The exchange of biogeochemically important gases across the air–sea interface. In *The Major Biogeochemical Cycles and their Interactions* (eds B. Bolin and R.B. Cook) SCOPE Rep. No. 21, pp. 411–426. Wiley, Chichester.

Mackenzie A.S. (1984) Applications of biological markers in petroleum geochemistry. In *Advances in Petroleum Geochemistry, Vol. 1* (eds J. Brooks and D. Welte) pp. 115–214. Academic Press, New York.

Mackenzie A.S., Beaumont C. and McKenzie D.P. (1984) Estimation of the kinetics of geochemical reactions with geophysical models of sedimentary basins and applications. *Org. Geochem.* **6**, 875–884.

Mackenzie A.S., Brassell S.C., Eglinton G. and Maxwell J.R. (1982) Chemical fossils: the geological fate of steroids. *Science* **217**, 491–504.

Mackenzie A.S., Leythaeuser D., Muller P., Quigley T.M. and Radke M. (1988) The movement of hydrocarbons in shales. *Nature* **331**, 63–65.

Mackenzie A.S. and McKenzie D. (1983) Isomerization and aromatization of hydrocarbons in sedimentary basins formed by extension. *Geol. Soc. Mag.* **120**, 417–528.

Mackenzie A.S. and Quigley T.M. (1988) Principles of geochemical prospect appraisal. *Am. Assoc. Pet. Geol. Bull.* **72**, 399–415.

Mango F.D. (1990) The origin of light hydrocarbons in petroleum: a kinetic test of the steady-state catalytic hypothesis. *Geochim. Cosmochim. Acta* **54**, 1315–1323.

Marlowe I.T., Brassell S.C., Eglinton G. and Green J.C. (1990) Long-chain alkenones and alkyl alkenoates and the fossil coccolith record of marine sediments. *Chem. Geol.* **88**, 349–375.

McCartney J.T. and Teichmüller M. (1972) Classification of coals according to degree of coalification by reflectance of the vitrinite component. *Fuel* **51**, 64–68.

McKirdy D.M., Cox R.E., Volkman J.K. and Howell V.J. (1986) Botryococcane in a new class of Australian non-marine crude oils. *Nature* **320**, 57–59.

Mello M.R., Telnaes N., Gaglianone P.C., Chicarelli M.I., Brassell S.C. and Maxwell J.R. (1988) Organic geochemical characterisation of depositional palaeoenvironments of source rocks and oils in Brazilian marginal basins. *Org. Geochem.* **13**, 31–45.

Miller G.H., Jull A.J.T., Linick T., Sutherland D., Sejrup H.P., Brigham J.K., Bowen D.Q. and Mangerud J. (1987) Racemization-derived late Devensian temperature reduction in Scotland. *Nature* **326**, 593–595.

Moldowan J.M. (1984) C_{30}-steranes, novel markers for marine petroleums and sedimentary rocks. *Geochim. Cosmochim. Acta* **48**, 2767–2768.

Moldowan J.M., Fago F.J., Carlson R.M.K., Young D.C., Van Duyne G., Clardy J., Schoell M., Pillinger C.T. and Watts D.S. (1991) Rearranged hopanes in sediments and petroleum. *Geochim. Cosmochim. Acta* **55**, 3333–53.

Moldowan J.M., Lee C.Y., Watt D.S., Jeganathan A., Slougui N.-E. and Gallegos E.J. (1991) Analysis and occurrence of C_{26}-steranes in petroleum and source rocks. *Geochim. Cosmochim. Acta* **55**, 1065–1081.

Moldowan J.M. and Seifert W.K. (1979) Head-to-head linked isoprenoid hydrocarbons in petroleum. *Science* **204**, 169–171.

Moldowan J.M. and Seifert W.K. (1980) First discovery of Botryococcane in petroleum. *J. Chem. Soc., Chem. Commun.*, 912–914.

Mopper K. and Degens E.T. (1979) Organic carbon in the ocean: nature and cycling. In *The Global Carbon Cycle* (eds B. Bolin, E.T. Degens, S. Kempe and P. Ketner) SCOPE Rep. No. 13, pp. 293–316. Wiley, Chichester.

Murchison D.G. and Raymond A.C. (1989) Igneous activity and organic maturation in the Midland Valley of Scotland. *Int. J. Coal Geol.* **14**, 47–82.

Murray J.W. and Grundmanis V. (1980) Oxygen consumption in pelagic marine sediments. *Science* **209**, 1527–1530.

Neal A.C., Prahl F.G., Eglinton G., O'Hara S.C.M. and Corner E.D.S. (1986) Lipid changes during a planktonic feeding sequence involving unicellular algae, *Elminius* nauplii and adult *Calanus*. *J. Mar. Biol. Assoc. U.K.* **66**, 1–13.

NERC (1989) *Our Future World — Global Environmental Research*. Abridged report to Advisory Council on Science and Technology by Natural Environment Research Council, Swindon.

Nip M., Tegelaar E.W., Brinkhuis H., de Leeuw J.W., Schenck P.A. and Holloway P.J. (1986) Analysis of modern and fossil plant cuticles by Curie point Py-GC and Curie point Py-GC-MS: Recognition of a new, highly aliphatic and resistant biopolymer. *Org. Geochem.* **10**, 769–778.

Orr W.L. (1986) Kerogen/asphaltene/sulphur relationships in sulphur-rich Monterey oils. *Org. Geochem.* **10**, 499–516.

Ourisson G., Albrecht P. and Rohmer M. (1979) The hopanoids. *Pure Appl. Chem.* **51**, 709–729.

Parkes R.J. (1987) Analysis of microbial communities within sediments using biomarkers. In *Ecology of Microbial Communities*. Cambridge University Press, Cambridge.

Parkes R.J., Cragg B.A., Fry J.C., Herbert R.A. and Wimpenny J.W.T. (1990) Bacterial biomass and activity in deep sediment layers from the Peru margin. *Phil. Trans. R. Soc. Lond. A* **331**, 139–153.

Peakman T.M. and Maxwell J.R. (1988) Early diagenetic pathways of steroid alkenes. *Org. Geochem.* **13**, 583–592.

Peakman T.M., ten Haven H.L., Rechka J.R., de Leeuw J.W. and Maxwell J.R. (1989) Occurrence of (20R)- and (20S)-$\Delta^{8,14}$ and $\Delta^{14}5\alpha(H)$-sterenes and the origin of $5\alpha(H),14\beta(H),17\beta(H)$-steranes in an immature sediment. *Geochim. Cosmochim. Acta* **53**, 2001–2009.

Pedersen T.F. and Calvert S.E. (1990) Anoxia vs. productivity: what controls the formation of organic-carbon-rich sediments and sedimentary rocks? *Am. Assoc. Pet. Geol. Bull.* **74**, 454–466.

Perry G.J., Volkman J.K., Johns R.B. and Bavor H.J. (1979) Fatty acids of bacterial origin in contemporary marine sediments. *Geochim. Cosmochim. Acta* **43**, 1715–1725.

Phillippi G.T. (1975) The deep sub-surface temperature controlled origin of the gaseous and gasoline-range hydrocarbons of petroleum. *Geochim. Cosmochim. Acta* **39**, 1353–1373.

Phillippi G.T. (1981) Correlation of crude oils with their oil source formation, using high resolution GLC C_6–C_7 component analysis. *Geochim. Cosmochim. Acta* **45**, 1495–1513.

Powell T.G. (1986) Petroleum geochemistry and depositional setting of lacustrine source rocks. *Mar. Pet. Geol.* **3**, 200–219.

Prahl F.G. and Carpenter R. (1983) Polycyclic aromatic hydrocarbons (PAH)-phase associations in Washington coastal sediment. *Geochim. Cosmochim. Acta* **47**, 1013–1023.

Prahl F.G., Eglinton G., Corner E.D.S., O'Hara S.C.M. and Forsberg T.E.V. (1984) Changes in plant lipids during passage through the gut of *Calanus*. *J. Mar. Biol. Assoc. U.K.* **64**, 317–334.

Prahl F.G. and Wakeham S.G. (1987) Calibration of unsaturation patterns in long-chain ketone compositions for palaeotemperature assessment. *Nature* **330**, 367–376.

Prather M.J. and Watson R.T. (1990) Stratospheric ozone depletion and future levels of atmospheric chlorine and bromine. *Nature* **344**, 729–734.

Püttmann W. and Villar H. (1987) Occurrence and geochemical significance of 1,2,5,6-tetramethylnaphthalene. *Geochim. Cosmochim. Acta* **51**, 3023–3029.

Quirk M.M., Wardroper A.M.K., Wheatley R.E. and Maxwell J.R. (1984) Extended hopanoids in peat environments. *Chem. Geol.* **42**, 25–43.

Radke M. (1987) Organic geochemistry of aromatic hydrocarbons. In *Advances in Petroleum Geochemistry, Vol. 2* (eds J. Brooks and D. Welte) pp. 141–207. Academic Press, New York.

Radke M. and Welte D.H. (1983) The methylphenanthrene index (MPI): a maturity parameter based on aromatic hydrocarbons. In *Advances in Organic Geochemistry 1981* (eds M. Bjorøy et al.) pp. 504–512. Wiley, Chichester.

Readman J.W., Mantoura R.F.C. and Rhead M.M. (1984) The physico–chemical speciation of polycyclic aromatic hydrocarbons (PAH) in aquatic systems. *Fresenius Z. Anal. Chem.* **319**, 126–131.

Repeta D.J. (1989) Carotenoid diagenesis in recent marine sediments: II. Degradation of fucoxanthin to loliolide. *Geochim. Cosmochim. Acta* **53**, 699–707.

Repeta D.J. and Gagosian R.B. (1982) Carotenoid transformation in coastal marine waters. *Nature* **295**, 51–54.

Repeta D.J. and Gagosian R.B. (1984) Transformation reactions and recycling of carotenoids and chlorins in the Peru upwelling region (15°S, 75°W). *Geochim. Cosmochim. Acta* **48**, 1265–1277.

Repeta D.J., Simpson D.J., Jørgensen B.B. and Jannasch H.W. (1989) Evidence for anoxygenic photosynthesis from the distribution of bacteriochlorophylls in the Black Sea. *Nature* 342, 69–72.

Revsbech N.P., Sørensen J. and Blackburn T.H. (1980) Distribution of oxygen in marine sediments measured with microelectrodes. *Limnol. Oceanogr.* 25, 403–411.

Reynolds C.S. (1984) Phytoplankton periodicity: the interactions of form, function and environmental variability. *Freshwater Biol.* 14, 111–142.

Reynolds C.S., Wiseman S.W., Godfrey B.M. and Butterwick C. (1983) Some effects of artificial mixing on the dynamics of phytoplankton populations in large limnetic enclosures. *J. Plankton Res.* 5, 203–234.

Rieley G., Collier R.J., Jones D.M., Eglinton G., Eakin P.A. and Fallick A.E. (1991) Sources of sedimentary lipids deduced from stable carbon-isotope analyses of individual compounds. *Nature* 352, 425–427.

Ries-Kautt M. and Albrecht P. (1989) Hopane-derived triterpenoids in soils. *Chem. Geol.* 76, 143–151.

Robin P.L. (1975) Caractérisation des kérogenes et de leur évolution par spectroscopie infrarouge. Thesis. University of Louvain.

Rodhe H. (1990) A comparison of the contribution of various gases to the Greenhouse Effect. *Science* 248, 1217–1219.

Rothschild L.J. and Mancinelli R.L. (1990) Model of carbon fixation in microbial mats from 3500 Myr ago to the present. *Nature* 345, 710–712.

Schidlowski M. (1988) A 3800-million-year isotopic record of life from carbon in sedimentary rocks. *Nature* 333, 313–318.

Schmitter J.M., Sucrow W. and Arpino P.J. (1982) Occurrence of novel tetracyclic geochemical markers: 8,14-seco-hopanes in a Nigerian crude oil. *Geochim. Cosmochim. Acta* 46, 2345–2350.

Schnitzer M. (1978) Humic substances: chemistry and reactions. *Dev. Soil Sci.* 8, 1–64.

Schoell M. (1984b) Stable isotopes in petroleum research. In *Advances in Petroleum Geochemistry, Vol. 1* (eds J. Brooks and D. Welte) pp. 215–245. Academic Press, New York.

Schoell M. (1984b) Stable isotopes in petroleum research. In *Advances in Petroleum Geochemistry, Vol. 1* (eds. J. Brooks and D. Welte) pp. 215–245. Academic Press, New York.

Schoell M. (1988) Multiple origins of methane in the Earth. *Chem. Geol.* 71, 1–10.

Selley R.C. (1985) *Elements of Petroleum Geology*. Freeman, New York.

Simoneit B.R.T. (1977) Diterpenoid compounds and other lipids in deep-sea sediments and their geochemical significance. *Geochim. Cosmochim. Acta* 41, 463–476.

Simoneit B.R.T. (1986) Cyclic terpenoids of the geosphere. In *Biological Markers in the Sedimentary Record* (ed. R.B. Johns) pp. 43–99. Elsevier, Amsterdam.

Sinninghe Damsté J.J., Rijpstra W.I.C., Kock-van Dalen A.C., de Leeuw J.W. and Schenck P.A. (1989) Quenching of labile functionalised lipids by inorganic sulphur species: evidence for the formation of sedimentary organic sulphur compounds at the early stages of diagenesis. *Geochim. Cosmochim. Acta* 53, 1343–1355.

Solomon S., Garcia R.R., Rowland F.S. and Wuebbles D.J. (1986) On the depletion of Antarctic ozone. *Nature* 321, 755–758.

Solomons T.W.G. (1988) *Organic Chemistry*. Wiley, New York.

Stainforth J.G. and Reinders J.E.A. (1990) Primary migration of hydrocarbons by diffusion through organic matter networks, and its effect on oil and gas generation. *Org. Geochem.* 16, 61–74.

Staplin F.L. (1969) Sedimentary organic matter, organic metamorphism, and oil and gas occurrence. *Can. Pet. Geol. Bull.* 17, 47–66.

Strachan M.G., Alexander R. and Kagi R.I. (1988) Trimethylnaphthalenes in crude oils and sediments: effects of source and maturity. *Geochim. Cosmochim. Acta* 52, 1255–1264.

Suess E. (1980) Particulate organic carbon flux in the oceans – surface productivity and oxygen utilization. *Nature* 288, 260–263.

Summerhayes C.P. (1983) Sedimentation of organic matter in upwelling regimes. In *Coastal Upwelling – Its Sediment Record. Part B: Sedimentary Records of Ancient Coastal Upwelling* (eds J. Thiede and E. Suess) pp. 29–72. Plenum Press, New York.

Summons R.E. and Powell T.G. (1987) Identification of aryl isoprenoids in source rocks and crude oils: biological markers for the green sulphur bacteria. *Geochim. Cosmochim. Acta* 51, 557–556.

Tan Y.L. and Heit M. (1981) Biogenic and abiogenic polynuclear aromatic hydrocarbons in sediments from two remote Adirondack lakes. *Geochim. Cosmochim. Acta* **45**, 2267–2279.

Tegelaar E.W., de Leeuw J.W., Derenne S. and Largeau C. (1989) A reappraisal of kerogen formation. *Geochim. Cosmochim. Acta* **53**, 3103–3106.

Tegelaar E.W., Kerp H., Visscher H., Schenck P.A. and de Leeuw J.W. (1991) Bias of the paleobotanical record as a consequence of variations in the chemical composition of higher vascular plant cuticles. *Palaeobiology* **17**, 133–144.

Teichmüller M. and Teichmüller R. (1968) Geological aspects of coal metamorphisms. In *Coal and Coal-Bearing Strata* (eds D.G. Murchison and T.S. Westoll) pp. 233–267. Oliver and Boyd, Edinburgh.

ten Haven H.L., de Leeuw J.W., Rullkötter J. and Sinninghe Damsté J.J. (1987) Restricted utility of the pristane/phytane ratio as a palaeoenvironmental indicator. *Nature* **330**, 641–643.

ten Haven H.L., de Leeuw J.W. and Schenck P.A. (1985) Organic geochemical studies of a Messinian evaporitic basin, northern Apennines (Italy) I: hydrocarbon biological markers for a hypersaline environment. *Geochim. Cosmochim. Acta* **49**, 2181–2191.

ten Haven H.L., Rohmer M., Rullkötter J. and Bisseret P. (1989) Tetrahymanol, the most likely precursor of gammacerane, occurs ubiquitously in marine sediments. *Geochim. Cosmochim Acta* **53**, 3073–3079.

ten Haven H.L. and Rullkötter J. (1988) The diagenetic fate of taraxer-14-ene and oleanane isomers. *Geochim. Cosmochim. Acta* **52**, 2543–2548.

Thomson S. and Eglinton G. (1978) Composition and sources of pollutant hydrocarbons in the Severn Estuary. *Mar. Pollut. Bull.* **9**, 133–136.

Tissot B. (1979) Effects on prolific petroleum source rocks and major coal deposits caused by sea-level changes. *Nature* **277**, 463–465.

Tissot B. and Espitalié J. (1975) L'évolution thermique de la matière organique des sédiments: application d'une simulation mathématique. *Rev. Inst. Fr. Pét.* **30**, 743–777.

Tissot B., Espitalié J., Deroo G., Tempere C. and Jonathan D. (1974) Origin and migration of hydrocarbons in the Eastern Sahara (Algeria). In *Advances in Organic Geochemistry 1973* (eds B. Tissot and F. Bienner) pp. 315–334. Editions Technip, Paris.

Tissot B.P. and Welte D.H. (1984) *Petroleum Formation and Occurrence*. Springer-Verlag, Berlin.

Towe K.M. (1990) Aerobic respiration in the Archaean. *Nature* **348**, 54–56.

Ungerer P. (1990) State of the art of research in kinetic modelling of oil formation and expulsion. *Org. Geochem.* **16**, 1–25.

Venkatesan M.I. (1989) Tetrahymanol: Its widespread occurrence and geochemical significance. *Geochim. Cosmochim. Acta* **53**, 3095–3101.

Venrick E.L. (1982) Phytoplankton in an oligotrophic ocean: observations and questions. *Ecol. Monogr.* **52**, 129–154.

Volkman J.K. (1986) A review of sterol markers for marine and terrigenous organic matter. *Org. Geochem.* **9**, 83–99.

Volkman J.K. and Johns R.B. (1977) The geochemical significance of positional isomers of unsaturated acids from an intertidal zone sediment. *Nature* **267**, 693–694.

Vollhardt K.P.C. (1987) *Organic Chemistry*. Freeman, New York.

Wakeham S.G. (1989) Reduction of sterols to stanols in particulate matter at oxic–anoxic boundaries in sea water. *Nature* **342**, 787–790.

Wakeham S.G., Schaffner C. and Giger W. (1980) Polycyclic aromatic hydrocarbons in Recent lake sediments. I. Compounds having anthropogenic origins. *Geochim. Cosmochim. Acta* **44**, 403–413.

Waples D.W. (1980) Time and temperature in petroleum formation: application of Lopatin's method to petroleum exploration. *Am. Assoc. Pet. Geol. Bull.* **64**, 916–926.

Waples D.W. (1984) Thermal models for oil generation. In *Advances in Petroleum Geochemistry, Vol. 1* (eds J. Brooks and D. Welte) pp. 7–67. Academic Press, New York.

Wardroper A.M.K., Hoffmann C.F., Maxwell J.R., Barwise A.J.G., Goodwin N.S. and Park P.J.D. (1984) Crude oil biodegradation under simulated and natural conditions – II. Aromatic steroid hydrocarbons. *Org. Geochem.* **6**, 605–617.

White C.M. and **Lee M.L.** (1980) Identification and geochemical significance of some aromatic components of coal. *Geochim. Cosmochim. Acta* **44**, 1825–1832.

Whiticar M.J. (1990) A geochemical perspective of natural gas and atmospheric methane. *Org. Geochem.* **16**, 531–547.

Whittaker R.H. and **Likens G.E.** (1975) The biosphere and man. In *Primary Productivity of the Biosphere* (eds H. Lieth and R.H. Whittaker) *Ecol. Stud.* **14**, 305–328. Springer-Verlag, Berlin.

Woese C.R. and **Wolfe R.S.** (1985) *The Bacteria – A Treatise on Structure and Function, Vol. VIII, Archaebacteria* (series ed. I.C. Gunsalus). Academic Press, New York.

Woese C.R., Kandler O. and **Wheelis M.L.** (1990) Towards a natural system of organisms: proposal for the domains Archaea, Bacteria and Eucarya. *Proc. Natl. Acad. Sci.* **87**, 4576–4579.

Wolff G.A., Lamb N.A. and **Maxwell J.R.** (1986a) The origin and fate of 4-methyl steroid hydrocarbons. I. Diagenesis of 4-methyl sterenes. *Geochim. Cosmochim. Acta* **50**, 335–342.

Wolff G.A., Lamb N.A. and **Maxwell J.R.** (1986b) The origin and fate of 4-methyl steroids – II. Dehydration of stanols and occurrence of C_{30} 4-methyl steranes. *Org. Geochem.* **10**, 965–974.

Wood D.A. (1988) Relationship between thermal maturity indices calculated using Arrhenius equation and Lopatin method: Implications for petroleum exploration. *Bull. Am. Assoc. Pet. Geol.* **72**, 115–134.

Woodwell G.M., Wurster C.F. and **Isaacson P.A.** (1967) DDT residues in an east coast estuary: a case of biological concentration of a persistent insecticide. *Science* **156**, 821–824.

Youngblood W.W. and **Blumer M.** (1975) Polycyclic aromatic hydrocarbons in the environment: homologous series in soils and recent marine sediments. *Geochim. Cosmochim. Acta* **39**, 1303–1314.

Further reading

General

Abercrombie M., Hickman C.J. and Johnson M.L. (1984) *The Penguin Dictionary of Biology*. Penguin Books, London.

Bates R.L. and Jackson J.A. (eds) (1987) *Glossary of Geology*. American Geological Institute, Falls Church, VA. (An updated but more costly version of Whitten and Brooks (1973), below.)

Sharp D.W.A. (1990) *Penguin Dictionary of Chemistry*. Penguin Books, London.

Whitten D.G.A. and Brooks J.R.V. (1973) *A Dictionary of Geology*. Penguin Books, London.

Chapter 1

Bolin B., Degens E.T., Kempe S. and Ketner P. (eds) (1979) *The Global Carbon Cycle*, SCOPE Rep. No. 13. Wiley, Chichester.

Bradbury I. (1991) *The Biosphere*. Belhaven Press, London.

Dineley D. (1975) *Earth's Voyage Through Time*. Paladin, St Albans. (Unfortunately no longer in press, but worth searching out.)

Gass I.G., Smith P.J. and Wilson R.C.L. (1972) *Understanding the Earth*. Artemis Press, London.

Thomas B. (1981) *The Evolution of Plants and Flowers*. Peter Lowe, London.

Ziegler B. (1983) *Introduction to Palaeobiology*. Ellis Horwood, Chichester.

Chapter 2

Bloomfield M.M. (1987) *Chemistry and the Living Organism*. Wiley, New York. (A good introductory chemistry text.)

Britton G. (1983) *The Biochemistry of Natural Pigments*. Cambridge University Press, Cambridge.

Harwood J.L. and Russell N.J. (1984) *Lipids in Plants and Microbes*. Allen and Unwin, London.

MacGregor E.A. and Greenwood C.T. (1980) *Polymers in Nature*. Wiley, Chichester.

Rose S. (1991) *The Chemistry of Life*. Penguin Books, London.

Stumpf P.K. and Conn E.E. (eds) (1980) *Biochemistry of Plants, Vol. 4, Lipids*. Academic Press, New York.

Chapter 3

Barnes R.S.K. and **Hughes R.N.** (1990) *An Introduction to Marine Ecology.* Blackwell Scientific, Oxford.

Bolin B. and **Cook R.B.** (eds) (1983) *The Major Biogeochemical Cycles and their Interactions.* SCOPE Rep. No. 21. Wiley, Chichester.

Campbell R. (1977) *Microbial Ecology.* Blackwell, Oxford.

Harvey J.G. (1982) *Atmosphere and Ocean: Our Fluid Environments.* Artemis Press, London.

Krumbein W.E. (ed.) (1983) *Microbial Geochemistry.* Blackwell Scientific, Oxford.

Reading H.G. (ed.) (1986) *Sedimentary Environments and Facies.* Blackwell Scientific, Oxford.

Tucker M.E. (1991) *Sedimentary Petrology, an Introduction.* Blackwell Scientific, Oxford.

Chapter 4

Durand B. (ed.) (1984) *Kerogen, Insoluble Organic Matter from Sedimentary Rocks.* Editions Technip, Paris.

Hayes M.H.B., MacCarthy P., Malcolm R.L. and **Swift R.S.** (eds) (1989) *Humic Substances II: In Search of Structure.* Wiley, Chichester.

Murchison D.G. and **Westoll T.S.** (eds) (1968) *Coal and Coal-Bearing Strata.* Oliver and Boyd, Edinburgh. (Quite old, but still a standard, particularly in association with Scott, 1987, below.)

Schobert H.H. (1987) *Coal – The Energy Source of the Past and Future.* Am. Chem. Soc., Washington D.C.

Scott A.C. (ed.) (1987) *Coal and Coal-Bearing Strata: Recent Advances.* Blackwell Scientific, Oxford.

Chapter 5

Brooks J. and **Fleet A.J.** (eds) (1987) *Marine Petroleum Source Rocks.* Geol. Soc. Spec. Publn. No. 26. Blackwell Scientific, Oxford.

England W.A. and **Fleet A.J.** (1991) *Petroleum Migration.* Geol. Soc. Spec. Publn. No. 59. Blackwell Scientific, Oxford.

Fleet A.J., Kelts K. and **Talbot M.R.** (eds) (1988) *Lacustrine Petroleum Source Rocks.* Geol. Soc. Spec. Publn. No. 40. Blackwell Scientific, Oxford.

North F.K. (1985) *Petroleum Geology.* Allen and Unwin, London. (A deeper coverage than Selley, 1985, below.)

Selley R.C. (1985) *Elements of Petroleum Geology.* Freeman, New York.

Tissot B.P. and **Welte D.H.** (1984) *Petroleum Formation and Occurrence.* Springer-Verlag, Berlin.

Chapters 6 and 7

Cranwell P.A. (1982) Lipids of aquatic sediments and sedimentary particulates. *Prog. Lipid Res.* **21**, 271–308.

Engel M.H. and **Macko S.A.** (eds) (1992) *Organic Geochemistry.* Plenum Press, New York.

Johns R.B. (ed.) (1986) *Biological Markers in the Sedimentary Record.* Elsevier, Amsterdam.

Mackenzie A.S. (1984) Applications of biological markers in petroleum geochemistry. In *Advances in Petroleum Geochemistry, Vol. 1* (eds J. Brooks and D. Welte) pp. 115–214. Academic Press, New York.

Moldowan J.M., Albrecht P. and Philp R.P. (1992) *Biological Markers in Sediments and Petroleum*. Prentice-Hall, Englewood Cliffs, New Jersey.

Radke M. (1987) Organic geochemistry of aromatic hydrocarbons. In *Advances in Petroleum Geochemistry, Vol. 2* (eds J. Brooks and D. Welte) pp. 141–207. Academic Press, New York.

Smith R.M. (1988) *Gas and liquid chromatography in Analytical Chemistry*. Wiley, Chichester.

Williams D.H. and Fleming I. (1987) *Spectroscopic Methods in Organic Chemistry*. McGraw-Hill, London. (A useful introduction to MS, IR and NMR techniques cited in these chapters and also in Chapter 4.)

Chapter 8

Benn F.R. and McAuliffe C.A. (eds) (1975) *Chemistry and Pollution*. Macmillan, London.

Harrison R.M., De Mora S.J., Rapsomanikis S. and Johnston W.R. (1991) *Introductory Chemistry for the Environmental Sciences*. Cambridge University Press, Cambridge.

Kurtz D.A. (ed.) (1990) *Long Range Transport of Pesticides*. Lewis, Chesea, USA.

Leggett J. (ed.) (1990) *Global Warming – The Greenpeace Report*. Oxford University Press, Oxford.

O'Neill P. (1985) *Environmental Chemistry*. Allen and Unwin, London.

Index

abietane, 180
abietic acid, 47, 48, 140, 180
abiogenic chemical evolution, 7, 8
absorption, **232**
    of hydrocarbons by sedimentary
        material, 197, 121, 131
abyssal plain, 69, **89**
acetate, 46, 79, 80, 81, 151, 152, 157
Acetobacterium, 79
acetogenesis, 79
acetosyringone, 163
acetovanillone, 163
acids, 22, 23, **35** *(see also* fatty acids)
    alkanoic, 39
    alkenoic, 39
acidity, 70–1
    bacterial activity, 81, 87, 101
    carbonate deposition, 86
    humic substance structure, 97, 234
    leaching of soils, 95
    metal complexation, 197–8, 238
    nutrient availability, 70
acid rain, 135
acritarchs, **17**, 143, 144
actinomycetes, 33, 78, 94
action spectrum, **55**, 56
activation energy, **37**, 167, 209
adenine, 38, 39
adenosine diphosphate (ADP), 10, 11,
    37
adenosine triphosphate (ATP), 10, 11,
    33, 38, 39, 41, 46, 70
ADP (see adenosine diphosphate)
adsorption, **232**
    of light by plants (*see* photosynthesis
        and pigments)
    of oil components during migration,
        140, 141
    of pollutants by particulate material,
        228, 230, 232, 236
    of solar energy by the Earth, 217–19
advection, **168**, 231
aerobes, 31, 33, 41, 64
    obligate, 11, **78**
agar, 34
agathalene, 185, 186
agathic acid, 47, 48
alanine, 34, 36, 43, 94
alcohols, 23, 30, 39, 46, 79
    fatty, 43, 45, 154
    steroidal (*see* stanols, stenols and
        sterols)
aldehydes, 23, 28, 30, 45

aldose, 27, 28, 30, 161
algae,
    benthic, 18, 55, 231
    blue-green (*see* cyanobacteria)
    brown, 14, 34, 53, 55, 154, 155, 158,
        176
    chemical components, 31, 33, 34, 40,
        41, 43, 47, 49, 53, 55, 56, 58, 98,
        108, 109, 120, 154–62, 187, 190,
        192, 232
    classification, 8
    contribution to sedimentary organic
        matter, 85, 86, 99, 100, 108,
        112–14, 120, 192, 194
    evolution, 13–14
    freshwater, 114
    green, 13, 14, 17, 34, 55, 68, 76, 143,
        144, 154, 155, 157, 158, 160
    in hypersaline environments, 194
    macroscopic (*see* multicellular)
    multicellular, 2, 14, 18, 55, 64, 157
    photosynthesis and isotopic
        composition, 164–5
    productivity, 224
    red, 14, 34, 53, 55, 58, 158
    succession, 76
    unicellular, 2, 12, 13, 18, 64, 154,
        157, 160
algaenan, **108**, 109
alginic acid, 32, 34
alginite, 99, 100, 109, 113, 115, 146
aliphatic structures, **23**, 26 (*see also*
        alkanes and alkenes)
    in coal, 102, 138
    in humic material, 96–8
    in kerogen, 108, 109, 112–14, 118,
        119
alkali, 35
alkalinity, 71
alkanes, **23**, 166, 182, 195
    GC-analysis of, 211, 212, 214, 215
    isotopic signatures, 151, 193–94
    in organisms 45, 154, 165, 172,
        190–2, 232
    in petroleum and bitumen, 107,
        114–15, 119, 122, 136–8, 141, 145,
        190–3, 197, 198, 202–3, 211,
        214–15, 231
    in sedimentary material, 97–8, 138,
        140, 166, 182, 190–3, 195, 198,
        214–15
alkanols, 43, 154 (*see also* alcohols)
alkenes, **24**, 46, 47, 166, 192, 203

allochemicals, **238**
allochthonous inputs, **62**, 84, 85, 88, 91,
        99, 112, 114, 194
alloisoleucine, 187
amino acids, **34**, 36, 70
    metabolism/degradation, 46, 79, 93–4
    occurrence and function, 33, 34–9,
        58, 154
    in sedimentary organic matter, 96, 98,
        110
    stereoisomerism, 26, 34, 187
    structure, 34–6
ammonium, 70, 79, 80, 110
amorphinite, 115
amylopectin, 31, 32, 33
amylose, 31, 32, 33
α-amyrin, 47, 49, 183, 184
β-amyrin, 47, 49, 140, 181–2, 185, 186,
        193, 200
Anabaena, 80, 225
anaerobes, 8, 31, 33
    facultative, 78, 79
    obligate, 11, 78
angiosperms, 7, **15**, 18, 56, 154, 155,
        161, 186, 193
    non-woody, 161, 162
    woody, 161, 162
anomerism, 29, 30
anomers, **30**
anthocyanins, 60
anthracene, 228, 229
anthracite, **101**, 102, 103, 104, 106, 144,
        146, 147, 151, 204
    semi-, 103
anthraquinones, **61**
anthrasteroids, 178
anthropogenic inputs, 4, **217**, 218–25,
        227, 230, 232, 237
anticline, 133, 134
API gravity, 135
Arabian Gulf, 230, 231
arabinose, 29, 33, 34, 161, 162, 166
arachidic acid, 40
arachidonic acid, 40
archaebacteria, **9**, 34, 43, 111, 155, 191,
        194, 196
arginine, 36
aromatic structures, 23, **24**, 26 (*see also*
        hydrocarbons, aromatic)
    in coal and peat, 100–2, 106
    in humic material, 95–8
    in kerogen, 112–18, 119, 122
    nitrogen-containing, 23, 100, 136,

aromatic structures *continued*
 137, 138, 169, 172–3, 201–2
 oxygen-containing, 23, 24, 25, 56, 58,
  60, 96, 97, 98, 100, 101, 103, 110,
  115, 136, 137, 138, 154, 162
 sulphur-containing, 23, 100, 111, 114,
  136, 137, 138, 210
aromatisation
 of biomarkers, 166, 172–3, 177–86,
  201, 204, 206, 208–10
 during coalification, 104–6, 138, 140
 during kerogen formation, 110–11
 as a maturity indicator, 201, 204, 206,
  209, 210
 during petroleum generation,
  122–3
Arrhenius
 constant, 126, 209
 equation, 208
arthropods, 33
ascomycetes, 171–2
Ascophyllum, 70
asparagine, 36
aspartic acid, 36
asphalt, 107, 228
asphaltenes, 106, **107**, 108, 116, 121,
 123, 135, 140, 141, 145, 151, 197,
 230
 precipitation, 107, 140, 141
 pyrolysis, 211
astaxanthin, 54, 56, 175
atmosphere
 as carbon reservoir, 2, 4–5, 21,
  141–2, 217–18
 evolution, 6–7, 12–13, 92
 stratification, 226
 trace gases, 220–4
ATP (see adenosine triphosphate)
autochthonous inputs, **62**, 64, 84, 86,
 88, 91, 99, 114, 120, 194
autotrophs, 3, 8, 16, 164, 165
axial substituents, **30**, 31, 50
Azotobacter, 80

Bacillariophyceae (*see* diatoms)
Bacillus, 33, 78, 79
bacteria,
 acetogenic, 79, 80
 aerobic, 78, 80, 81, 141
 anaerobic, 43, 78, 79, 80, 81
 autotrophic, 17
 chemical composition, 31–4, 43–5,
  47, 49, 52, 55, 57–8, 61, 62,
  155–9, 161–2, 165, 191
 chemosynthetic, 65, 80, 91, 164, 194
 classification, 8–9, 13–14
 competition between anaerobic
  heterotrophs, 81, 195
 contribution to sedimentary organic
  matter, 17–18, 81, 101, 109, 113,
  114, 120, 154, 192–3, 194, 195
 degradation of organic matter, 73,
  77–80, 81, 82, 86, 89–90, 93–4,
  101, 106–7, 113, 114, 120, 168,
  170, 186, 190, 192, 231

evolution, 13, 18
 Gram-negative, 33, 34, 43, 159
 Gram-positive, 9, **33**, 34, 43, 159
 green non-sulphur, 9
 green sulphur, 8, 58, 80, 91, 165, 174
 halophilic, 9, 43, 81, 194, 196, 197
 heterotrophic, 17, 78, 155
 metabolism, 42, 80, 81, 171–2 (*see*
  also respiration)
 methanogenic (*see* methanogens)
 methylotrophic (*see* methylotrophs)
 nitrate-reducing (*see* nitrate reducers)
 nitrifying, 80–1
 nitrogen-fixing, 80
 photosynthetic, **11**, 13, 21, 55, 56, 58,
  65, 80, 87, 91, 155, 158, 164, 165
 purple non-sulphur, 9, 11
 purple sulphur, 8, 9, 80, 165, 174
 red photosynthetic, 165
 sedimentary communities, 80–1, 153
 substrates for anaerobic heterotrophs,
  79–80
 sulphate-reducing (*see* sulphate
  reducers)
 sulphur-oxidising, 80
 thermoacidophilic, 9, 43, 81, 155, 191
bacteriochlorophylls, 57–8, **58**, 91, 155,
 173
bacteriohopanepolyols, 47, 49, 182, 183
bacterioruberin, 155, 156
banded iron formations, 7, 12
barrier sand bars/islands, 85, 134
base, **35**, 42
basidiomycetes, 94
Basin,
 Douala, 113, 119
 East Shetland, 208
 Gippsland, 120, 132
 Guaymas, 125, 230
 North American, 236
 North Sea, 209
 Panonian, 125, 208, 209
 Paris, 113, 119, 125
 Songliao, 145
 Uinta, 113, 119
 Williston, 131
basins, **83**, 83–4, 85, 91, 124, 125, 208,
 209, 210
 Brazilian marginal, 198–9
 classification, 83–4
 enclosed, 90–1
 extensional, 122, 209, 210
 formation, 84
 marginal marine (paralic), 83
 oceanic, 76, 89, 144
 restricted, 83, 89, 90, 92
 saline lake, 86
 salt-pan, 86
 silled, 83, 90–1
beaches, 85
Beggiatoa, 80
benthic organisms, **3**, 14, 18, 55, 70, 76,
 231
benzanthracene, 229
benzene, 23, 24, 61, 116, 137, 138

alkyl-, 136, 137, 138, 141, 174
benzofluoranthene, 229
benzohopanes, 183, 186, 214
benzoperylene, 228, 229, 230
benzopyrene, 228, 229, 230
benzothiophene, 136, 137
betulin, 49
bicarbonate, 4, 19, 20, 21, 78
bile acids, 53
biodegradation,
 of biomarkers, 211
 of oil in-reservoir, 141, 145, 213
 of organic matter, 2–5, 11, 63, 73,
  77–83, 86, 89–90, 93–4, 98, 101,
  106–7, 108, 110, 113, 115, 120,
  121, 165–6, 168, 190, 225, 235
 of spilt oil, 231
biomacromolecules, 107, 110
 degradation, 101
 resistant, 102, 107, 108, 109, 110
biomarkers, **109**, 110, 121
 acyclic isoprenoidal, 191–2
 analysis, 211–16
 in asphaltenes, 211
 biodegradation, 141, 211, 231
 in coals, 139–40
 as depositional environment
  indicators, 194–9
 diagenesis, 166–86
 incorporation in kerogen, 139, 197,
  203
 isotopic signatures, 165, 193–4
 as maturity indicators, 149, 199–202,
  203–5
 in oil source rock correlation, 210–11
 in oils, 138–40, 202, 231
 as source indicators, 153–5, 157–60,
  190–4
 in source rocks and bitumen, 148,
  151, 202
 sulphur incorporation, 111–12
biopolymers (*see* biomacromolecules)
bioturbation, **78**, 86
bitter principles, 47
bitumen, 101, 106, **107**, 109, 110, 114,
 135, 145
 abundance, 150
 composition, 21, 107, 140–1, 147,
  148, 151, 192, 202
blooms, 55, 72, 73, 74, 76, 82, 86, 224,
 225
bond,
 breakage in petroleum generation,
  116, 122, 123, 124, 127, 129, 150
 conjugated, **24**, 41, 53
 covalent, **23**, 24
 double 23, 24, 40, 47, 50, 110, 111,
  158, 166, 178, 180–2, 190, 192
 glycosidic, **31**
 peptide (*see* peptide linkage)
 vibration, 218
botryococcane, 192, 194
botryococcenes, 191, 192
Botryococcus, 109, 114, 115, 192
brassicasterol, 53, 160

bromophenols, 238
bryophytes, 8, 14, 87, 159
butane, 138
butyrate, 79

β-cadinene, 48
Calanus, 168, 170
calcite, 1, 84, 86, 87, 188
Calvin cycle, **11**
campesterol, 53, 159
camphor, 48
cap rock, 124, 134
capillary pressure, 130, 133, 134
carbohydrates, 3, 9, 10, 11, 12, 22, **28**,
    28–34, 79
    composition, 28–31
    diagenesis, 41, 46, 79, 93–4, 96, 98,
        101, 102, 108, 165–6
    occurrence and function, 31–4, 37,
        39, 43, 61, 154
    as source indicators, 161–2, 165–6
carbon cycle, 1–6, 7, 19, 79–80
    biochemical subcycle, 5, 5–6
    fluxes, **1**, 1–5, 170, 217
    geochemical subcycle, 5, 5–6
    human influence on 217–25
    reservoirs, **1**, 1–5, 19, 217
carbon dioxide,
    anthropogenic sources, 217, 218, 219,
        220, 230, 227
    geochemical production, 101–2, 106,
        116
    in natural gas, 138
    from oxidation of CO and CH_4 in
        atmosphere, 223
    in photosynthesis and respiration,
        2–5, 10–12, 19–21, 33, 41, 63–4,
        78–82, 86, 101, 106, 110, 118, 121,
        151–2, 162, 164, 220
    role in global warming, 217–220
    variation in atmospheric levels, 7, 92,
        217–19
carbon monoxide, 4, 7, 223
carbon preference index (CPI), 203
carbonate, 1, 4, 5, 18–21, 147, 187
    compensation depth (CCD), 187, **188**
    equilibrium with bicarbonate, 188,
        218
    reservoir rocks, 134
    source rocks, 147
carbonisation of polynomorphs, 148
carboxylation, 20, 162, 164
carcinogens, 228, 232, 234
carnivores, 3, 13, 16
β-carotane, 174–5, 195, 198, 199
β-carotene, 54, 56, 174, 175, 195
carotenes, **53**, 175
carotenoids, 46, **53**, 55, 56, 69, 91, 155,
    174, 192, 195
    diagenesis, 170, 171, 174–5
    epoxide containing, 174, 175, 195
carrier rock, 129, 133, 134
carvone, 25, 26
catagenesis, **93**
    and biomarkers, 109, 139–140, 166,

    171, 172–3, 177–8, 189, 190, 193,
        197, 199–205
    indicators of, 146–50, 202–10
    of kerogen, 116–19, 121–23, 131,
        137, 139–40
catalysis, 93, 100, 203
catalyst, 35, **37**, 227
CCD (*see* carbonate compensation
    depth)
cedrene, 48
cell membranes, 31, 33, **42**, 155, 199
    bacterial, 43, 44, 47, 155, 157
    plant, 43, 52, 166
cell walls, 17, **31**, 43, 93
    bacterial, 33, 34, 44, 45, 108, 109,
        165
    fungal, 45
    plant, 33, 45, 108, 109
cellulose, **31**, 32, 33, 37, 58, 61, 94, 101,
    102, 104, 165, 166
cerinite, 99, 100
CFCs (*see* chlorofluorocarbons)
chair (*see* conformation, chair)
channel sands, 134
charcoal, 88, 89, 100, 238
chemosynthesis, 10, **11**, 79, 91
chemotrophs, 3, 81
chiral centre, **25**, 26, 28, 30, 34, 47,
    166–7, 176, 181, 183, 187, 190,
    199, 201
chitin, 32, **33**
chlorbactene, 174
Chlorobium (*see* bacteria, green
    sulphur)
chlorofluorocarbons (CFCs), **220**,
    220–1, 227
chlorophyllides, 169, 170
chlorophylls, 6, 10, 11, 55, **56**, 56–8,
    69, 70, 114, 140, 155, 203
    compositional variation in
        organisms, 47, 58, 155, 172, 194,
        195
    diagenesis, 111, 169–70, 172–4,
        195–6
Chlorophyta (*see* algae, green)
chloroplasts, 43, 155
cholestane, 50
cholestanol, 50, 51, 176
cholesterol, 46, 50–2, 53, 159, 160, 168
cholic acid, 53
choline, 42
Chromatium (*see* bacteria, purple
    sulphur)
chromatogram, **213**, 214–15, 231–2
    mass (*see* single ion)
    single ion, 214, 216
chromatography, 107, 115, 160, 211–16,
    231–2
chrysene, 137, 228
    methyl-substituted, 183, 184
chrysanthemic acid, 47, 48
Chrysophyceae, 76
ciliates, 9, 77, 195
cinnamyl lignin constituents, 162, 163,
    164

citric acid cycle (*see* Krebs cycle)
Cladium (*see* grass, saw-)
clarain, 100
clarite, 100
classification,
    of bitumens and pyrobitumens, 107
    of coals, 98–9, 103
    of humic substances, 95
    of kerogen, 113–115
    of oils, 137–8, 144–5
    of organisms, 8–9
    of plankton, 13, 14, 76
    of sedimentary basins, 83
    of source rocks, 131–2
clay, 77, 78, 232
    catalysed reactions, 129, 157, 166,
        176, 178
    Kimmeridge, 130, 131
    minerals, 232
Clostridium, 33, 78, 79, 80, 94
cloud condensation nuclei, 224
clubmosses, 15
coal, 1, 2, 87, 95, 98–106, 107, 220
    bituminous, **101**, 103–5, 106, 138,
        144, 146, 149, 150, 151, 201, 204
    boghead, 99, 104, 113, 120
    brown, 94, 95, 98, 99, **101**, 102, 103,
        104, 105, 113, 144, 146, 147, 151,
        186, 190, 204, 238
    cannel, 99, 104, 113, 120
    classification, 98–9, 103
    composition, 99–100
    detrital, 88
    formation, 6, 7, 87–9, 100–6, 108,
        115, 120, 144
    hard, 101, 103
    humic, 98, 99, 100, 101, 103, 104,
        105, 120, 131, 138, 201
    hydrocarbons, 109, 120, 132, 136,
        138, 146, 193, 205, 222, 230
    lithotypes, 100
    macerals (*see* macerals)
    microlithotypes, 100
    oil-prone, 120
    petrology, 99–100, 147
    rank, **100**, 101, 102, 103, 106,
        148–51, 205
    reserves, 3, 15, 16, 98, 132, 141–4,
        146
    sapropelic, 98, **99**, 104, 113, 120
    structural evolution, 103–6, 116
    swamps and forests, 6, 15, 16, 64, 94,
        115
    temperature of formation, 102
    volatiles, 102, 103, 148, 150
    water content, 102, 103, 151
coalification, **100**, 102, 103, 104, 105,
    120, 138, 148–9, 166, 193, 201,
    204
    biochemical stage, **100**, 101–2
    geochemical stage, **100**, 102–3, 105
coals,
    Asian, 143
    Australian, 120, 132, 143
    European, 98, 132, 143

coals *continued*
  Indonesian, 132, 143
  North American, 132, 143
  Westphalian, 98
coccolithophores, **13**, 18, 72, 76, 77,
    143, 144
  chemical composition, 4, 111, 158,
    160, 162, 187, 224
co-elution, **205**, 214
coenzyme A, 38, 39, 41
  acetyl, 31, 33, 41, 46, 164
  malonyl, 41
coenzymes Q (*see* ubiquinones)
collagen, 37
collinite, 99
columbin, 47, 48
comets, 8, 15
trans-communic acid, 47, 48
compensation,
  depth, 69
  light intensity, 69
condensate, 118, **122**, 131, 132
condensation, **31**, 35, 36, 47, 58, 59, 94,
    98, 101, 102, 104, 106, 110
conessine, 53
configuration, **25**, 27, 28
  absolute, **26**, 27
  in acyclic isoprenoids, 199
  in amino acids, 34
  in carbohydrates, 30–3
  in steroids, 52–3, 159–60, 166,
    176–9, 190, 199–200, 214, 215
  in terpenoids, 47–8, 166, 180–3, 190,
    200–1, 215
  in unsaturated fatty acids, 40, 157–9
conformation, **30**, 31
  chair, **30**, 50, 51
  envelope, 30, 31
congeners (*see* PCBs)
conifers, 15, 47, 60, 101, 107, 161
coniferyl alcohol, 58, 59, 60
continental
  margin, **89**, 144
  rise, 69, 89
  shelf, 69, **89**, 90, 92, 132
  slope, 69, 82, **89**, 90, 92
convergence, **66**, 74
Cooksonia, 15
copepods, **14**, 77, 159, 170
  lipid content, 61–2, 159, 168
coprophagy, 168
Coriolis effect, 66, 82
coronene, 228, 229, 230
correlation of petroleums and source
    rocks, 151–2, 202, 210–11
Corynebacterium, 78
coumaric acid, 163
coumaryl alcohol, 58, 59
CPI (*see* carbon preference index)
cracking,
  of petroleum components, 122–3,
    128, 129, 131–2, 138, 152, 190,
    201–2
  temperature of, 128, 131, 132
  zone, 122

Cretaceous anoxic event, 91–2
Cretaceous/Tertiary boundary, 15
crustaceans, 14, 33, 53, 159, 168
Cryptophyceae, 76
cutan, **45**, 107, 108, 109, 110
cuticles, 100, 102, 104, 108–9, 114, 120
cutin, **45**, 108, 114, 159, 171
cutinite, 99, 100, 109
cyanadin chloride, 60
cyanobacteria, 3, 8, **9**, 11, 13, 17, 164,
    76, 80, 143, 144, 164, 165, 224–5
  chemical composition, 43, 56, 58,
    109, 155, 158, 159, 160, 225
cyanobacterial mats, 10, 18, 21, 62, 85,
    87, 90, 194
Cyanophyta (*see* cyanobacteria)
cycads, 15
cycloartenol, 50, 52
cyclohexane, 137
  alkyl-, 136, 174
cyclopentane, 137
  alkyl-, 136
cypress, 87, 88, 89
cysteine, 34, 36
cytochromes, 37, 58, 173

DDA, 233–4
DDD, 233–4
DDE, 233–4
DDT, 217, 233–4, 235, 236
deamination, 59, 60
de-asphalting of oil (*see* asphaltenes,
    precipitation)
decarboxylation, 41, 45, 46, 52, 105,
    154, 166, 172–3, 190, 195
decomposers, **3**, 63, 70, 78, 80, 81, 86,
    87, 93, 98, 101
decomposition of organic matter, (*see*
    biodegradation)
deforestation, 217, 218, 222
defunctionalisation, 101, 106, 109, 110,
    166, 172, 174, 179, 181, 190
dehydration, 41, 58, 105, 166, 178, 195,
    200
dehydroabietane, 180
dehydroabietic acid, 180
dehydroabietin, 180
dehydrogenation, 41, 58, 110, 116, 166,
    179
Delta,
  Mahakam, 87, 88, 90, 132, 194, 203,
    208
  Niger, 194
  Nile, 132
demethylation, 105, 205
denitrification (*see* nitrate reduction)
deoxyphylloerythroetio porphyrins (*see*
    porphyrins, DPEP)
depolymerisation, 31, 78, 93–4, 101,
    106–7, (*see also* hydrolysis)
depositional environments, 63, 148, 197,
    198–9, 201
  aerobic, **6**, 82, 92, 98, 172, 176, 195
  anaerobic, **6**, 63, 82, 83, 86, 88, 91,

    92, 102, 113, 114, 144, 170, 171,
    172, 173, 222
  anoxic, **6** (*see also* anaerobic)
  aquatic, 8, 9, 17, 62, 63, 64, 65, 94,
    106, 108, 224, 231
  bog, marsh and swamp, 65, 87, 88,
    90, 144, 145, 222
  brackish, 114
  coastal, 16, 70, 75, 145, 197, 225, 228
  continental shelf and shallow seas, 65,
    75, 83, 89–90
  deltaic, 4, 83, 85, 86, 88, 90, 194
  deserts, 194
  dysaerobic, **6**, 78
  epeiric, 89
  estuarine, 4, 64, 65, 70, 76, 83, 159,
    160
  eutrophic, **75**, 76
  freshwater, 3, 4, 15, 64, 83, 222
  hypersaline, 9, 87, 197
  lacustrine, 3, 9, 16, 65, 82, 83, 84–7,
    114, 145, 159, 160, 198–199, 222,
    228
  lagoonal, 83, 85, 88, 113, 145
  land-fill, 222
  marine, 3, 17, 64, 65, 81, 82, 83, 88,
    89–92, 114, 159, 160, 197, 198–9,
    222, 225, 236
  molecular distinction of, 194–9
  oligotraphic, **74**, 75, 76, 83
  oxic, **6** (see also aerobic)
  paralic, 83, **88**
  sabkha, **86**, 90
  saline, 15, 199
  subtoxic, **6** (*see also* dysaerobic)
  terrestrial, 65, 94, 98, 159, 160
  tundra, 223
Desulfobacter, 80, 157, 159
Desulfobulbus, 157, 158
Desulfovibrio, 80, 81, 158
deposit feeders, 53
desmosterol, 53
detritivores, **3**, 63, 78, 82
detritus, **3**
  adsorption of pollutants, 233
  preservation/degradation, 4–5, 11,
    70, 71, 73, 75, 77, 78–81, 82, 90,
    91, 107, 120, 121, 224–5
  sedimentary inputs 13, 17, 77, 83, 89,
    106, 113–15, 120, 157, 168, 194
  sinking rate, 77, 168
deuterium, 151, 152
diadinochrome, 175
diadinoxanthin, 45, 56, 174, 175
diagenesis, **93**, 93–4, 190, 198, 204
  and coal, 99, 100–2, 103, 120
  and humic material, 96, 98
  and kerogen, 106–12, 113, 115–18,
    121
  maceral changes, 148–50
  at molecular level, 109–10, 138–9,
    153–4, 165–86, 189, 192–3,
    195–7, 199, 200, 202
  in sediment, 170–86, 196
  in water column, 168–70, 195

diahopanes, 193
diasteranes, 139, 140, 141, 177, **178**, 214, 215
    4-methyl, 179
diasterenes, **176**, 179
diastereomers, 25, 26, **28**, 30, 212
diasteroids, 50
diatoms, **13**, 18, 64, 68, 72, 76, 77, 86, 87, 143, 144, 225
    chemical composition, 53, 55, 61, 70, 158, 160, 170
diatoxanthin, 54, 55, 174, 175
dibenzofuran, 137
dibenzothiophene, 136, 137
    alkyl-, 138
2,2-dichloro-di-(*p*-chlorophenyl)- acetic acid (*see* DDA)
2,2-dichloro-2,2-di-(*p*-chlorophenyl)- ethane (*see* DDD)
2,2-dichloro-2,2-di-(*p*-chlorophenyl)- ethylene (*see* DDE)
diffusion, 78, 130, 131, 164, 237
diglycerides, 39
dihydroactinidiolide, 175
dihydrophytol (see phytanol)
dihydroxyacetone phosphate, 29
dimethylnapthalene index (DNI), 204, 207, 208
dinoflagellates, **13**, 18, 64, 68, 76, 77, 143, 144
    chemical composition, 53, 55, 61, 154, 158, 160, 178, 192
Dinophyceae (*see* dinoflagellates)
dinosterol, 53, 160
dioxins, 236
dipeptides, 35
diploptene, 47, 49, 182, 183, 184
disaccharides, **31**
dissolved organic carbon (DOC), 2, 4, 75
diterpenoids, **46**, 47, 48, 109, 140, 214
    diagenesis, 179–81, 192, 201
divergence, **66**, 67, 74, 75, 92
diversity of species, 15, 64, 87
DMS (*see* sulphide, dimethyl)
DNI (*see* dimethylnapthalene index)
DOC (*see* dissolved organic carbon)
Doldrums, 66
dolomite, 1, 7
drimane, 182, 183, 192
drimenol, 48, 182, 183, 185, 186, 192
dry deposition, **236**
durain, 100
durite, 100

E_h (*see* redox potential)
electromagnetic spectrum, 55, 218
elemental composition,
    of coal, 100, 102–5, 151
    of humic material, 95–7, 238
    of kerogen, 112–18, 148
    of oil, 135
ellagic acid, 60
elution, **212**, 215
Emiliania huxley i, 111, 162, 187, 188

emoldin, 60, 61
enantiomers, **25**, 26, 27, 28, 166
    dextrorotatory, 26, 27
    laevorotatory, 26, 27
Enteromorpha, 55
envelope (*see* conformation, envelope)
enzymes, 11, 34, **37**, 41, 43, 45, 46, 47, 50, 52, 58, 64, 69, 93, 158, 170
EOP (*see* even-over-odd predominance)
epiabietic acid, 180
epilimnion, 67, **68**, 73, 74, 225
epimerisation, **166**, 167, 187
epimers, **28**, 166, 167, 175, 177, 178, 179, 181, 182, 200
epiphytes, **76**
equatorial substituents, **30**, 31
ergosterol, 52, 53, 149
erythrose-4-phosphate, 59
esters, 45, 47, 56, 60, 97, 113
    glyceryl, 39, 43, 44
    hydrolysis of, 35, 153–154
    steryl, 39, **45**, 154, 159, 170
    wax, 39, **43**, 61, 62, 154, 157, 170
ethane, 138, 222
ethanol, 79, 135
ether lipids, 42–43, **43**, 47
ethers, 23, 43
    biphytanyl, 43, 44, 155, 191
    phytanyl, 43, 44, 109, 194, 196, 197
etioporphyrins (see porphyrins, etio)
eubacteria, **9**, 33, 158 (*see also* bacteria)
eudesmol, 48
Euglenophyceae, 76
eukaryotes, 7, 8, 9, 13
euphausiids, 170
euphotic zone (see photic zone)
eustatic sea level, 89
eutrophication, **224**, 224–5
evaporites (*see* sediments, evaporitic)
even-over-odd predominance (EOP), **154**, 157, 198
evolution,
    of Earth's surface, 6–8
    of life, 6–16, 21, 62
exinite, 99, **100**, 102, 103, 104, 112–16, 120, 126, 147, 149
    fluorescence, 99, 112, 150

facies, **84**, 87, 92, 207
farnesol, 46, 47, 48, 57
fats, 33, 34, **39**, 39–42
fatty acids, **39**, 39–42, 46, 52, 56, 154, 155
    biosynthesis, 33, 34, 41, 158
    branched, 155–7, 158–9
    cyclopropyl, 156, 159
    α,ψ-di-, 159, 172
    diagenesis, 96–8, 110, 168, 170–2, 190
    hydroxy, 45, 157, 159, 171, 190
    shorthand notation scheme, 40
    as source indicators, 157–9
    unsaturated, 40, 157–8, 168, 170
    volatile, 79

fermentation, 12, 78, **79**, 80, 151, 152
ferns, 15
filter feeders, 231
fires,
    peat, 88
    vegetation, 87, 88, 99, 100, 222, 228, 230
flagellates, 8, 9, 76
flame ionisation detector, 147, 211, 213, 214, 215
Flavobacterium, 9, 78
flavonoids, 60
fluoranthene, 229
food chain, 3, 228, 230, 233, 234
    detrital, **3**, 17, 82
    grazing, **3**, 16, 17
foraminiferans, **13**, 14, 18, 77, 188, 189
forests, 64, 88, 89, 90
formate, 79, 80
fossil fuels, 2, 5, 135, 220, 221
    calorific value, 103, 135, 151
    combustion, 2, 4, 138, 217, 218, 220, 222, 224, 225, 227–8
    geographical distribution, 143–6
    temporal distribution, 141–4
fossil record, 17–18
freidelin, 49
fructose, 28, 29, 30, 31, 33
fucoids, 55
fucose, 29, 161, 167
fucosterol, 53, 192
fucoxanthin, 54, 55, 170, 171, 174, 195
fucoxanthinol, 170–1, 174
fulvic acids, **95**, 96, 97, 106
fumaroles, 9
functional groups, **22**, 23
    in amino acids and peptides, 34, 35
    in carbohydrates, 28–30
    in coal, 100–105
    in humic material, 95–7, 238
    in kerogen, 106, 112–18
    in lignins and tannins, 58–60
    in lipids, 39, 43–6, 53, 166, 172, 192
fungi,
    chemical composition, 31, 33, 34, 47, 50, 53, 55, 56, 61, 101, 155, 157–9
    classification, 8, 9, 171
    contribution to sedimentary organic matter, 17, 99, 100, 101, 120
    degradation of organic matter, 3, 81, 93–4, 101, 106–7, 120, 168, 172, 190
    evolution, 14, 18
furanoses, 28, 30, 31
furan, 23, 28, 96, 137
fusain, 100
fusinite, 99, 100, 109, 115, 228
    semi-, 99, 100, 228

galactosamine, 28, 29
galactose, 29, 33, 34, 161, 162, 166
galacturonic acid, 29, 32, 33, 166
gallic acid, 60
gammacerane, 191, 195, 198

gas (natural) 1, 2, 107, 114, 124, 136, 222
  carrier, 212
  chromatography, 107, 115, 160, 202, 205, 211–212, 231–2
  chromatography/mass spectrometry, 205, 211–16, 229
  composition, 122, 127, 138
  condensate, 118, **122**, 127, 131, 132, 149, 204, 207
  generation, 88, 100, 114, 115, 119, 121, 122, 128–32, 147, 150, 152, 202
  generation temperature, 121, 122, 128–9, 131, 132
  migration, 133–4, 141, 144
  reserves, 142, 144, 146
  solubility in oil, 122, 130–1, 134, 152
  trace, 4, 220–7
  wet, 118, 121, **122**, 127, 149, 204, 207
genetic potential, 112, 115, **147**, 148
geochemical fossils, 109, 112 (*see also* biomarkers)
geochromatography, 141
geomacromolecules, 94, 106, 109
geopolymers (*see* geomacromolecules)
geothermal gradient, 103, 118, **124**, 125, 126, 178, 208
  and molecular maturity parameters, 208–11
geranylgeranyl, 46, 57
gibberellic acid, 47, 48
gibberellins, 47
ginkgos, 15
global warming, 217–25
  potential of gases, **221**, 223, 227
Globigerina, 77, 189
Glossopteris, 16
glucosamine, 28, 29
  N-acetyl, 33, 34
glucose,
  degradation, 12, 79, 101, 165, 166
  occurrence and function, 33–4, 41, 43, 58–9, 60, 101, 161, 166
  synthesis and structure, 10, 28–31
glucuronic acid, 29
glutamic acid, 34, 36
glutamine, 36
glyceraldehyde, 25, 26, 27
  phosphate, 29
glycerides, **39**, 39–43
glycerol, 33, 39, 41, 42, 43, 44
glycine, 34, 36, 94
glycogen, 31
glycolipids, 39, 42–3, **43**
glycolysis, **31**, 33
glycoproteins, 34, 37
glycosides, **61**, 159
glycosidic bond (*see* bond, glycosidic)
Gondwanaland, 15, 16
graben, 124
graphite, 1, 106, 116, 121, 122, 128, 129, 149
grass, 88, 161

cord-, 64
eel-, 64
saw-, 87
sea-, 70
greenhouse effect, 217–20, **218**, 224
gymnosperms, **15**, 16, 18, 56, 154, 161
  non-woody, 161, 162
  woody, 161, 162
gypsum, 7, 8, 87
gyres, **66**, 67, 74

haems, **58**, 173
Halobacterium, 34
halocarbons, 225–7, **227**
halocline, **67**, 86
Haptophyceae, 76, 158, 160
heat flow, 124, 125, 128, 208, 209
helium, 6, 138, 212
hemicellulose, 33, 94, 101, 161, 165, 166
herbicides, 234
herbivores, 3, 16, 17, 60, 62, 64, 73, 77, 94, 170, 195
heteroatoms, **22**, 96, 113, 116
heterogrophs, 3, 8, 13, 16, 93
heteropolysaccharides, **31**, 33
γ-hexachlorocylcohexane (*see* lindane)
hexane, 24, 25, 39, 212
  methylated, 203
  ratio to methylcyclopentane, 202
hexaprenol, 191, 192
hexoses, 28, 94
HI (*see* hydrogen index)
hibaene, 47, 48
higher plants, 8, **14**
  chemical composition, 22, 31–4, 37, 39–45, 47–53, 55, 56, 57–61, 62, 97, 101, 103, 112, 138, 140, 154–5, 157–62, 165, 170–1, 186, 190, 203, 231
  contribution to sedimentary organic matter, 6, 16–17, 62, 63, 82, 85, 86, 87, 96–9, 101, 106, 107, 112, 114–15, 120, 138, 140, 146, 148, 157, 159–60, 165, 171, 186, 190, 192, 194, 203
  decomposition, 86, 87, 90, 94
  productivity, 64–5, 69, 70, 223
histidine, 36
homopolysaccharides, **31**, 33, 34
hopanes, 47, 110, 112, 120, 139, 140, 141, **183**, 201
  analysis, 205, 214–16
  demethylated, 191, 193, 199
  environmental/source indicators, 193–4, 195
  methylated, 50, 193, 214
  'rearranged' (*see* diahopanes and neohopanes)
  C-ring opened (*see* secohopanes)
  stereoisomerism, 200–1, 204
hopanoic acids, 182, 183
hopanoids, **47**
  aromatic, 183–4, 186 (*see also* benzohopanes)
  diagenesis, 111–12, 182–3, 185–6

occurrence and function, 47, 49, 52, 109, 155, 182
C-ring opened (*see* secohopanoids)
structural notation scheme, 50–1
hopenes, **183**, 184
hormones, 37, 53
horsetails, 15
Howarth structure, **28**
humic acids, 95, 95–7, 106, 238
humic material, **95**, 95–8, 99, 106, 110, 112
  classification and occurrence, 97–8
  composition and structure, 95–8, 109
  formation, 95, 98, 101
  freshwater, 95, 96, 97
  marine, 95, 98
  metal complexation, 97, 234, 238
  pollutant association, 233, 234, 238
  terrestrial, 95, 98, 106
humic substances (*see* humic material)
humification, 98, 99
humin, 94, **95**, 98
huminite, **99**, 102, 103, 104, 148
hydrocarbons, 1, 22, 23, 24
  aliphatic (*see* alkanes and alkenes)
  aromatic, 136–8, 139, 140, 141, 145, 151, 166, 174, 180, 183–186, 192, 201, 202, 204, 205–10, 213, 214, 227–30, 232
  biomarker, 47, 53, 110, 138–40, 148, 166, 174, 176–86, 190–7, 198–205, 206, 208–16
  chromatographic analysis, 211–16
  gaseous, 107, 118, 122, 130, 131, 134, 137, 138, 146, 152, 2;2, 222
  gasoline range, 202
  generation, 106, 116, 118, 119, 120–2, 123, 124, 127, 131, 150
  liquid, 107, 118, 120, 129, 130, 131, 134, 137, 147, 202, 212, 230
  in petroleum and bitumen, 107, 108, 114, 120, 129–30, 135, 136–41, 145, 146, 147, 148, 150, 151, 202–11, 230–2
  saturated (*see* alkanes)
  as source/environmental indicators, 45, 115, 168, 190–7, 198–9
  thermal alteration, 122, 123, 199–210
  viscosity, 130
hydrochlorofluorocarbons (HCFCs), 227
hydrodynamic flow, 133, 141
hydrofluorocarbons (HFCs), 227
hydrogen index (HI), **148**
hydrogenation, 110, 166, 170, 176, 181, 192 (*see also* reduction)
hydrolysis, 30, **35**, 41, 79, 94, 96, 98, 108, 153, 154, 161, 162, 170, 172, 176, 196
hydrothermal
  activity, 222, 230
  vents, 125, 152
hydroxyacetophenone, 163
hydroxybenzaldehyde, 163

hydroxybenzoic acid, 163
hydroxycinnamate, 59, 60
hydroxyhopanone, 50, 51
p-hydroxyl lignin constituents, 162, 163, 165
hydroxylation, 59, 60
hypersalinity, **64**, 90, 194–5
hypolimnion, 67, **68**, 73, 74, 78

indene, 23
indenopyrene, 229
intertinite, **99**, 100, 102, 103, 104, 112, 113, 115, 120, 122, 126, 146, 147, 149
   reflectance, 99, 150
insecticides, 47, 217, 233, 234
insolation, **65**, 66, 67, 73, 218, 219, 224
isoarborinol, 49
isoleucine, 36, 155
   racemisation, 187
isololiolide, 175
isomerism, 29, 111, **166**, 167, 176, 178, 182, 190
   in amino acids, 187
   at C-6 and C-10 in pristane, 196, 199, 204, 209
   at C-17 and C-21 in hopanoids, 201, 206
   at C-22 in hopanoids, 200, 204, 206, 209, 210
   at C-14 and C-17 in steroids, 199–200, 204, 206
   at C-20 in steroids, 178, 199, 204–6, 208, 209, 210
   of C=C bonds, 111, 166, 176, 178, 181, 183, 190, 200
   cis/trans convention, 26, 27, 28, 40
   E/Z convention, 26, 27, 40
   geometric 26, 27, **28**, 40, 157
   of methyl groups, 186, 193, 201, 205–8
   optical, **25**, 26, 28
isopentenyl pyrophosphate (*see* isoprene unit)
isoprene unit, **46**
isoprenoids, 111 (*see also* terpenoids)
   acyclic, **46**, 114, 139, 140, 214
isorenieratane, 174
isorenieratene, 54, 91, 174
isotopes, **20**
   carbon, 18–21, 193–4
   oxygen, 187, 188
isotopic fractionation, **20**, 123–4
   carbon, 148, 151–2, 162–5, 193–4, 198, 222–3
   hydrogen, 151, 152
   oxygen, 188
   sulphur, 151

juvenile volatiles, 7

kaurene, 47, 48
kelp, 55, 70
keratin, 37, 94

kerogen, 94, **106**, 106–18
   amorphous, **109**, 113, 120
   classification, 113–15, 128–9, 148, 150, 198
   composition, 98, 100, 112, 116–18, 123, 127, 139, 148, 150, 151, 152, 154, 172, 174, 196, 203
   formation, 106–9, 154, 190, 197, 203
   humic, 112
   hydrogen-rich, 99
   inert, **128**, 129
   labile, **128**, 129, 131, 132, 152
   petroleum generation, 119–33, 149, 152, 192, 193, 196–7, 202–3, 231
   pyrolysis, 112, 114, 115, 152, 203
   reactive, **128**, 129, 131
   refractory, **128**, 132, 152
   sapropelic, 112
   structure and evolution, 106, 115–17, 118, 119, 122, 123–4, 125, 126, 147
   sulphur-rich, 111, 115, 135, 137
   type I, **113**, 113–14, 115, 116, 118, 119, 120, 123, 148, 150, 198
   type II, **113**, 114, 115, 116, 117, 118, 119, 123, 148, 198, 199
   type III, **113**, 114–15, 116, 118, 119, 120, 148, 207
   type IV, **113**, 115, 119
   type II-S, **114**, 123
   volatiles, 106
ketones, 23, 28, 43
   long-chain III, 155, 187–9
ketoses, 28, 30, 33
kinetic,
   isotope effect, **20**, 223
   models of hydrocarbon generation, 125–9
Krebs cycle, **31**, 33, 34, 41

lactate, 79, 157
lactobacillic acid, 155
Lake,
   Baikal, 74
   Chad, 85
   Kivu, 85
   Maracaibo, 91
   Salt, 85
   Tanganyika, 74
lakes, 75, 83, 84–7, 90, 192
   amictic, 68
   cold monomictic, 68
   dimictic, **67**, 68, 76
   eutrophic, 78, 83
   freshwater, 194, 198
   glacial, 67, 85
   holomictic, 74
   hydrologically closed, **85**, 86–7
   hydrologically open, **85**, 85–6
   hypersaline, 195
   meromictic, **74**
   polymictic, **67**, 74
   saline, 86, 194, 195, 198, 199
   soda, 81
   stratified, 73–4

temperate, 73, 74
   tropical, 74, 76
   volcanic, 85
   warm monomictic, **68**, 74
Laminaria, 70
lamnosterol, 45, 50, 52
Laurasia, 15, 16
lauric acid, 40
leaching, 5, 101
lecithin, 42
Lepidodendron, 15
leucine, 36, 155
lichens, 8, 61
light,
   intensity and photosynthesis, 68–70
   wavelength and photosynthesis, 55
lignin, 58, 58–60
   constituent source indicators, 162–4
   contribution to sedimentary organic matter, 62, 96, 98, 101, 102, 104, 109, 110, 115, 138
   diagenesis, 93, 94, 165–6, 170
   formation, 58–60
   occurrence and function, 22, 37, 58, 61, 154
   structure, 59
lignite 86, 90, **94**, 101, 103 (*see also* coal, brown)
limestone, 1, 6, 84
limonene, 48
lindane, 234
linoleic acid, 40
linolenic acid, 40
lipids, 22, **39**, 39–58
   bound, **153**, 154
   compositional variation in organisms, 61–2, 154–7
   contribution to sedimentary organic matter, 98, 100, 106, 108–10, 138, 193
   diagenesis, 79, 94, 168–86, 189
   free, **153**, 154
   occurrence and function, 33, 39–58, 61, 62, 93, 153, 162, 188, 192
lipopolysaccharides, **33**, 43, 159
lipoproteins, 39, **52**
liptinite, 100, 104, **112**, 113, 114, 115, 116, 120, 122, 128, 147, 149
   fluorescence, 112, 150
loliolide, 171, 175
long-chain ketones (*see* ketones, long-chain)
lupane, 47
lupeol, 47, 49
lutein, 175
lycopane, 173, 191
lycopene, 54, 56, 175, 191
Lyngba, 87
lysine, 36
lyxose, 29, 161

macerals, **99**, 100, 106, 108, 109, 112–15
   microscopic analysis, 99–100, 112, 147–50

Macrocystis, 70
macroplankton, 13, 14, 77
Maillard reaction, 98
maltenes, **107**
mangroves, 64, 87, 90
    swamps, 64, 65, 76, 83
mannose, 29, 33, 34, 161, 162, 166
mannuronic acid, 32, 34
marine,
    regressions, 89, 143, 144
    transgressions, **89**, 92, 143, 144, 198
mass: charge ratio, 213
mass spectrometry, 20, 211–16
maturation, **100**
    of coal and kerogen, 104, 111, 118,
        124, 125, 131, 144, 146, 148–51,
        193, 202–11
    of petroleum, 131, 140, 141, 152, 190,
        201
maturity indicators, 148–51, 208, 216
    factors affecting, 208
    molecular, 149, 150–51, 199–211
    optical, 204, 205
    ranges, 204, 205
    temperature dependence, 199, 208–11
megaplankton, 13, 14
melanoidins, **98**
menthol, 47, 48
mesopause, 226
mesosphere, 226
metagenesis, **116**, 118, 119, 121, 148,
    149, 204
metal chelation, 197
    by coal, 100
    depositional conditions, 197–8
    by humic material, 97, 238
    by NSO compounds in oils, 135, 197
    by porphyrins, 171–3, 197
metazoa, 7, **13**
methane,
    anthropogenic, 221–2
    atmospheric, 2, 4, 7, 221–3
    biogenic 4, 12, 79, 80, 81, 101, 106,
        121, 122, 151–2, 194, 202, 221–3
    hydrates, 222, **223**
    oxidation, 6, 79, 81, 220, 223
    role in global warming, 221–223
    sinks for, 223
    sources and isotopic signatures,
        123–4, 151–2, 221–3
    sulphonate, 224
    thermally generated 102, 115, 118,
        119–20, 122, 123, 138, 146, 151–2,
        202, 222–3
Methanobacillus, 80
Methanococcus, 80
methanogenesis, 79–81, 121, 122, 151,
    174, 195, 222
methanogens, 9, 13, 43, 58, 78, **80**, 81,
    155, 165, 191, 194, 195, 196, 197,
    222
methanol, 79, 80
methionine, 36
methyl group migration (*see* isomerism
    of methyl groups)

1,2-methyl shift, **186**
methylated amines, 79, 80
methylation, 59, 60, 205
Methylomonas, 33, 81, 194
methylotrophs, 79, **81**, 194, 222
methylphenantrhene index (MPI), 204,
    205, 207
methyltrophy, 79
mevalonate, 46
micrinite, 99, 100
microbial mats, (*see* cyanobacterial
    mats)
Micrococcus, 79
Microcoleus, 87
microplankton, 13, 14, 77
mid-ocean ridges, 69, 89, 92, 125, 144,
    152
migration of petroleum, 120, 129–33,
    140, 150, 151, 152, 202, 208, 211
    distance, 133, 151, 210
    mechanisms of primary, 129–31
    primary, **129**, 129–32
    secondary, 132–3, 140
mineralisation, **78**, 79, 94, 110 (*see also*
    biodegradation of organic matter)
mobile phase, 212
molecular ions, 212
    fragmentation, 212, 213, 214, 216
molluscs, 33, 158, 187
Monera, 8
monoaromatic steroidal
    hydrocarbons, 110, 139, 140, 166,
        177, 178, 201, 204, 214
    aromatisation, 201, 204, 206, 208,
        209, 210
Monochrysis lutheri, 176
monoglycerides, 39
monosaccharides, **28**, 28–34, 58, 93
    source indicators, 161–2, 165, 166
    stereoisomerism, 27, 29–31, 32
monoterpenoids, **46**, 47, 48, 179
moretanes, **201**
mosses, 8, 14, 87, 159
MPI (*see* methylphenanthrene index)
muramic acid,
    N-acetyl, 32, 33
murein, 32, **33**, 43, 45, 110
mussels, 231, 232
mutarotation (*see* anomerism)
mycolic acid, 44, 45
myrcene, 48
myristic acid, 40
Mytillus edulis, 94, 231

NAD (*see* nicotiniamide adenine
    dinucleotide)
NADP (*see* nicotinamide adenine
    dinucleotide phosphate)
nanoplankton, **13**, 14, 76, 144
naphthalene, **61**, 114, 137, 138
    alkyl-, 136, 137, 138
    alkylpherhydro-, 136, 137
    methyl substituted, 185–6, 193, 207
naphthenes, 107, 115–16, **136**, 138, 202
naphtheno-aromatics, 136, 137

neoabietic acid, 180
neohopanes, 50, 51, 140, 193, 215
nickel: vanadium ratio in oils, 197–9
nicotinamide, 38
nicotinamide adenine dinucleotide
    (NAD), 38, 39
    phosphate (NADP), 10, 11, 38–9, 46
nitrate, 70, 78–81, 110, 224, 225, 227
    reducers, 78, **79**, 80, 81
    reduction, 78, 79, 81
nitrification, 79, **80**
nitrite, 79, 80
Nitrobacter, 81
nitrogen,
    cycle, 79–80
    fixation, 79, 80
    oxides (NOx), 135, 227
Nitrosomonas, 81
nitrous oxide (N_2O), 220, 221, 227
Nocardia, 45, 231
norhopanes, 193, 194, 214
Nostoc, 80
NSO compounds in oil, 107, 135, 137,
    138
nuclear magnetic resonance (NMR), 96
nucleic acids, 8, 9, 37, 161
nucleotides, **37**, 38, 39, 161
nutrients, 65, **70**, 70–6, 82, 83, 90, 92,
    144, 155
    biointermediate, 71
    biolimiting, **70**, 71, 72, 220, 225
    eutrophication, 224–5
    phytoplankton blooms, 72–4, 224

Ocean,
    Atlantic, 66, 72, 73, 88, 92, 188, 198
    Indian, 66
    Pacific, 66, 72, 73, 92, 188
    Tethyan, 92
oceans,
    anoxic events, 91
    carbon cycle, 65–7, 71–3
    deep circulation, 65, 66, 82, 91, 92
    latitudinal variation in
        phytoplankton and productivity,
        15, 71–3, 76–7
    pollutants, 218, 228, 233
    surface currents, 66, 67, 74, 219
    thermal stratification, 65–7, 71–3
odd-over-even predominance (OEP),
    120, 138, 140, **154**, 172, 190, 198,
    203
OEP (*see* odd-over-even predominance)
OI (*see* oxygen index)
    aromatic-asphaltic, 138, 141, 145
    aromatic-intermediate, 137, 138, 141
    aromatic-naphthenic, 138, 141
    biodegradation, 140, 141, 231, 232
    classification, 137–8, 144–5
    composition, 135–41, 152, 186, 192,
        194, 202, 209
    conventional, 142, **144**, 144–5
    correlation with source rocks, 210–11
    de-asphalting, 140, 141
    density, 135

distillation fractions, 135–6
essential, 47
generation, 93, 114, 118, 120, 121, 122–5, 127–33, 134, 202, 208
heavy, 134, 135, 141, 142, 144, **145**
light, 134, 135, 141
migration of, 120, 133–4
naphthenic, 138
paraffinic, 137, 138
paraffinic–naphthenic, 137, 138, 141, 145
potential, 109, 113, 114, 115, 116, 118, 120, 147
pour point, 135
reserves, 142, 143–5, 146
shale, 6, 84, 99, 107, 114, **145**
shows, 147
solubility in gas, 122, 131
source rocks, 22, 125, 129, 130, 131, 143–5, 151
specific gravity, 133, 135
spills, 230–2
stringers, 130, 132
sulphur-rich, 123
temperature of generation, 121, 123, 124, 128–9, 130, 131, 149
thermal alteration, 128–9, 131, 140, 141, 144
water washing, 140, 141
waxy, 120
weathering, 230
window, **122**, 127, 128, 150, 201, 202, 204, 205, 207, 208
oils,
African, 143, 194, 198
Australian, 143, 192
Central Asian, 143
Chinese, 195
Indonesian, 143, 192, 194
Middle Eastern, 143, 144, 198
North American/Canadian, 143, 145
North Sea, 143
Russians, 143, 145
South American, 143, 145, 197
Okefenokee swamp, 88–9
okenone, 54, 56, 175
oleanane, 47, 140, 181–2, 193, 194, 200
isomerism, 181
oleanenes, 181, 182
oleic acid, 40, 157
oligosaccharides, **31**
onoceranes, 183
onocerins, 49, 183
orogenic events, 125
oxaloacetic acid, 162
oxidation, **6**, 39, 70–1 (*see also* respiration)
biosynthetic, 39, 52
of DMS in atmosphere, 224
metabolic/diagenetic, 12, 39, 41, 42, 45, 81, 88, 159, 171–3, 174–5
of methane in atmosphere, 223
oxychlorin-p$_6$, 171
oxygen
atmospheric levels, 7, 12

index (OI), **148**
minimum layer, **82**, 83, 89, 92
ozone, 223, 225–7
depletion, 225–7
layer, 7, 12, 14, 226
role in global warming, 220, 221

PAHs (*see* polycyclic aromatic hydrocarbons)
palaeotemperature measurement, 187–9
palmitate, 41
palmitic acid, 40
palmitoleic acid, 40, 157
paraffins, 101, 107, 115, **136**, 138, 202
particulate organic carbon (POC), 4 (*see also* detritus)
PCBs (*see* polychlorinated biphenyls)
PDB (*see* Peedee belemnite)
peat, 2, 84, 86, 98, 99, **101**, 204
composition, 101, 102, 104, 150, 151, 182
erosion, 88
formation, 3, 6, 82, 87–9, 101
swamps/bogs, 62, 87–9, 99, 101, 173
peatification, **100**, 101, 102, 103, 104
pectins, 32, **33**, 162, 165, 166
PEE (*see* petroleum expulsion efficiency)
Peedee belemnite (PDB), 20
pelagic organisms, **3**
pentamethyleicosane, 191, 198
pentane, 122, 138
methylated, 203
pentoses, 28, 37, 94, 161
peptide, 32, 33, 37, 93
linkage, **35**
peptidoglycans, 33, 34
peridinin, 54, 55, 174
permeability, **129**, 130, 132, 133, 134
Persian Gulf, 90
pesticides, 234
perylene, 229
petroleum, **107**
buoyant flow, 133, 134
composition, 47, 107, 122, 127–9, 131, 135–41, 144
expulsion efficiency (PEE), **131**, 131–2, 133
generation, 50, 88, 93, 100, 114, 115, 116, 118, 119–25, 127–33, 134, 147, 152, 202, 204, 208, 221
generation and effect of temperature and time, 121–2, 123, 124–9, 130–2, 146, 149
generation index (PGI), **131**, 132
migration, 129–133, 134, 141, 144
potential, 71, 109, 113, 114, 115, 116, 110, 120, 146–8
reserves, 142, 143–5, 146
reservoirs and traps (*see* reservoir rocks)
source rocks (*see* source rocks)
PFI (*see* pristane formation index)
PGA (*see* phosphoglyceric acid)
PGI (*see* petroleum generation index)

pH (*see* acidity)
phaeophorbides, 169, 170, 172, 173
Phaeophyta (*see* algae, brown)
phaeophytins, 169, 170, 172
phenanthrene, 138, 205, 229
alkyl-, 136, 137, 138
methyl substituted, 180, 205, 207
phenolic compounds, 23, 24, 25, 56, 58, 60, 96, 97, 98, 100, 101, 103, 110, 115, 154, 162
phenylalanine, 36, 58, 59, 60
pheromones, 45, 47
phorbides, 169, 172
phosphate, 42, 43, 45, 46, 70, 79, 80, 110, 224, 225
phosphatides (*see* phospholipids)
phosphoglyceric acid (PGA), 10, 11, 162
phospholipids, 39, **42**, 42–3, 52, 155, 157
phosphorite, 90
photic zone, **69**, 71, 72, 73, 77, 82, 91, 168, 170, 225
photorespiration, 11, 64
photosynthesis, **10**, 10–11, 29, 39, 55–6, 58, 63–4, 75, 79, 86, 91, 170, 218
and carbon cycle, 2–6
C3 path, 11, 20, 162
C4 path, 162, 164
dark stage, 10, 11, 64, 68
and evolution of life, 6–16
and isotopic fractionation, 18–21, 162–5
and light levels, 68–70, 73, 230
light stage, 10, 11
phototrophs, **3**, 6–16, 55, 64–5, 70, 73, 80, 81, 154, 155
phycobilins, 58
phycocyanobilins, 57, 58
phycothrobilins, 57, 58
phyllocladane, 180
isomerism, 181
phytadienes, 111
phytane, 136, 137, 139, 140, 193, 194, 195, 196, 197, 215
phytanol, 47, 195, 196
phytenic acid, 195, 196
phytochrome, 57, 58
phytol, 43, **47**, 48, 56, 57, 114, 195–6
diagenesis, 195–7
phytoplankton, **13**, 64
blooms, 72, 73, 74, 76, 82, 225
chemical composition, 31–4, 39–40, 43, 47–8, 53, 54–8, 61, 154–5, 156–62, 170, 172, 176, 195, 223
classification, 13, 76
contribution to sedimentary organic matter, 4, 6, 17, 62, 63, 77, 85, 86, 90, 91, 112–14, 144, 145, 153–4, 168
evolution, 7, 13, 17, 18, 143
population variations, 15, 72, 76–7, 224

phytoplankton *continued*
  predation by zooplankton, 3, 13, 16,
    73, 77, 168, 169, 170, 172
  productivity, 2, 3, 4, 11, 12, 16,
    68–76, 77, 82, 83, 143, 170, 218,
    224, 230
  senescence, 169, 170, 172
  succession, 76–7, 225, 238
phytosterols, 53
picene, 140
  methyl-substituted, 139, 140, 183,
    184
pigments,
  accessory, **55**, 56, 58, 69, 155
  carotenoid, 53–6, 91, 155, 170, 171,
    174–5, 195
  diagenesis, 170, 171–5, 195, 196
  flavonoid, 60–1
  hydroxy-aromatic, 60–1
  photosynthetic, 39, 58
  tetrapyrrole, 55, **56**, 56–8, 91, 155,
    170, 172–4, 196
pimanthrene, 139, 140
pimaric acid, 47, 48, 140, 180
α-pinene, 48
plankton, 4, **13**, 22, 62, 81, 86, 112, 161,
    233, 234
  classification, 13, 14
plants,
  aquatic, 2, 4, 6, 14 (*see also* algae,
    phytoplankton, mangroves and
    grass)
  C3, **11**, 64, 162, 165
  C4, 11, **162**, 165
  CAM, 11, **162**, 165
  evolution, 7, 9, 14–15, 17–18
  flowering (*see* angiosperms)
  higher (*see* higher plants)
  lower, 14
  terrestrial, 2, 4, 6, 7 (*see also* higher
    plants and bryophytes)
  vascular, **14** (*see* higher plants)
plasmalogens, 43, 44
POC (*see* particulate organic carbon)
polar NSO compounds (*see* NSO
    compounds in oil)
pollen, 56, 94, 100, 101, 102, 104, 108,
    112, 114, 120, 148
pollutants, 218, 226, 238
  anthropogenic gases, 217–4, 225–7
  anthropogenic nutrients, 224–5
  chloro-aromatics, 217, 232–7
  hydrocarbons, 227–2
  interaction with humic material, 233,
    234, 238
  particulate association, 228, 230, 232,
    236
  pathways and fluxes, 223, 237
polyamides (*see* polypeptides)
polychaete worms, 78
polychlorinated biphenyls (PCBs), 233,
    234, 234–7
  congeners, 234–5
polycyclic aromatic hydrocarbons
    (PAHs), 137, 138, **227**

peri-condensed, 228, 230
pyrolytic, **228**, 230, 232
sedimentary distributions, 227–30
polyhydroxyflavonol, 60
polymerisation, 101
polypeptides, 34, **36**, 37
polysaccharides, 30, **31**, 32–4, 43, 45,
    93, 96, 101, 108, 110, 161, 166
polyterpenoids, 109
pore waters, 5, 6, 78, 80, 129, 130, 133,
    140, 141, 197
porosity, 102, 124, **129**, 130, 132, 133,
    134
porphyrins, **56**, 139, 194
  catagenesis, 201–2
  cracking, 201, 202
  diagenesis, 170, 172–4
  DPEP, 169, 172–4
  DPEP : etio ratio, 201–202, 204
  enrichment of short-chain
    components, 201–2
  etio, 172–4
  free-base, 169, 172
  iron, 172–3
  metallated, 169, 172, 173
  nickel, 172, 197, 199, 201
  vanadyl, 172, 197, 199, 201, 202
preservation of sedimentary organic
    matter, 2–6, 22, 63, 64, 77–83,
    84–92, 93–4, 101, 108–10, 144,
    195, 225
primary production, 2–6, 12, 13, 16, 20,
    21, 55, 63, 64, 77, 82, 89, 90, 91,
    92, 93, 101, 106, 144, 168, 218,
    220, 224, 230
  factors affecting, 55, 63–77, 224, 230
  gross, **2**
  net, 2, 3, 5, 16, 64
  spatial variation in oceans, 74–6
pristane, 136, 137, 139, 140, 193, 194,
    196, 203, 215
  formation index (PFI), 203, 204, 209,
    210
  meso-, 199
  stereoisomerism, 196, 199, 204
  ratio to phytane, 195, 197, 198
pristenes, 195, 196, 203
production index (*see* transformation
    ratio)
prokaryotes, 7, **8**, 13
proline, 36
propane, 38, 222
propionate, 79, 157
proteins, 22, **34**, 34–9, 110
  composition, 34–6
  diagenesis, 79, 93, 94, 108, 110
  occurrence and function, 41, 42, 43,
    61
Protista, 8
protozoa, **13**, 77, 195
Pseudomonas, 33, 78, 79
psilophytes, 15
pteridophytes, 8, 15, 18
purine, 38, 39
pycnocline, 67

pyran, 23, 28
pyranoses, 28, 30, 31
pyrene, 228, 229
pyridine, 23, 137
pyrimidine, 38, 39
pyrite, 100
pyrobitumen, 107
pyrophaeophorbides, 169, 172
pyrrole, 23, 56, 137
pyruvate, 79
pyruvic acid, 12, 31, 33, 41
  phenyl, 59
phosphoenol, 59, 162

quinones, 23, 60, 61, 96

racemic mixture, **26**, 166
racemisation, **166**
radiation balance of Earth, 218, 219,
    220, 221, 223
radicals,
  free, **55**
  hydroxyl, 223
radiolarians, **13**, 18, 77
rank (*see* coal rank)
rate constant, **126**
RDP (*see* ribulose diphosphate)
reactions,
  addition, 111
  cracking, 122, 123, 128, 129, 131,
    132, 138, 152, 190, 201–2
  cyclisation, 122
  disproportionation, 122
  equilibrium, 4, 167, 188, 200, 201,
    209
  in hydrocarbon generation, 122–4
  Maillard, 98
  photochemical, 223, 226–8, 231
  rates of, 37, 126, 208, 209
  redox, **70**
  reversible, 166, 167
  ring-opening, 122, 183
red beds, 7, 12
redox,
  indicators, 195–8
  potential, 70–1, 197, 198, 234
  reaction, **70**
reduction, 6, 39, 70–1
  biosynthetic, 41, 43, 46, 59–60
  metabolic/diagenetic, 110, 151–2,
    166, 170, 176, 178, 181–3, 190,
    191, 192, 195, 196, 200
  nitrate (*see* nitrate reduction)
  sulphate (*see* sulphate reduction)
reeds, 87
reefs, 65, 83, 134
reservoir rocks, 129, 133–4, **134**, 140,
    141, 144, 150, 151
residence time, 5, 77, 82, 221, 227, 237
resin acids, 47, 155
resinite, 99, 100, 109, 115
resins,
  in oil, 103, **107**, 108, 116, 121, 123,
    135, 140, 141, 145, 151

plant, 47, 99, 100, 101, 102, 107, 114,
    120, 140, 155, 179, 180
respiration, 2, 4, 16, 58, 61, 69, 218 (*see
    also* oxidation)
  aerobic, 12, 33, 41, 77–8, 80, 91 (*see
      also* methylotrophy)
  anaerobic, 33, 78–80, 81, 174 (*see
      also* acetogenesis, fermentation,
      methanogenesis, nitrate reduction
      and sulphate reduction)
retene, 139, 140, 180
  1,2,3,4-tetrahydro-, 180
retention time, 212
retinol (*see* vitamin A)
rhamnose, 29, 34, 161, 167
Rhizobium, 80
Rhizophora, 64, 87
rhizopods, 77
Rhodophyta (*see* algae, red)
Rhodospirillum, 11
ribonucleic acid (RNA), 8, 9, 37, 161
ribose, 27, 30, 34, 38, 43, 161
ribulose diphosphate (RDP), 10, 11, 20,
    162
River,
    Amazon, 90
    Mississippi, 90
RNA (*see* ribonucleic acid)
Rock-Eval, 147, 148, 150
rubber, 45, 46

S_1, S_2 and S_3 peaks (*see* Rock-Eval)
sabkha (see depositional environments,
    sabkha)
salinity, 4, 61, **64**, 67, 81, 86, 90, 92,
    198
salt,
    domes, 124, 134
    lakes, 86, 194, 195, 198, 199
    marshes, 64, 65, 76, 83
sandarocopimaric acid, 180
saponification, **35**
sapropels, 63 (*see also* kerogen,
    sapropelic and coals, sapropelic)
saturated compounds, **23** (*see also*
    aliphatic structures)
saturates, 107, **136**
sclerotinite, 99, 100
Sea,
    Adriatic, 255
    Azov, 91
    Baltic, 91
    Black, 17, 63, 91
    Caspian, 17
    Marmara, 91
    Mediterranean, 90, 225
    North, 75, 130, 131, 208, 231
    Red, 90
    Sargasso, 237
sea,
    floor spreading, 12, 92, 125, 144
    level changes, 89, 92, 143, 144, 187,
        188, 219
    surface temperature (SST), 187, 188,
        189

seamounts, 92
8,14-secohopanes, 182, **183**, 186
8,14-secohopanoids, 50
    monoaromatic, 185, 186, 213, 214
8,14-secotriterpanes, **183**, 192, 214
secotriterpenoids,
    aromatic, 186, 193
sedges, 87, 88
sediments,
    abiogenic, **4**, 86
    ancient, 22, 116, **171**, 190–211
    anoxic, 13, 171, 175, 195, 199, 222
    argillaceous, 111
    authegenic, 84
    biogenic, **4**, 6, 21, 84, 86
    and carbon cycles, 2–6, 178–21
    carbonate, 80, 85, 86, 90, 111, 134,
        145, 147, 187, 199
    chemical, **84**, 86
    clastic, **84**, 86, 111, 134, 147
    coaly, 186
    deep-sea, 94, 223
    density flows, 86
    detrital, **84** (*see* clastic)
    evaporitic, 84, 86, 111, 124, 198, 199
    fluvial, 4, 86, 88, 90
    freshwater, 80, 198, 199
    lacustrine, 85–6, 94, 95, 174, 195,
        198, 199
    laminated, 85, 86, 89
    major contributing organisms, 16–18
    marine, 4, 79, 94, 95, 97, 170, 174,
        195, 197
    mass flows, 86
    organic-rich, 5, 6, 18–19, 63, 74, 77,
        82, 83, 84, 86, 89, 90, 91, 92, 99,
        125, 144, 194, 195
    pelagic, 78, 80, 85, 86, 90, 196
    Recent, **153**, 153–89, 190, 192, 195,
        202
    shelf, 78, 81
    siliceous, 85, 111
    turbidity flows, 86, 90, 92
senescence, 169, **170**, 172
serine, 36
sesquiterpenoids, **46**, 47, 48, 109, 179,
    183, 186, 192, 214
sesterpenoids, 46
Severn Estuary, 228
Shale, 17, 86, 115, 124, 134, 145, 147,
    214, 215
    Bakken, 131
    black, 91
    Green River, 84, 113, 114
    Messel, 108, 193, 194
    organic-rich, 92, 124
shikimic acid, 58, 59
silicate, 70, 225
silicoflagellates, **13**, 18, 70, 76, 143, 144
simonellite, 180
sinapyl alcohol, 58, 59, 60
β-sitosterol, 52, 53, 159, 160, 192
SML (*see* surface mixed layer)
soil,
    and carbon cycle, 2, 3, 5

and chloroaromatic pollutants, 233,
    234, 235, 236
decomposition of organic matter
    in, 3, 81, 82, 222, 223
hopanoids, 182
humics, 94, 95, 97, 98, 106
source indicators, 193
    isotopic, 148, 151–2, 162–5, 193–4,
        198, 222–3
    molecular, 148, 190–4
    optical, 147–8
source rock, 129, 130, 131, 132, 133,
    135, 137, 140–1, 142–52, 192, 195,
    197, 205
    amount of organic matter, 146–7
    classification, 131–132, 147
    evaluation, 146–52
    lacustrine, 145
    marine, 145
    maturity of organic matter, 148–51
    type of organic matter, 146, 147–8
Spartina (*see* grass, cord-)
spermatophytes, 8, 15
Sphagnum, 87, 159, 182
spirosterenes, 176, 200
    4-methyl, 178, 179
spores, 15, 56, 94, 99, 100, 101, 102,
    104, 108, 112, 114, 148, 155
sporinite, 99, 100, 109, 115
sporopollenin, **56**, 109
squalane, 191, 198
squalene, 46, 47, 50, 52, 191
    -2,3-oxide, 52
SST (*see* sea surface temperature)
stanols, **53**, 168, 170, 176, 177, 178, 179
    5α(H) and 5β(H), 176
    4-methyl, 178, 179
stanones, 176
starch, 31, 33, 93
stationary phase, 211, 212, 214
steady state, 1, 4, 217
    deviations, 4, 217
stearic acid, 40
stenols, **53**, 166, 168, 170, 176, 178, 200
steradienes, 177, **178**
steranes, 110, 166, **176**, 177, 192, 194,
    199, 201, 204, 211, 215, 231
    4-desmethyl (*see* regular)
    4-methyl, 140, 179, 192, 199, 214, 215
    rearranged (*see* diasteranes)
    regular, 139, 140, 141, 192, 199, 214
    as source indicators, 192, 210–11
    stereoisomerism, 181, 199–200, 204
sterenes, 166, **176**, 176–9, 200
    isomerism, 176–9, 200
    4-methyl, 178, 179
    rearranged (*see* diasterenes and
        spirosterenes)
stereoisomerism, **25**, 25–8, 30, 166–7,
    177–9, 180–3, 199–201, 204
steric hindrance, **166**, 167, 205, 207
steroids, 39, 46, **47**, 50, 52–3, 139, 140,
    211
    aromatic, 110, 139, 140, 141, 166, 177,
        179, 201, 204, 206, 208–10, 214

steroids *continued*
  diagenesis, 110, 166, 168, 172, 175–9, 201
  4-methyl, 139, 140, 201
  rearranged (*see* diasteroids)
  ring-A degraded, 176
  short-chain enrichment, 201, 204, 205, 206
  structural notation scheme, 50–1
sterols, 45, **52**, 154 (*see also* stanols and stenols)
  diagenesis, 110, 166, 168, 172, 176–7
  4-methyl, 160, 178, 179, 192
  regular, 178
  source/environment indicators, 52–3, 156, 159–60, 192
stigmasterol, 50, 52, 53, 159, 160, 192
stratification,
  of atmosphere, 226
  of lakes, 67, 68, 85, 86, 224
  of oceans, 67, 92
  of water column, 65–8, 71, 75, 85, 86, 90, 224
stratopause, 226
stratosphere, 225, **226**
stromatolites, 9, 17, 21, 62
subduction zones, 89
suberan, **45**, 109
suberin, **45**, 159, 171
suberinite, 99, 109
submarine fans, 134
succession, **76**, 77, 87, 224, 225, 238
sucrose, 31, 32
sugars, 26, **31**, 37, 43, 61, 79, 93, 94, 96, 98, 110, 161
sulphate, 8, 43, 64, 78, 79, 80, 81, 110, 197
  reducers, 78, 79, **80**, 81, 110, 157, 158, 194
  reduction, 79, 80, 81, 91, 173, 195
sulphide, 79, 80, 100, 110, 111, 114
  dimethyl (DMS), 223–4
  hydrogen, 8, 11, 79, 80, 81, 91, 98, 101, 111, 116, 118, 138
sulphur,
  cycle, 79–80
  dioxide, 135, 224
  forms/availability in sedimentary environments, 197–8, **199**
  incorporation into organic matter, 95, 97, 98, 100, 109, 110–12, 114
  volcanic emissions, 224
surface mixed layer (SML), 67, **236**, 237
suspension feeders, 78
swamps/bogs, 83, 90, 144
syringaldehyde, 163
syringic acid, 163
syringyl lignin constituents, 162, 163, 164, 165

2,4,5-T, 233, 234
T$_{max}$ (*see* Rock-Eval)
TAI (*see* thermal alteration index)

tannins, 58, **60**, 61, 98, 109
taraxerane, 181, 182
taraxerene, 181, 182
taraxerol, 49, 181, 182
tar, 100
tar sands, 144, **145**
  Athabasca, 133, 145
tasmanites, 56, 114, 115
Taxodium (*see* cypress)
taxonomic rank, 8, 9
TCDD (*see* 2,3,7,8-tetrachlorodibenzo-*p*-dioxin)
tectonic activity, **12**, 83, 84, 85, 89, 124, 133, 144
teichoic acids, 41, 43
telinite, 99
teratogens, 232
termites, 222, 223
terpanes, 141, 215, 231
  pentacyclic, 139, 214, 215
  tetracyclic, 139, 140, 214, 215
  tricyclic, 139, 140, 141, 192, 214, 215
terpenoids, 33, 34, **45**, 45–56, 139, 155, 192, 211
  classification, 45–46
  diagenesis, 179–186
α-terpineol, 48
test, 4, 80, 84, 153, 187
2,3,7,8-tetrachlorodibenzo-*p*-dioxin (TCDD), 233, 234, 236
Tetrahedron, 108
tetrahymenol, 191, 195
tetrasaccharides, 31
tetraterpenoids, **46**, 53–6 (*see also* carotenoids)
Thalassia (*see* grass, sea-)
thermal,
  alteration index (TAI), **148**, 149, 204
  conductivity, 124, 126
  conductivity detector, 147
thermocline, **65**, 67, 68, 72, 73, 74, 92
  diurnal, 67, 74
  permanent, 65, **66**, 67, 69, 71, 73
  seasonal, **67**, 69, 72, 73, 84
thiacycloalkanes, 111
Thiobacillus, 79, 80
thiols, 23, 100, 111
Thiothrix, 80
thiophenes, 23, 111, 114, 135, 137, 138
threonine, 36
time-temperature index (TTI), 125–8, 149, 204
tocopherols, 191k 196
Torbanite, 99, 114, 115
toxicants, **230**
toxins, 238
Trade Winds, 66, 67, 92, 228
transamination, **34**, 58, 59
transfer efficiency, **16**
transformation ratio, **150**
traps, 133–4
  sediment, 168, 170, 237
  stratigraphic, 133, **134**
  structural, **133**, 134

triaromatic steroidal hydrocarbons, 139, 140, 177, 201, 214
  short-chain enrichment, 201, 204, 205, 206
1,1,1-trichloro-2,2-di(*p*-chlorophenyl)-ethane (DDT), 233
2,4,5-trichlorophenoxy acetic acid (*see* 2,4,5-T)
triglycerides, 39, 41, 42, 61, 62, 155, 157, 170
triose phosphate, 10, 11, 29, 33
triphenylene, 229
trisaccharides, 31
22,29,30-trisnorhopane (Tm), 50, 51, 140, 215
25,28,30-trisnorhopane (Ts), 50, 51, 140, 193, 215
triterpanes, 183, 192, 211, 214, 215 (*see also* hopanes)
  ring-C degraded (*see* 8,14-secotriterpanes)
  stereoisomerism, 200–1, 204
triterpenoids, **46**, 47, 49–51
  acyclic, 47, 52
  aromatic, 183–6, 201
  bacterial (*see* hopanoids)
  diagenesis, 181–6
  higher plant, 47, 49, 140, 155, 181–3, 193
  pentacyclic, 47, 49, (*see also* triterpenoids, higher plant)
  ring-A degraded, 183–184
  ring-C degraded (*see* secotriterpenoids)
  structural notation scheme for pentacyclics, 50–1
  tetracyclic, 47 (*see also* steroids)
trophic level, **16**
tropopause, 226
troposphere, 225, **226**
tryptophan, 36
TTI (*see* time-temperature index)
turbidites, 85, **86**, 90, 92, 196
tyrosine, 36

U$^K_{37}$, 187, 188, 189
ubiquinones, 60, 61
UCM (*see* unresolved complex mixture)
ultrananoplankton, **13**, 14
Ulva, 55
unresolved complex mixture (UCM), 231, 232
unsaturated compounds, **24**, 26, 28, 41, 52, 53, 111
upwelling, 12, 65, 66, 67, 74, 75, 82, 89, 90, 92
  Peru, 75, 82, 170
uranium minerals, 90
uronic acids, **29**
ursane, 47

vaccenic acid, 157
valine, 36, 155, 187

van Krevelen diagram, 103, 104, 113,
    115, 116, 118, 148, 150
vanillic acid, 163
vanillin, 163
vanillyl lignin constituents, 162, 163,
    164, 165
varves, **86**
vitamins, 77
    A, 48, 55
    E, (*see* tocopherols)
vitrain, 100
vitrinite, **99**, 100, 112, 113, 114, 116
    chemical composition, 102, 103, 104,
        105, 109, 115, 119, 120
    reflectance, 99, 103, 147, **148**,
        148–50, 151, 203–5, 207–8
    reworked, 149, 150
    structure, 105

thermal evolution, 102–4, 116, 119,
    120, 128, 146
vitrite, 100

water lilies, 88, 89
waxes, 39, **43**, 43–5, 90, 98, 101, 102,
    104, 115, 119, 120, 138, 140
    higher plant, 43, 61, 90, 98, 101, 102,
        104, 114, 115, 119, 120, 121, 138,
        140, 154, 165, 190, 203
    leaf cuticular, 43, 61, 115, 121, 154,
        165, 190, 203
    microbial, 45
westerlies, 67
wet deposition, **236**
wood, 2, 58, 81, 87, 99, 104, 115, 217,
    228, 230

Xanthophyceae, 76
xanthophylls, **53**, 175
xenobiotic compounds, **217**, 232
xylem, 58
xylose, 29, 33, 34, 161, 166
yeasts, **172**

zeaxanthin, 175
zooplankton, 3, **13**, 72, 222
    chemical composition, 56, 62, 155,
        159–60, 161, 168, 196
    contribution to sedimentary organic
        matter, 13–14, 16, 17, 73, 114
    faecal pellet sinking rate, 77, 168
    grazing of phytoplankton, 13, 16, 62,
        73, 77, 168–72, 223
    migration, 168
Zostera (*see* grass, eel-)
zwitterion, 34, 36